25 Springer Series in Chemical Physics
Edited by Robert Gomer

Springer Series in Chemical Physics

Editors: V. I. Goldanskii   R. Gomer   F. P. Schäfer   J. P. Toennies

Ion Formation from Organic Solids

Proceedings of the Second International Conference
Münster, Fed. Rep. of Germany
September 7 – 9, 1982

Editor: A. Benninghoven

With 170 Figures

Springer-Verlag
Berlin Heidelberg New York Tokyo 1983

Professor Dr. Alfred Benninghoven

Physikalisches Institut der Universität Münster, Domagkstraße 75
D-4400 Münster, Fed. Rep. of Germany

Series Editors

Professor Vitalii I. Goldanskii

Institute of Chemical Physics
Academy of Sciences
Vorobyevskoye Chaussee 2-b
Moscow V-334, USSR

Professor Dr. Fritz Peter Schäfer

Max-Planck-Institut für
Biophysikalische Chemie
D-3400 Göttingen-Nikolausberg
Fed. Rep. of Germany

Professor Robert Gomer

The James Franck Institute
The University of Chicago
5640 Ellis Avenue
Chicago, IL 60637, USA

Professor Dr. J. Peter Toennies

Max-Planck-Institut für Strömungsforschung
Böttingerstraße 6-8
D-3400 Göttingen
Fed. Rep. of Germany

ISBN 978-3-642-87150-4   ISBN 978-3-642-87148-1 (eBook)
DOI 10.1007/978-3-642-87148-1

Library of Congress Cataloging in Publication Data. Main entry under title: Ion formation from organic solids. (Springer series in chemical physics ; v. 25). Includes index. 1. Secondary ion emission–Congresses. 2. Secondary ion mass spectrometry–Congresses. 3. Organic compounds-Spectra-Congresses. 4. Field desorption mass spectrometry–Congresses. I. Benninghoven, A. II. Series. QC702.7.E4157 1983   547.1'3722 83-4714

© by Springer-Verlag Berlin Heidelberg 1983
Softcover reprint of the hardcover 1st edition 1983

2153/3130-543210

Preface

The Second International Conference on Ion Formation from Organic Solids (IFOS II) was held at the University of Münster, Federal Republic of Germany, from September 7 to 9, 1982. The subject of the conference was the rapidly developing field of ion formation from involatile, thermally labile organic compounds. Rapid progress has been made in this field in the last few years, mainly because of the discovery of unexpected new ionization processes such as sputtering and laser-induced desorption.

The aim of the conference was twofold: to acquire a basic understanding of these "soft" ionization processes on the one hand, and to examine their present and future analytical applications on the other. We sought to bring together scientists working in fundamental as well as applied research. The participants represented such widely varied fields as pure and applied physics and chemistry, biochemistry, nuclear and solid-state physics, medicine, and pharmacology.

These proceedings contain all of the papers presented at the conference. Six review papers cover the fundamentals of different ionization processes. The authors of these reviews were asked to give up-to-date surveys including characteristics of spectra, the influence of excitation parameters, tentative models for ion formation processes, and assessments of their analytical applications. These reviews are followed by 26 contributed papers dealing with more specialized aspects of the ionization processes and their analytical applications.

This volume is the first comprehensive treatment of all recently developed ionization techniques for involatile, especially organic, material. It is the hope of the editor that this volume will not only be helpful to those seeking a better understanding of ion formation processes, but will also meet the needs of the rapidly increasing number of laboratories applying these new developments in mass spectrometry as an analytical tool.

I am grateful to the Deutsche Forschungsgemeinschaft, Bonn, and the Office of Naval Research, Washington, D.C., for their generous financial support, and to the University of Münster for its hospitality. The conference was prepared in close cooperation with Dr. F.E. Saalfeld of the ONR, and I would also like to thank Dr. R.J. Colton, ONR, Dr. W. Sichtermann, Münster, and my secretary Miss I. Bekemeier for their assistance. The success of the conference was largely due to their efforts.

Münster, Fed. Rep. of Germany   *A. Benninghoven*
January 1983

Contents

Part 4   Laser Induced Ion Formation

Part 5 Other Ion Formation Processes

Part 1

Field Desorption

1.1 Principles of Field Desorption Mass Spectrometry (Review)

F.W. Röllgen

Institut für Physikalische Chemie der Universität Bonn,
D-5300 Bonn, Fed. Rep. of Germany

1. Introduction

Field desorption mass spectrometry (FD MS) is widely applied for soft ionization of thermally labile compounds. It was introduced in 1969 by BECKEY [1]
in order to avoid introduction of sample molecules into the vapor phase
prior to their ionization. With this technique (Fig.1) the sample is deposited from solution onto the surface of a field anode, usually a thin 10 μm
W-wire covered with field—enhancing microneedles. Ionization and desorption
of molecules is achieved by applying a high voltage of about 10 kV between
the wire and a 2-3 mm distant counterelectrode while resistively heating the
emitter wire. The mass spectra typically exhibit molecular ions such as $M^{+\cdot}$,
$(M + H)^+$ and $(M + Na)^+$. No or at most only a few fragment ions are formed.

The method was termed FD MS [1] because molecules were assumed to be ionized exclusively in adsorbed states. This is in contrast to field ionization
mass spectrometry (FI MS) [2] where molecules are supplied from the gas
phase onto the emitter and are ionized by FI in the gas phase or in adsorbed
layers.[+]

In the past FD MS has been applied in almost every field of analytical
chemistry mostly for molecular weight determination and to a lesser degree
for structure elucidation. These applications have been reviewed by several
authors [2-5].

The development of FD MS since its introduction has mostly been a sequence of experimental optimizations resulting from its analytical applications. Less attention has been given to understanding the mechanism of ion
formation.

For a long time ion formation during field desorption (FD) of organic
compounds was considered to occur only by field induced electron tunneling,
i.e.,FI at the tips of field-enhancing microneedles on activated emitters
[3]. These ion forming processes do not differ in principle from those by
which molecules supplied from the gas phase are ionized under FI MS conditions. Accordingly, field desorption of organic molecules was discussed
within the framework of the FD theory developed by GOMER et al. [6] to explain the removal of mono- or submonolayer adsorbates from metal surfaces
taking, however, in addition reaction and desorption mechanisms into account which had been discovered under conditions of FI and pulsed field
desorption mass spectrometry experiments [7].

[+]
   In the following FI is used to denote any ion formation by field—induced
   electron tunneling processes.

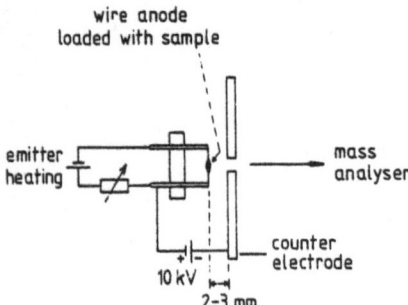

wire anode
loaded with sample

emitter
heating

mass
analyser

counter
electrode

10 kV

2-3 mm

Fig.1. Schematic representation of a FD ion source

At odds with this model of FD are a number of observations [8-12], in particular with ionic compounds which cannot be explained by field desorption through field—induced electron tunneling. They point to a different emission process occurring at lower threshold field strengths in which ions are extracted from condensed layers via a field—induced desolvation mechanism [13-15]. This desorption mechanism alone governs the production of non-radicalic molecular ions from thermally labile compounds.

In the following the present state of both the experimental method and the theory of FD MS will be reviewed. In addition to the conventional positive ion FD MS the recently introduced negative ion field desorption mass spectrometry (NFD MS) will also be discussed. Applications of FD MS in biochemistry, medicine and environmental research will not be considered.

2. Experimental Techniques and Methods

Introduction of the wire emitter by BECKEY [16] was an important prerequisite for the development of FD MS of organic compounds because in retrospect single tips are not suited for this mode of ionization. Furthermore, improvements in the production of activated emitters, i.e., thin wires covered with field—enhancing microneedles were of particular importance for analytical applications of the FD method. Up to now only activated wire emitters have found widespread applications as field anodes. The diameter of the wire is usually around 10 μm and the length of the microneedles is typically between 20 μm and 50 μm. Carbon microneedles grown at 1500 K from benzonitrile [17] or indene [18] are most frequently applied. The activation period is considerably shorter for indene but still requires a few hours. In addition silicon microneedles [19] are also used in routine FD work. Various types of activated emitters and their preparation techniques have been reviewed by BECKEY [20]. In general the field enhancement achieved with the microneedles is sufficient for FI of organic molecules (range of field strength: 10^9 to 10^{10} V/m).

Smooth 10 μm W-wires without field—enhancing microneedles have so far been used in only a few analytical applications [e.g., 21,22]. A disadvantage of the smooth wire emitter is the absence of any FI signal from a background gas which can be used for focusing purposes prior to the emission of ions by FD. The low field strength ($< 10^9$ V/m) at the wire surface does not allow any FI to occur but is high enough for the extraction of preformed ions from condensed sample layers as will be discussed below. Accordingly, simpler mass spectra are obtained from thermally labile molecules with smooth wires than with activated emitters. This effect is shown

3

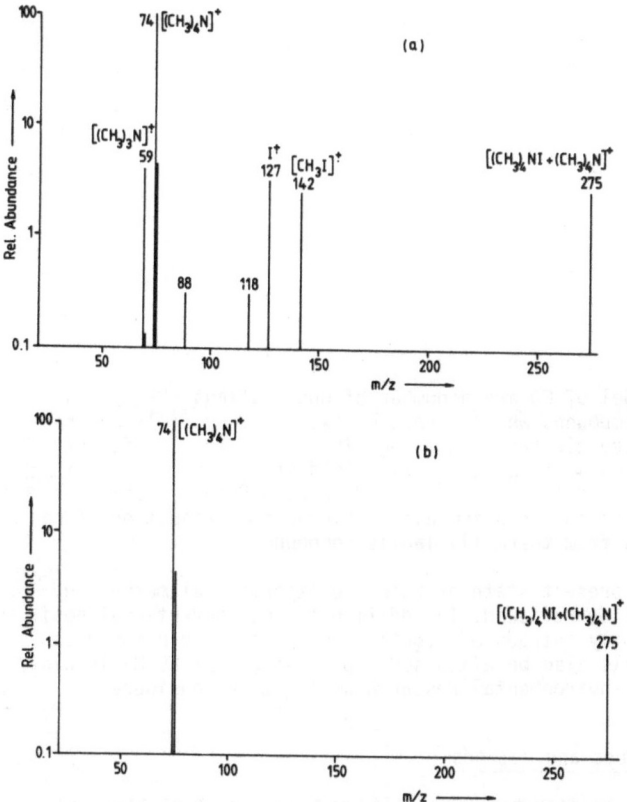

Fig.2. FD mass spectra of tetramethyl ammonium iodide obtained (a) by using a conventional activated emitter [23] and (b) by using a smooth 10 μm W-wire emitter [11]

in Fig.2 for the case of $(CH_3)_4NI$. The $(CH_3)_3N^{+\cdot}$, $I^{+\cdot}$ and $CH_3I^{+\cdot}$ ions are formed by FI of volatilization products on top of the field-enhancing microneedles, whereas the $(CH_3)_4N^{+\cdot}$ cation and the clusterion of the salt are desorbed directly from the salt layer.

Effect of different microneedle lengths on FD spectra is shown in Fig.3a and b with the example of sucrose. The contribution from FI processes at the tips of the field-enhancing microneedles to the mass spectrum decreases with decreasing length of the microneedles. This effect is, of course, also dependent on the amount of sample loaded onto the activated emitters. Without the addition of an acid or salt no spectrum of sucrose could be obtained by using a very smooth 10 μm W-wire [24]; the spectrum obtained by adding LiI is displayed in Fig.3c. Under certain experimental conditions no fragment ions are formed.

Samples are deposited on field anodes from solution either by dipping the emitter into the solution or by using a syringe technique [2]. The emitter loading can only be controlled with the latter procedure. In practice a

4

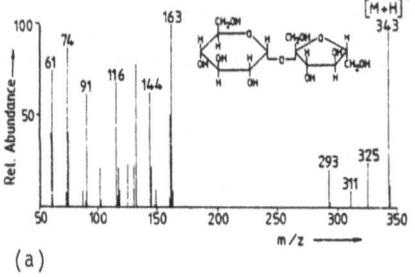

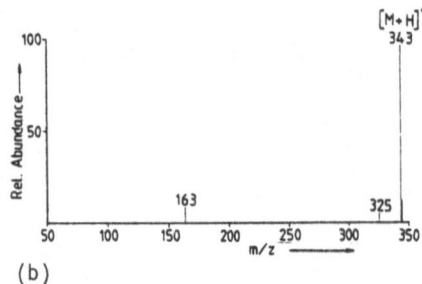

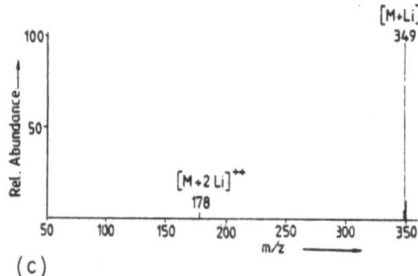

Fig.3. FD mass spectra of sucrose obtained (a) by using a conventional activated wire emitter [4], (b) by using a slightly activated wire [24], and (c) by using a smooth wire and adding LiI to the aqueous solution of the sample [25]

sample amount of 10^{-8} - 10^{-6}g is used for one measurement. However, under optimum experimental conditions 10^{-12} - 10^{-10}g are sufficient for recording a mass spectrum.

The emitter temperature is the most critical parameter in determining the appearance of a mass spectrum. The search for optimum desorption conditions, viz. a molecular ion emission of sufficient intensity and duration accompanied by a low level of fragmentation, is considerably facilitated if the emitter heating current is increased automatically at a preselected rate. Different types of emitter heating controllers have been constructed [20] and successfully applied to obtain more reproducible mass spectra.

Radiative heating is also used as a means to promote the desorption of ions. Both infrared (globar) [26] and laser radiation [27,28] have been used. Preferential absorption of radiation by the carbon microneedles [27] and by the sample itself [29], if non-activated metal wires are used, leads to a reduction of the noise level caused by the emission of charged particles [30]. Moreover, an increase in the ion yield compared to that obtained with resistive heating is observed.

Almost all types of mass analysers have been used in connection with FD ion sources, e.g., single and double focusing magnetic mass spectrometers, quadrupole mass spectrometers [31], and in one known instance an energy focusing time of flight (TOF) instrument [32]. It is important to note that single focusing mass spectrometers are not suited for analytical FD MS since a broad time varying energy distribution of the desorbed ions makes in general an unambiguous mass assignment above m/z 400 with unknown samples impossible. Investigations of pulsed FD of thermally labile sample molecules from smooth wire emitters have shown [33] that TOF mass analysis is also not an appropriate technique for FD MS since ions formed via the desolvation mechanism cannot be produced within a sufficiently small time interval (< 100 ns) to achieve an acceptable mass resolution. Furthermore, FD induced by short laser pulses has been attempted in some preliminary experiments using smooth wire emitters also without success [34]. Again this is probably due to the fact that short temperature pulses in the ns time range do not allow sufficient time for the slow field-induced desolvation mechanism to occur.

The FD method can be significantly improved for molecular weight determination and structure elucidation if certain additives are applied to sample solutions. Three kinds of additives can be distinguished with respect to different promotion effects.

1. Matrix compounds which promote desolvation of ions. The admixture of a number of polymeres and higher molecular weight compounds such as poly-vinyl alcohol [12,15], polyglycol, polyethers and emulsifying agents [35] was found to promote a constant ion emission over a broader temperature range and to decrease the threshold field strength for desolvation of ions. Similar results were also obtained for some small polyhydroxy compounds such as pentaerythritol [36], mono- and disaccharides, and the solvent dimethyl sulfoxide. The admixture of matrix compounds affects the viscosity of the sample solution and favors the electrohydrodynamic disintegration of the sample layer which results in the formation of ion emitting tips and fibers as will be discussed below. The analytical capabilities of matrix assisted FD MS have not yet been sufficiently explored.

2. Alkali salts which promote cationization. The addition of alkali salts to a sample gives rise to the ionization of polar molecules by alkali ion attachment as demonstrated in Fig.2c. Alkali ion attachment is the most important mode of ionization for thermally labile polyhydroxy compounds because the resulting molecular ions are particularly stable with respect to unimolecular decomposition [9,37]. Various alkali halides, lithium tetra-phenyl borates, and other salts have been used for cationization [9,14,38-41]. It has become obvious that not only the cation but also the anion affects the ionization efficiency of molecules. In practice alkali ions are already present as impurities in samples taken from crude extracts. Their concentration is usually high enough to account for the formation of $(M + Na)^+$ ions.

3. Acids which promote protonation and fragmentation. The addition of acids to weakly or nonacidic samples favors the formation of $(M + H)^+$ ions. This was first demonstrated by the admixture of HCl to a sample of adenine [17]. In general, however, more useful additives are tartaric acid [14] and toluene sulfonic acid [42] because they induce viscous behavior in the sample solution in the same way as the above mentioned matrix compounds. The addition of acids is not only important for the ionization of zwitter-ionic compounds but also for the promotion of fragmentation which provides additional structural information.

Pretreatment of the emitter by forming an acidic layer, for example, tar-taric acid [43] or 1-naphthalenylboronic acid [44] leads to a protonation of molecules just the same way as acid addition to the sample solution does. This technique also works with salts in the case of cationization.

The influence of various types of emitters and additives on the appearance of a mass spectrum and on the selectivity of the method for the detection of different components of a mixture has numerous practical implications for analytical applications of FD MS. Surprisingly, these potentially useful features of the FD method have been utilized only rarely. A detailed discussion of the analytical applications of FD MS has been given elsewhere [45].

6

3. Ionization Mechanisms

Before discussing ion formation mechanisms it may be instructive to consider the various types of molecular ions which appear in FD spectra. Some frequently observed molecular ions are listed in Table 1. The analytically most important ions are $M^{+\cdot}$, $(M + H)^+$ and $(M + Alkali)^+$ ions. In the case of salts (C^+X^-) the generally observed ions are C^+, $C_{n+1}X_n^+$ with n > 1, and less often $(CX + H)^+$ ions. C^{2+} and $C^{2+}X^-$ ions have been observed with metal halides and onium salts [46,47]. Of course, cation exchange with salts and the formation of salts from acids may also occur in a sample solution. The latter chemistry gives rise to the appearance of $(M - H + 2Alkali)^+$ species in FD spectra.

Table 1. Frequently observed molecular ions

Singly charged ions:

$M^{+\cdot}$ (typical FI product)

$(M + H)^+$ (formed by FI or the desolvation mechanism)

$(M + Alkali)^+$ (typical desolvated ion)

$(2M + H)^+$, $(2M + Alkali)^+$

Multiply charged ions:

M^{2+} (FI product, observed mainly with aromatic compounds)

$(M + nH)^{n+}$, $n \geq 2$ (e.g., peptides)

$(M + n\ Alkali)^{n+}$, $n \geq 2$ (e.g., oligosaccharides)

$(M + nH + m\ Alkali)^{(n + m)+}$ with $(n + m) \geq 2$

Three distinctly different mechanisms contribute to the production of gaseous ions in FD MS. The ionization of molecules by electron tunneling which takes place on the tips of the field enhancing microneedles of activated emitters and requires a high threshold field strength (> 10^9 V/m) may be denoted as a *FI mechanism*. Ions formed in condensed sample layers can, as already mentioned above, be extracted via a field-induced *desolvation mechanism*. In contrast to the FI mechanism the desolvation mechanism requires lower threshold field strengths for the production of gaseous ions. The third mechanism is a *thermal ionization mechanism* which is principally independent of external fields. In general this mechanism does not contribute to the emission of ions except in the cases of metal ions at emitter temperatures > 800 K and quaternary salts at temperatures > 380 K, and it will, therefore, not be discussed in any further detail.

1. *FI mechanism*. As is well known from FI MS there are a number of FI processes possible on the tips of field-enhancing microneedles. The most important ionization processes are due to the following two reactions:

$$M_{ads} \xrightarrow{-e} M^{+\cdot} \tag{1}$$

$$XH + M_{ads} \xrightarrow{-e} X + (M + H)^+ \ . \tag{2}$$

7

XH represents an acidic surface or another adsorbed molecule. The mechanism and the energetics of these and other FI processes involving the removal of electrons by tunneling have been extensively investigated [7]. It is interesting to note that under FI conditions $(M + H)^+$ ions are not always formed by reactions involving electron tunneling. High fields ($> 10^9$ V/m) also give rise to the creation of surface ions or field adsorbed ions which react with neutral molecules to form $(M + H)^+$ ions by ion-molecule reactions [7]. It is outside the scope of this paper, however, to discuss field ion formation by surface interaction in more detail.

For FI molecules deposited in layers on the shanks of or between the microneedles of activated emitters have to be supplied to the field enhancing tips via gas phase transport or surface diffusion. In the case of gas phase transport the close proximity of the evaporation sites to the ionization zones gives rise to a "direct exposure" mode of FI (DFI). As with direct exposure electron impact (DEI) and direct exposure chemical ionization (DCI) the sample temperature required for the ionization of low vapor pressure molecules is reduced.

2. *Desolvation mechanism.* The field-induced extraction of ions from electrolytic solutions or from salt layers is the most important mechanism by which gaseous molecular ions are formed from thermally labile or nonvolatile compounds. This mechanism is schematically shown in Fig.4 for the formation of $(M + Na)^+$ ions from a liquid sample layer containing NaI.

The desorption of desolvated ions is not directly achieved by the action of the high field on the sample layer. Detailed investigations, using combined optical microscopy and mass spectrometry, of the behavior of small droplets of sample solution deposited on 10 μm wire anodes and exposed to a high field revealed the following sequential steps in FD of highly polar molecules [48-50]:

i) The external field first causes a charging of the surface of the sample layer by solvated ions. Accordingly, negative ions are neutralized at the sample layer/metal anode interface. The neutralization reactions may lead to corrosion of the anode surface as observed, for example, with alkali halides, or to the evolution of gases. It is important that the sample layer has mobile ions to build up the surface charge.

ii) In the second stage the stress of the high field exerted on the charged layer creates electrohydrodynamic instabilities resulting in a disintegration of the sample layer. The disintegration, on the one hand, gives rise to the emission of charged droplets, particles and cluster ions; most of the sample layer is usually lost during this stage of decomposition. On the other hand, the disintegration also results in the formation of field-enhancing protuberances, e.g., sharp projections and fibers, which are stable, with respect to the foregoing decomposition processes, for time intervals ranging from seconds to minutes. It has to be emphasized that a liquid or viscous state of the sample is needed for the formation of these protruding features.

iii) In the third stage evaporation of the solvent from the field enhancing projections results in the formation of a nearly solid, amorphous state in, at least, the surface layer of the protrusions. For oligosaccharides, peptides and similar highly polar compounds not only small protrusions but even long projections transform into this apparent glassy state which has a low ionic conductivity.

8

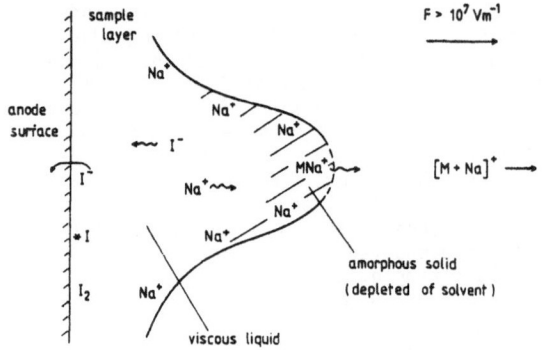

Fig.4. Schematic representation of ion formation by
desolvation

iv) In the last stage field—induced emission of desolvated molecular ions
   such as $(M + Na)^+$ or $(M + H)^+$, takes place from the amorphous or glassy
   state of the sample layer. The high kinetic energy deficits observed in
   the desorbed ions (up to 80 eV) and the rather small field enhancement
   caused by the long projections are consistent with ion emission from
   nearly solid projections of low ionic conductivity. Such a solid, amorphous state of the surface of the projections can also account for the
   fact that sometimes dimers, but in general no higher cluster ions, are
   observed in FD spectra. In the case of field ion emission from a liquid
   surface a broad distribution of cluster ions, i.e.,partially solvated
   ions, is typically obtained. This is well known in the instance of
   "electrohydrodynamic desorption" mass spectrometry [51], but it has
   also been observed in field ion emission from charged liquid droplets
   [52,53] and under conditions of low temperature FI of polar molecules
   [54].

The desolvation itself must proceed in several steps from the amorphous, molecularly very rough surface structure to overcome the high
solvation energy which is in the order of several eV.

The field strengths required for desolvation of molecular ions are not accurately known but should lie for most samples below the threshold field
strength for FI of the sample molecules, i.e.,below $3 \cdot 10^9$ V/m.

The main role of emitter temperature is to decrease the viscosity of the
sample layer and thus to increase the rate of disintegration of the layer.
The desolvation is at every stage a thermally activated process. It is important to note that the temperature required for the desolvation of molecular ions is significantly lower than the temperature required for "fast"
volatilization of the sample layer. For sucrose the emission of $[M + Na]^+$
ions was observed between 90 and 130^0 C, whereas an emitter temperature of
more than 175^0 C was needed for a field free evaporation of a sucrose droplet within a period of a few minutes [49].

The electrohydrodynamic disintegration processes involved in the production of ions account for a number of experimental difficulties in recording
FD spectra, viz. delayed onset of ion emission after an emitter temperature
increase, high intensity fluctuations, short desorption times and sensitive
dependence of the appearance of mass spectra on emitter type, sample loading
and temperature. It is obvious that sample deposited between the long microneedles of activated emitters, i.e., in areas of low field strength, does
not contribute to the emission of molecular ions via the desolvation mecha-

nism. The sample layer must be exposed to field strengths above the threshold of this ionization mechanism. For activated emitters a strong adhesion of the sample to the microneedle surface is important because there is no transport of the charged sample layer via a viscous flow from areas of low to high field strengths [30]. Evidence for the emission of ions from deposits near the apex of the microneedles was obtained from measurements of the energy distribution of $(M + Na)^+$ sucrose ions [50]. The mean kinetic energy deficit of these ions was found to be significantly smaller with activated emitters than with smooth wires. This is in accordance with smaller protusions and higher field strengths in the case of activated emitters. It seems to be probable that for sample layers deposited near the apex of microneedles, i.e., in areas of high field strength, the electrohydrodynamic destruction is no longer a prerequisite for the emission of desolvated ions.

It is appropriate at this point to briefly discuss the formation of fragment ions observed in FD spectra. These species result (a) from thermal degradation of molecules and successive ionization of the resulting products by FI or the desolvation mechanism, (b) from reactions within the liquid sample layer, (c) from field-induced reactions in the space charge region of the sample layer during the course of desolvation of ions, (d) from FI processes involving surface interaction, (e) from field dissociation of molecular ions in the gas phase, and, of course, (f) from unimolecular decomposition of metastable ions. Typical fragment ions from (b) and (c) are products of solvolysis reactions [14]. So far the field-induced ion chemistry accompanying the desolvation of ions has been investigated for some negative ions [55] but no positive ions. It is highly probable that most of structurally significant fragment ions of a large, polar molecule result from reactions (c) rather than (a) or (b). It must also be noted that reactions (a) to (d) may also give rise to the production of those ions appearing in the spectrum at higher molecular weights than the molecular ion group.

4. Field Desorption Mass Spectrometry of Negative Ions (NFD MS)

In the past field electron emission has been the main obstacle to the development of NFD MS since the detection of negative ions is difficult or even impossible above the threshold of field electron emission. The disturbing influence of electron emission on the formation of negative field ions can be attributed to several effects, such as surface heating, sputtering of the surface by impinging positive ions formed near the surface by electron impact, and destruction of the surface during the course of a discharge between the cathode and the counterelectrode. The first FD mass spectra obtained with reversed polarity for the detection of negative ions [56] provide evidence for ion formation under conditions of a discharge.

More recent experiments, however, demonstrate that NFD MS is possible even below the onset field strength of field electron emission [57,58]. With this method the compounds to be investigated are mixed with an aqueous solution of polyethylene oxide 4000. The sample mixture is deposited onto smooth wire emitters or activated emitters [59]. The desorption of negative ions is achieved by applying a negative potential to the emitter cathode with respect to the counterelectrode and by heating the emitter. The applied potential is between -4kV and -8kV and depends on the onset of field electron emission.

NFD mass spectra of a number of compounds, in particular of salts and acids, have been obtained [58-60]. With salts the formation of cluster ions is analogous to that observed in the positive ion FD mode. Typically, no fragment ions of polyethylene oxide are detected. With acids (M - H)$^-$ ions are formed. So far, doubly charged negative ions have only been observed with some disulfonic acids [59].

NFD mass spectra of some sulfonic acids and peptides have also been obtained under conditions of strong electron emission [61]. The absence of any fragmentation in these instances suggests that the electrons are emitted from microneedles not covered by the sample. This conclusion is supported by the observation that NFD spectra of some carbonic and sulfonic acids [58] and of salts from biological extracts [62] can be recorded below the onset of electron emission even without the use of a polymer matrix.

Ionization of molecules by halogen and NO_3^- ion attachment has been reported [58,59]. The method is analogous to cationization by alkali ion attachment. The anionization is easily achieved by adding salts such as LiCl to the samples. An example is given in Fig.5.

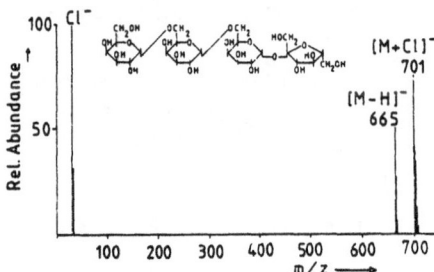

Fig.5. NFD mass spectrum of stachyose mixed with polyethylene oxide 4000 and LiCl

Ion formation in the described mode of NFD MS is based exclusively on the desolvation mechanism discussed above. The desolvation of negative ions from the viscous electrolytic solution, prepared by admixing a high molecular weight polymer with the aqueous sample solutions, is possible at field strengths below the onset of electron emission. Since the electron work functions of electrolytic solutions are higher by one or several eV than those of metals or graphite, the threshold field strength for field electron emission from smooth wires or activated emitters increases if they are covered by an electrolytic solution.

The formation of M$^-$· ions by FD have been observed with quinones [61] having unusually high electron affinities. These ions are not typical for NFD MS, and are probably formed by an FI mechanism discussed in Ref. [62].

An important feature of NFD MS for analytical applications is its insensitivity to most nonacidic compounds. Hence, a high selectivity exists for the detection of acids and salts from mixtures with nonacidic components as has been demonstrated [60,62,63].

Acknowledgement

The author is grateful to the Deutsche Forschungsgemeinschaft, the Ministerium für Wissenschaft und Forschung des Landes Nordrhein-Westfalen, and the Fonds der Deutschen Chemischen Industrie for their continuous financial support.

References

1   H.D.Beckey, Int.J.Mass Spectrom.Ion Phys. $\underline{2}$, 500 (1969).
2   H.D.Beckey, Principles of Field Ionization and Field Desorption Mass Spectrometry, Oxford, Pergamon Press 1977.
3   H.D.Beckey and H.-R.Schulten, Angew.Chem.Int.Ed. 14, 403 (1977).
4   H.R.Schulten, Int.J.Mass Spectrom.Ion Phys. $\underline{32}$, 97 (1979).
5   G.W.Wood, Tetrahedron $\underline{38}$, 1125 (1982); Mass Spectrom.Rev. $\underline{1}$, 63 (1982).
6   R.Gomer, Field Emission and Field Ionization, Havard University Press, Cambridge 1961.
7   F.W.Röllgen and H.D.Beckey, Surface Sci. $\underline{23}$, 69 (1970); Ber.Bunsenges. Physik.Chem. $\underline{75}$, 988 (1971); Int.J.Mass Spectrom.Ion Phys. $\underline{12}$, 465 (1973); Z.Naturforsch. $\underline{29a}$, 230 (1974); H.J.Heinen, F.W.Röllgen and H.D. Beckey, Z.Naturforsch. $\underline{29a}$, 773 (1974).
8   D.F.Barofsky and E. Barofsky, Int.J.Mass Spectrom.Ion Phys. $\underline{14}$, 3 (1974).
9   F.W.Röllgen and H.-R.Schulten, Org. Mass Spectrom. $\underline{10}$, 660 (1975); Z.Naturforsch. $\underline{30a}$, 1685 (1975).
10 U.Giessmann and F.W.Röllgen, Org. Mass Spectrom. $\underline{11}$, 1094 (1976).
11 F.W.Röllgen, U.Giessmann, H.J.Heinen and S.J.Reddy, Int.J.Mass Spectrom.Ion Phys. $\underline{24}$, 235 (1977).
12 M.Anbar and G.A.St.John, Anal.Chem. $\underline{48}$, 198 (1976).
13 F.W.Röllgen, U.Giessmann and H.J.Heinen, Z.Naturforsch.31a, 1729 (1976)
14 F.W.Röllgen, H.J.Heinen and U.Giessmann, Naturwiss. $\underline{64}$, 222 (1977).
15 H.J.Heinen, U.Giessmann and F.W.Röllgen, Org.Mass Spectrom. $\underline{12}$, 710 (1977).
16 H.D.Beckey, Z.Instrumkde., $\underline{71}$, 51 (1963).
17 H.-R.Schulten and H.D.Beckey, $\underline{6}$, 885 (1972).
18 M.Rabrenovic, T.Ast and V.Kramer, Int.J.Mass Spectrom.Ion Phys. $\underline{37}$, 297 (1981).
19 T.Matsuo, H.Matsuda and I.Katakuse, Anal.Chem. $\underline{51}$, 69 (1979).
20 H.D.Beckey, J.Phys. E $\underline{12}$, 72 (1979).
21 S.Goenechea, K.J.Goebel and H.J.Heinen, Z.Anal.Chem. $\underline{290}$, 110 (1978).
22 W.Frick, E.Barofsky, G.D.Daves, D.F.Barofsky, D.Chang and K.Folkers, J.Am.Chem.Soc. $\underline{100}$, 6221 (1978).
23 H.-R.Schulten and F.W.Röllgen, Org. Mass Spectrom. $\underline{10}$, 649 (1975).
24 U.Giessmann, H.J.Heinen and F.W.Röllgen, Org. Mass Spectrom. $\underline{14}$, 177 (1979).
25 F.W.Röllgen, U.Giessmann and H.-R.Schulten, Adv.Mass Spectrom. $\underline{7}$, 1419 (1978).
26 H.U.Winkler and H.D.Beckey, Org.Mass Spectrom. $\underline{7}$, 1007 (1973).
27 H.-R.Schulten, W.D.Lehmann and D.Haaks, Org.Mass Spectrom. $\underline{13}$, 361 (1978).
28 H.-R.Schulten, P.B.Monkhouse and R.Müller, Anal.Chem. $\underline{54}$, 654 (1982).
29 R.Stoll and F.W.Röllgen; unpublished experiments.
30 U.Giessmann, R.Stoll and F.W.Röllgen, Adv.Mass Spectrom. $\underline{8}$, 1047 (1980).
31 see H.J.Heinen in Dynamic Mass Spectrometry, Vol.5, Eds. D.Price and J.F.J.Todd, Heyden London 1977.
32 R.W.Odom and S.E.Buttrill, 28th ASMS meeting, New York 1980.
33 F.W.Röllgen, unpublished experiments.
34 Experiments done in cooperation with Prof.F.Hillenkamp, Frankfurt.
35 F.W.Röllgen and K.H.Ott, Z.Naturforsch. $\underline{33a}$, 736 (1978).
36 G.W.Wood and W.F.Sun, Biomed.Mass Spectrom. $\underline{9}$, 72 (1982).
37 U.Giessmann and F.W.Röllgen, Org.Mass Spectrom. $\underline{11}$, 1094 (1976).
38 H.J.Veith, Angew.Chem.Int.Ed. $\underline{15}$, 696 (1976).
39 H.J.Veith, Tetrahedron $\underline{33}$, 2825 (1977).
40 J.C.Prome and G.Puzo, Org.Mass Spectrom. $\underline{12}$, 28 (1977).
41 J.C.Prome and G.Puzo, Israel J.Chem. $\underline{17}$, 172 (1978).

42 T.Keough and A.J.DeStefano, Anal.Chem. 53, 25 (1981).
43 I.Katakuse, T.Matsuo, H.Matsuda, Y.Shimonishi and Y.Izumi, Mass Spec-
    troskopy (Jpn) 27, 127 (1979).
44 T.L.Youngless, M.M.Bursey, Int.J.Mass Spectrom.Ion Phys., 34, 9 (1980).
45 F.W.Röllgen, Trends in Anal.Chem. 1, 304 (1982).
46 G.W.Wood, J.M.Mcintosh and P.Y.Lau, J.Org.Chem. 40, 636 (1975).
47 M.C.Sammons, M.M.Bursey and C.K.White, Anal.Chem. 47, 1165 (1975).
48 U.Giessmann and F.W.Röllgen, Int.J.Mass Spectrom. Ion Phys. 38, 267
    (1981).
49 S.S.Wong and F.W.Röllgen, 29th Proc.Int.Field Emission Symposium Eds.
    H.Norden and H.Andrien, Göteborg 1982.
50 S.S.Wong, U.Giessmann, M.Karas and F.W.Röllgen, Int.J.Mass Spectrom.
    Ion Phys., in preparation.
51 B.P.Stimpson, D.S.Simons and C.A.Evans, J.Phys.Chem. 82, 660 (1978).
52 B.A.Thomson and J.V.Iribarne, J.Chem.Phys. 71, 4451 (1979).
53 M.L.Vestal, this volume.
54 H.H.Gierlich and F.W.Röllgen, Int.J.Mass Spectrom. Ion Phys. 29, 125
    (1978).
55 P.Dähling, K.H.Ott, F.W.Röllgen, J.J.Zwinselman, R.H.Fokkens and N.M.
    M.Nibbering, Int.J.Mass Spectrom. Ion Phys., in press.
56 M.Anbar and G.A.St.John, J.Am.Chem.Soc. 97, 7196 (1975).
57 K.H.Ott, F.W.Röllgen, G.F.Mes, J.van der Greef and N.M.M.Nibbering,
    Proc. 26th Int.Field Emission Symposium, Berlin 1979.
58 K.H.Ott, F.W.Röllgen, J.J.Zwinselman, R.H.Fokkens and N.M.M.Nibbering
    Org.Mass Spectrom. 15, 419 (1980).
59 J.J.Zwinselman, R.H.Fokkens, N.M.M.Nibbering, K.H.Ott and F.W.Röllgen
    Biomed. Mass Spectrom., 8, 312 (1981).
60 K.H.Ott, F.W.Röllgen, J.J.Zwinselman, R.H.Fokkens and N.M.M.Nibbering,
    Angew.Chem.Int.Ed. 20, 111 (1981).
61 P.Higuchi, E.Kubota, F.Kunihiro and Y.Itagaki, Adv.Mass Spectrom. 8,
    1061 (1980).
62 S.A.Carr, C.E.Costello, C.Orvig, A.Davison and K.Biemann, ASMS meeting
    Minneapolis 1981.
63 G.F.Mes, J.van der Greef, N.M.M.Nibbering, K.H.Ott and F.W.Röllgen,
    Int.J.Mass Spectrom.Ion Phys., 34, 295 (1980).
64 P.Dähling, F.W.Röllgen, J.J.Zwinselman, R.H.Fokkens and N.M.M.Nibbering
    Fresenius Z.Analyt.Chem. 312, 335 (1982).

1.2 Analytical Application of Field Desorption Mass Spectrometry (Review)

Hans-Rolf Schulten

Institute of Physical Chemistry, University of Bonn, Wegelerstraße 12
D-5300 Bonn, Fed. Rep. of Germany

Field desorption (FD) mass spectrometry (MS) is used to investigate almost any major class of organic and inorganic compounds in the mass range m/z 6 to m/z 4000 and above. The focal point is the application to biochemical, medical and environmental problems in which context comprehensive reviews on the analytical capacity of FD-MS have been published [1,2].

So far, the main targets are substances which cannot be evaporized undecomposed prior to ionization, which is generally due to their high polarity or thermal lability. The typical process of ion formation in FD leads to even electron species of type $[M+H]^+$ or/and $[M+cation]^+$. The attachment of singly or multiply charged cations (cationization) is observed for metallic and organic cations and leads to the most intense and commonly observed signals. Thus the formation of $[M]^{+\cdot}$ ions may be occurring in field ionization (FI) but this is not characteristic for FD-MS. This means, however, that the basic (and previously often described) approach to describe field desorption as an electronic process (electron tunneling, etc.) is only of limited value. Therefore, the evaluation of new models is desirable as there is a fundamental necessity to understand and predict the results of soft ionization of solids by FD-MS. It should be stated very clearly that this effort is not just academic but is also a prerequisite for analytical work with as yet unknown substances.

Owing to the complexity of the ionization process in FD-MS, obviously a straightforward explanation is not simple. A number of interdependent parameters are involved such as the chemical character and reactivity of the compound, its purity or more precisely, its impurities (e.g., alkali), composition and structure of the matrix involved, characteristics of the emitter (morphology and surface properties) - and all this underlying a simultaneous influence of thermal energy and the forces of the high electric field. It seems appropriate, therefore, to illustrate the basic facts and commonly observed effects in FD-MS from the viewpoint of an analytical chemist and to contribute from this (nonspeculative) side to a better understanding of the principles of the method.

As an estimate 200 groups worldwide are sporadically or routinely applying the FD method in concert with other analytical tools to solve their specific problems. As an example our group ran about 2500 samples in the last five years for more than 80 different customers. The other half of the instrument capacity has been employed in methodological developments. Of the wide-scattered areas of application the disciplines chemistry, biology, medicine, physics and environmental research set the framework (Fig.1). Since the analytical results of FD-MS have been described previously in detail, emphasis in this short survey is put on the observation of prevalent effects and problems in:

1. the identification of *drugs* and *drug metabolites*,
2. the off-line combination of high pressure liquid chromatography (HPLC) and FD-MS for the identification of *biocides*,
3. the molecular weight determination and structure elucidation of *natural products*, and
4. the isotope determination and ultratrace quantification of *metals*.

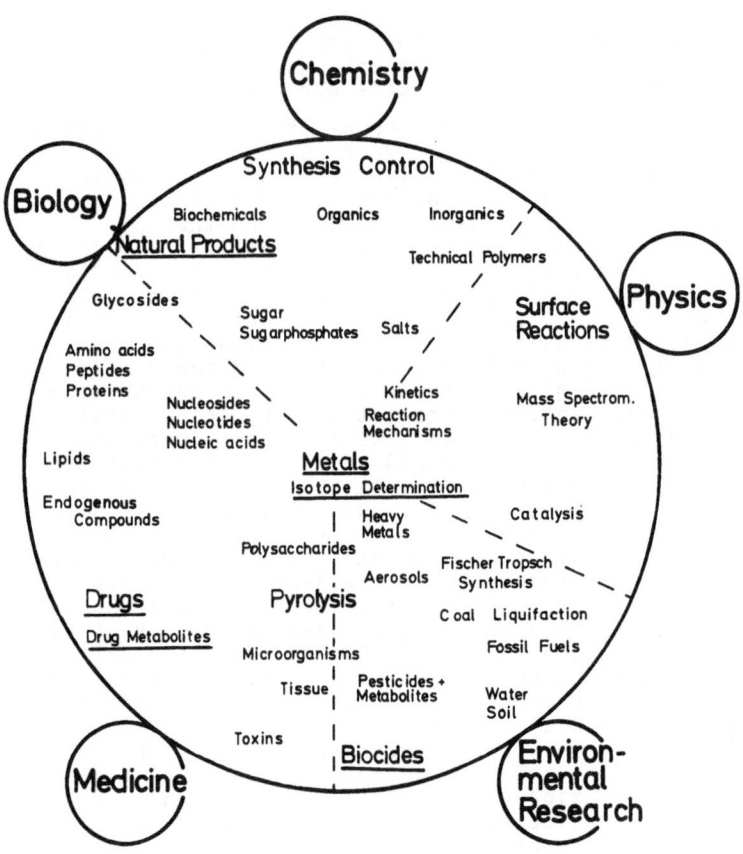

Fig. 1. Main areas of analytical and fundamental research in FD-MS. The topics which are treated in the following are underlined [2]

1. Identification of Drugs and Drug Metabolites

The mass range in which compounds have to be detected and identified by FD-MS has shifted gradually to m/z 1000 and above in
the recent years. This holds in particular for the investigations
of natural products such as oligoglycosides, lipids and peptides
and, beyond that, for technical polymers [2]. However, the majority of studies of drugs and their metabolic or nonmetabolic
products is still performed at lower m/z values ($\sim$ m/z 300).
This is due partly to their high thermal instability and chemical reactivity as well as the serious difficulties in isolating
these compounds in a pure state from their biological embodi-
ment.

In the field of cancer research a new metabolite has been
found in the metabolism of tamoxifen by rat liver microsomes [3]
and the structure of a new oxidized derivative of cyclophosphamide obtained from ozonolysis of O-3-butenyl N,N-bis
(2-chloroethyl)phosphorodiamidate has been established [4].
Further, in continuation of the development of FD-MS as a
quantitative method, cyclophosphamide levels in serum, urine
and spinal fluid have been estimated [5] and pilot assays of
two metabolites in physiological fluids could be achieved [6].

As an example of an FD study of the N-nitrosation product of
the parent drug cimetidine [7], Figure 2 shows the FD mass
spectrum of nitrosocimetidine. Cimetidine is applied in widespread use for disorders of the oesophagus, stomach and duo-
denum and is often used in long-term maintenance therapy.
Concern has been expressed that there may be a link between the
treatment with this drug and the appearance of gastric carcino-
ma. The *in vitro* product of the reaction of cimetidine in 2M
hydrochloric acid in the presence of excess sodium nitrite is
a mono-nitroso derivative, the structure of which has been
elucidated by application of FD-MS and H-NMR. As can be derived from Figure 2, FD gives intense signals for the protonated molecule and its isotopic satellites. Thus the essential
information of the molecular weight was obtained. There is,
however, a distinct signal at m/z 253 which is interpreted on
the basis of accurate mass measurements as the $[M+H]^+$ ion of
the parent drug cimetidine. Without additional information from
other analytical techniques it is not easy to decide whether a
small impurity of cimetidine is present in the originally submitted specimen or is generated in a field-and temperature-
promoted chemical process on the emitter surface. As the amount
of sample available allowed only one FD measurement, the first
detected spectrum is shown in Fig. 2. Due to inorganic salt
impurities in the sample, which generally increase the necessary emitter heating current (e.h.c.) for desorption, these
FD signals were recorded at approximately 15 mA e.h.c. It is
quite possible that at the corresponding sample temperature
elimination of NO occurs and (subsequent) protonation is responsible for the formation of the parent drug (mol.wt.= 252).
Indeed at higher heating currents (18-20 mA) a coincident drop
in the relative abundance of m/z 282 and increase of m/z 253
clearly indicate the occurrence of this process.

In summarizing this effect it should be stated that in sol-
ving authentic analytical problems by FD-MS, the question of an

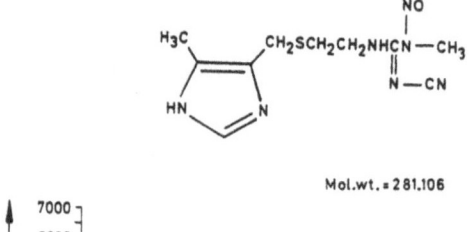

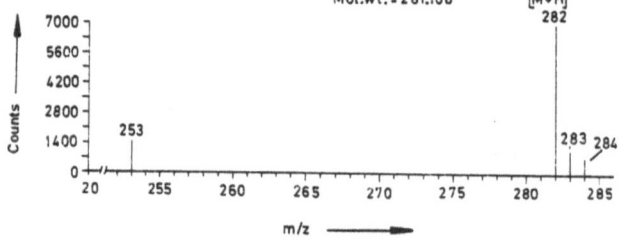

Fig. 2. FD mass spectrum of mono-nitrosocimetidine. Single magnetic scan at 15 mA emitter heating current which is close to
the best anode temperature (BAT) for this compound in a purified extract from a reaction mixture

accompanying impurity can be difficult to answer if this component can also be produced by 'chemistry on the emitter'. In cases
where sample composition and amount allow the recording of the
complete desorption process continuously, the application of
fractionated desorption [1] will clarify the role of a potential
chemical precursor.

If the salt concentration in the sample is even more enhanced,
the so-called salt effect is encountered. It is often observed,
that with increasing polarity of the compounds under FD investigation, for instance in the direction parent drug → drug metabolite, shorter desorption times, strong ion current fluctuations
and sparking between the emitter (+ 8kV) and the counter electrode (- 3kV) occur. Concomitantly the preparative difficulties
for purification, mainly the separation of inorganic and organic
salts from the highly polar sample substance increase considerably. This may be the cause of the short rise in ion emission
seen on the total ion monitor which results each time the e.h.c.
is raised, although no organic, high-molecular weight FD ions
are detected. Only after destruction of the organic substance
at the end of desorption of the sample mixture intense alkali
ions, typically m/z 23 [Na]$^+$, m/z 39 and 41 [K]$^+$ or salt clusters are found.

In Fig.3 a characteristic example of an FD spectrum is given
for a polar biochemical (phospholipid from human serum) which
contains a high amount of salts. Sodium, potassium chloride and
very intense alkali clusters of carboxylates (HCOONa, sodium
formiate ≙ 68 mu) are found which cover the mass range between
m/z 23 and m/z 363 like a network. Since no signal of the intact phospholipid is detected, the FD analysis is, in this
respect, a complete failure.

17

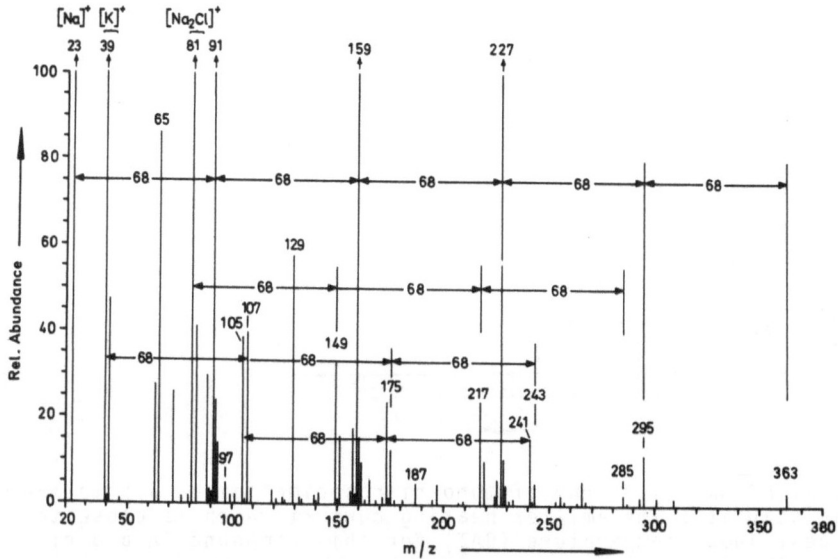

Fig. 3. FD mass spectrum of a prepurified sample which contains a high molecular weight phospholipid. Its detection is hampered by the salt effect and only alkali salt clusters are found

As shown schematically in Fig.4 there are two main reasons for the decomposition or loss of adsorbed organic molecules in FD-MS. First, the so-called salt effect is visualized in the left part of the figure. In general, the signals of inorganic cations are observed with high intensities, the organic molecules, however, encircled in the inorganic matrix decompose before desorption. Second, since the highest temperature in the sample layer is produced close to the emitter surface, the formation of gas bubbles at this site,for instance from adsorbed or chemically produced water (Fig.4, center), may lead to sputtering of the sample. This occurs after the high electric field has been applied and the adsorbed layer has remained stable during a slow increase in e.h.c. up to 15-25 mA. The sudden and eruptive burst of neutral and charged sample material is often accompanied by discharges which may cause loss of the whole sample and, what is more, the complete emitter wire. Even if only occasional sparking is observed this has a detrimental effect on the adjustment of the ion beam, particularly under high mass resolution conditions. Moreover, malfunction of the data system sometimes follows an intense spark. From these illustrations of pitfalls in FD-MS clearly the question arises: What are the possible remedies?

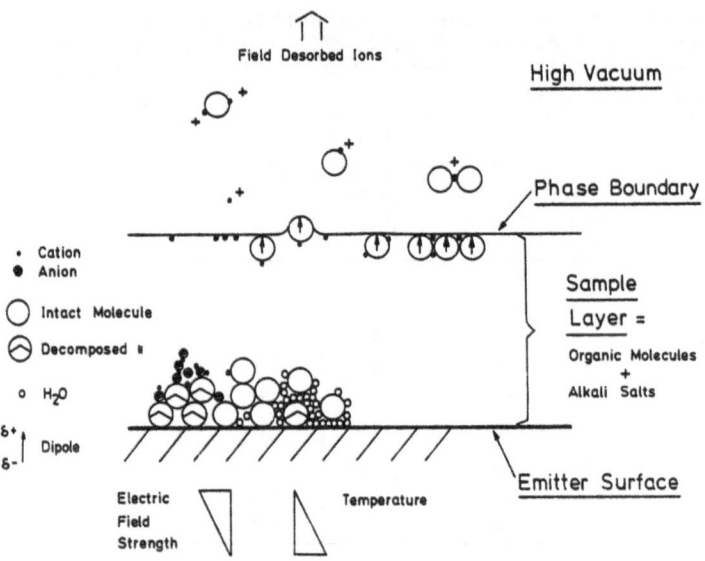

Fig. 4. Schematic representation of the main process in the field desorption of an analytical sample which is a mixture of polar, organic molecules and alkali salts [2]. The two large triangles on the bottom of the figure visualize the field and temperature gradients in the microneedle tip of the FD emitter and the adsorbed sample layer. If the adsorbed molecule is a polar polymer the cationized subunits and their oligomers with n= 1,2,3,4 ... are observed[24]. Formally in the production of these ions transfer of one hydrogen and cationization leads to cyclic structures whereas transfer of two hydrogens plus cation attachment results in singly and multiply charged ions.

2. Off-line Combination of High-Pressure Liquid Chromatography and FD-MS

In our opinion, the best solution to the salt effect is a pre-treatment of the sample by high-pressure liquid chromatography (HPLC) [32]. The off-line combination of these two methods has been proven successfully,e.g.,for studies of drugs, drug meta-bolites [6], natural products [8] and biocides [9].

In Table 1 the results of a longstanding cooperation with an environmental agency in the state of Northrhine-Westfalia are listed which were obtained in this manner. The concentrations of phenylurea herbicides in Rhine river water (salt content be-tween 450 and 800 mg l^{-1}) have been determined between 10 and 332 ppt by HPLC and, simultaneously, the reliable identification of the biocides has been performed by FD-MS. As an illustrative example of this procedure the total chromatogram of an extract from Rhine water from which the volatile as well as ionic com-ponents have been separated is shown in Fig.5a. The FD mass

TABLE 1 . Concentrations of phenylurea herbicides in Rhine
river water between August 1977 and September 1978,
in brackets ( ) the total number of investigated
samples is given [9]

Substance	Number of samples which were found to contain the herbicide		Concentration in Rhine river water	
			min. ng/l	max. ng/l
Buturon	1	(18)	-	76
Chloroxuron	0	(21)	-	-
Chlorotoluron	36	(36)	14	230
Diuron	33	(34)	10	196
Linuron	28	(28)	23	209
Methabenz-thiazuron	2	(10)	47	62
Metobromuron	13	(28)	10	332
Metoxuron	32	(35)	10	210
Monolinuron	12	(28)	10	290

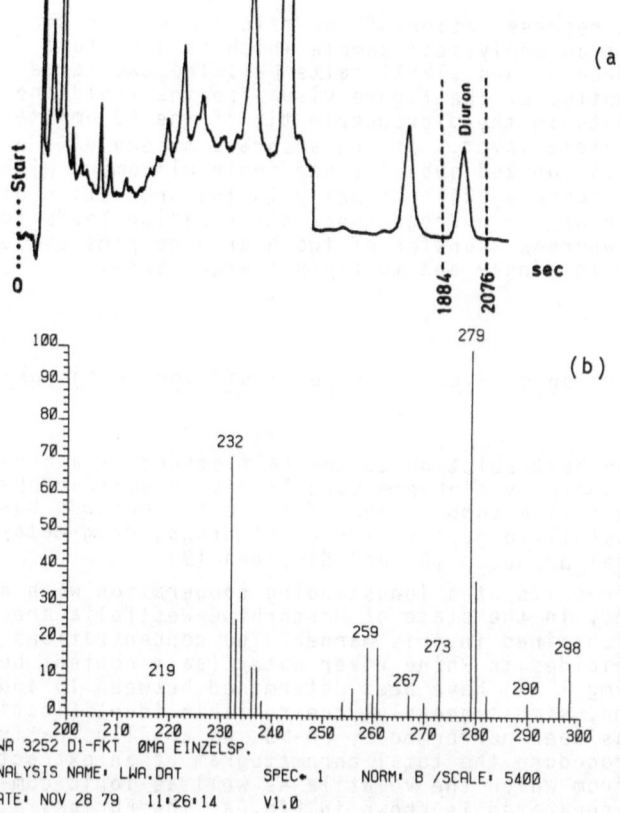

(a)

(b)

LWR 3252 D1-FKT ØMA EINZELSP.
ANALYSIS NAME: LWR.DAT   SPEC: 1   NORM: B /SCALE: 5400
DATE: NOV 28 79   11:26:14   V1.0

Fig. 5. a) HPLC-
chromatogram of a
fraction of an
extract from
Rhine water. b)
Single FD mass
spectrum of the
collected portion
of the HPLC ef-
fluent as indi-
cated in a) [9]

spectrum corresponding to the fraction of the eluate eluting between 1884 and 2076 sec. after injection of the sample (Fig.5b) allows unambiguous identification of the pesticide diuron.

The isolation, identification and determination of cyclo-phosphamide and two of its metabolites in urine of a patient suffering from multiple sclerosis has been reported [6]. The parent drug and the metabolites were extracted from the polar matrix and separated using reverse phase HPLC. For identification and quantification FD-MS was employed and showed a limit of detection which is about a factor of 4×10^3 to 10^5 times lower than that of a common variable UV-detector.

3. Molecular Weight Determination and Structure Elucidation of Natural Products

Primarily, FD-MS has been utilized to determine the molecular weights of natural products. Following the technical improvements of the methodology such as better sample handling and ionization efficiency, higher mass marking and stronger magnets, most of recent investigations have been aimed at the mass range above m/z 1000. To illustrate the scope of these analytical applications, two examples which represent extremes in sample behavior under FD conditions are given.

Because of the complexity of natural waxes (hundreds of substances are identified already) and of the relatively high molecular size of many of the components, their chemical investigation presents a challenge to the chemist. Whereas generally derivatization and chromatographic separation are used to characterize these mixtures of long-chain compounds, Figure 6 shows the molecular ion group of a component of a natural wax which was obtained by simply scraping the wax from the leaves

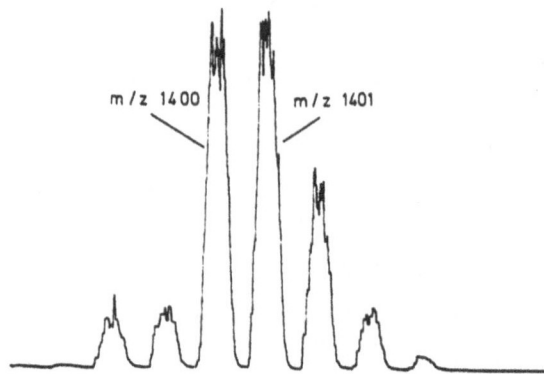

Fig. 6. FD mass spectrum of an untreated palm wax. At 20 mA e.h.c. 25 repetitive, magnetic scans were accumulated on a multichannel analyzer (range 16 K). Direct isotope determination revealed an elemental composition of approximately C_{96}-C_{98}, $H_{192-198}$ O_{2-4}

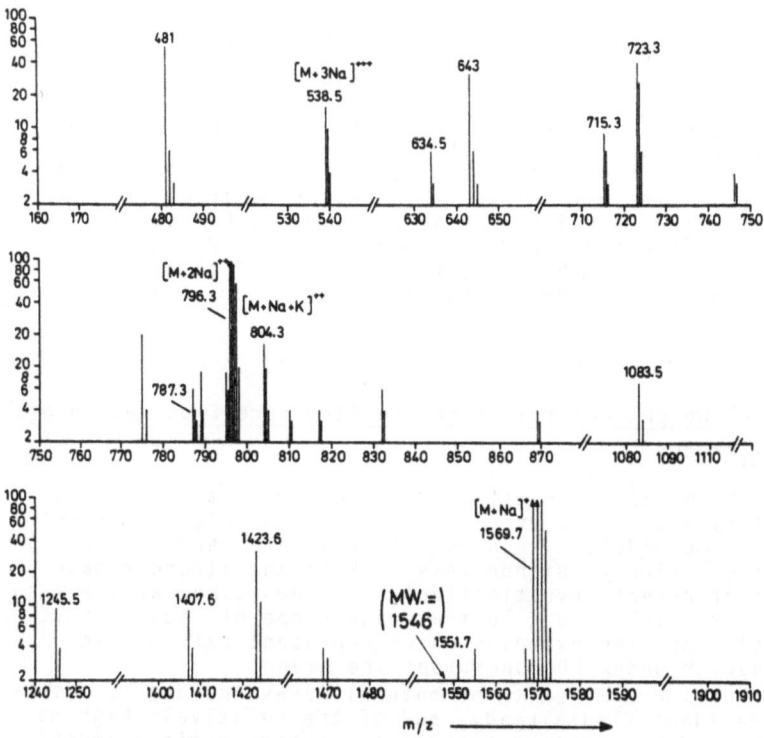

Fig. 7. FD mass spectrum of an oligoglycoside of yet unknown
structure. Recording (Nov. 28th, 1978) on gelatine-free (Iono-
met) photoplate, solvent methanol/water, mass range m/z 45 to
m/z 1950, exposure time 12 min., best observed ion emission
between 20 and 35 mA e.h.c., ordinate ≙ rel. blackening

of a palm tree and applying this sample directly to the emitter
[11]. Characteristic groups of signals were recorded in the mass
range from m/z 400 to m/z 2100. In particular the high-mass signals gave unique analytical data of the natural product which
cannot be obtained by other methods. Since representatives of
long straight-chain alcohols, acids and esters were identified
without any pretreatment in lipids of known structures [11],
one important feature of the desorption process appears to be a
'chromatographic' separation of the mixture components during
the rise in sample temperature. Subtilized one could speak of
the carbon surface and the adsorbed, immobile sample as the corresponding stationary phase. Obviously the intermolecular forces
(in the electric field) allow a gradual mobility of the sample
molecules and the result is a fractionated desorption. Typically
the individual substances desorb in accordance with the number
and polarity of their functional groups.

With regard to the analytical aspects, however, this means
that there is no strict borderline between FI and FD processes

and that essentially the chemical properties of the substances under investigation determine their ionization pathway. For the genuine wax components, ketones showed no and α,ω - alcohols almost exclusively [M+H]$^+$ ions, cationization by metal cations was not observed at all. This makes clear that these samples are settled 'more on the FI side'.

In contrast, the information obtained from an oligoglycosidic saponin*isolated from the flowers of a plant shows the predominance of two mechanisms: *cationization and acidic solvolysis*. All observed ions are explained by attachment of Na$^+$ or K$^+$ cations and the formation of fragments is in close analogy to solution chemistry [12]. To obtain the FD mass spectrum of this natural product (Fig.7), only microgram amounts of specimen were available which allowed only one photographic registration. Opposed to the wax sample, the collection of the saponin containing material is limited in its size and sampling to a certain season of the year. Moreover, its isolation consists of many, highly sophisticated separation and purification steps which may last altogether a year or more. With the intention of demonstrating the capacity and limitations of FD-MS for the analysis of natural products this example of a still unresolved problem is given. The question was, what is the prognosis of this compound without any prior knowledge of other analytical data *based on the FD results alone?*

(1) If the molecular weight does not exceed the recorded mass range, from the high abundance of the [M+Na]$^+$ion at m/z 1569.7 molecular weight of 1546 is derived. This assumption is supported by two facts. First, as commonly observed in the FD mass spectra of oligoglycosides and other polar biochemicals, multiply charged ions of type [M+2Na]$^{++}$ at m/z 796.4 and [M+3Na]$^{+++}$ at m/z 538.6 are found. Second, loss of small neutral molecules from the tentatively assigned, high-weight molecule is observed, in this case for [(M+Na)-H$_2$O]$^+$ at m/z 1551.7 and [(M+2Na)-H$_2$O]$^{++}$ at 787.3.

(2) The structural details can be described as follows: the aglycon has a nominal mass of 458 (m/z 481) and is linked to a glucose unit (m/z 643.3). It follows arabinose (m/z 775) and subsequently a (branched) rhamnose+glucose moiety (m/z 1083). Finally, the FD signals are consistent with three terminal glucose units which are bound to the molecule, e.g. [(M+Na)-162]$^+$ at m/z 1407.7, [(M+Na)-2x162]$^+$ at m/z 1245.6 and [(M+Na)-3x162] at m/z 1083.6. Thus the saponin should be a heptaglycoside containing five hexoses, one deoxyhexose and one pentose.

(3) Probably the specimen represents a mixture of more than two saponins. For instance, the ion at m/z 1423.6 (30% rel. abund.) appears not to be due to the loss of (terminal) rhamnose from the saponin described above, because the loss of 146 mass units is not observed in any other sequential fragment. Note that there is no signal at m/z 1261 for loss of rhamnose plus glucose and that the relative abundance of the doubly charged ion at m/z 723.3 is 40%.

---

* We are grateful to Prof. T.Kawasaki and Prof. T.Komori, Kyushu University, Department of Plant Chemistry, Fukuoka, Japan, for their kind permission to use these unpublished, preliminary data.

The main restrictions of this interpretation are that no structural details about the aglycon are obtained, the assignment of the sugar is only tentative and needs unequivocal confirmation by other analytical techniques as isomers (anomers) cannot be distinguished and finally, the information given on the position of the sugars in the oligoglycosidic chain is just of preliminary diagnostic value, but by no means yields reliable sugar sequencing.

Essentially the knowledge of two mechanisms, namely *cationization* and *acidic solvolysis* allows this prediction about molecular weight and structure. The attachment of metal cations according to the cation affinity of the organic molecule and the formation of stable, even electron ions appears to be a general feature which is observed in many other techniques for soft ionization of organic solids. The qualitative and, to some extent, quantitative analogy between the chemical behavior of these polar substances in solution and under FD conditions can serve as heuristic principle for interpretation of fragment ions [13]. For instance, KOMORI et al. [12] could demonstrate that identical products of oligoglycosidic natural products are formed in FD-MS and acidic aqueous solution. Whereas in a solvent the strength of the acid, duration of attack and reaction temperature play the major role, in FD these parameters can be simulated by adding proton donors to the sample, increasing the emitter temperature and, last not least, by catalytic effects. It appears that cleavage at the linking heteroatom accompanied by proton (hydrogen) transfer generates subunits which are intact chemical species formed in the adsorbed layer and desorbed by cationization. Thus the governing principle in analytical FD-MS is *chemistry*.

It is noteworthy that the thickness of the adsorbed sample layer can be quite small. As shown in Fig.8 coating of a 0.01 molar sugar solution in water onto the FD emitter, transfer into the ion source and short recording of the [M+Na]$^+$ ions, leaves small droplets of viscous sample on the tips of the microneedles (Fig.8a) and in the forking of the needle branches (Fig.8b). Clearly the promoting effect of the electric field strength on the formation and mobility of ions cannot be underestimated.

4. Isotope Determination and Ultratrace Determination of Metals

Together with the recent improvements in qualitative [13,14] and quantitative [15,26] interlaboratory reproducibility and the reliability of direct isotope determination by signal accumulation using a multichannel analyzer [16-18], the identification and quantification of metals by FD-MS is a rapidly developing analytical method [1,19,25,27,28,29].

Originally developed for soft ionization of organic compounds of low volatility, FD unexpectedly proved to be a powerful technique for the analysis of metals. In particular from biological and medical specimens trace (ppm) and ultratrace (ppb-ppt) quantification using stable isotope dilution and multichannel-analyzer recording is feasible without pretreatment of the

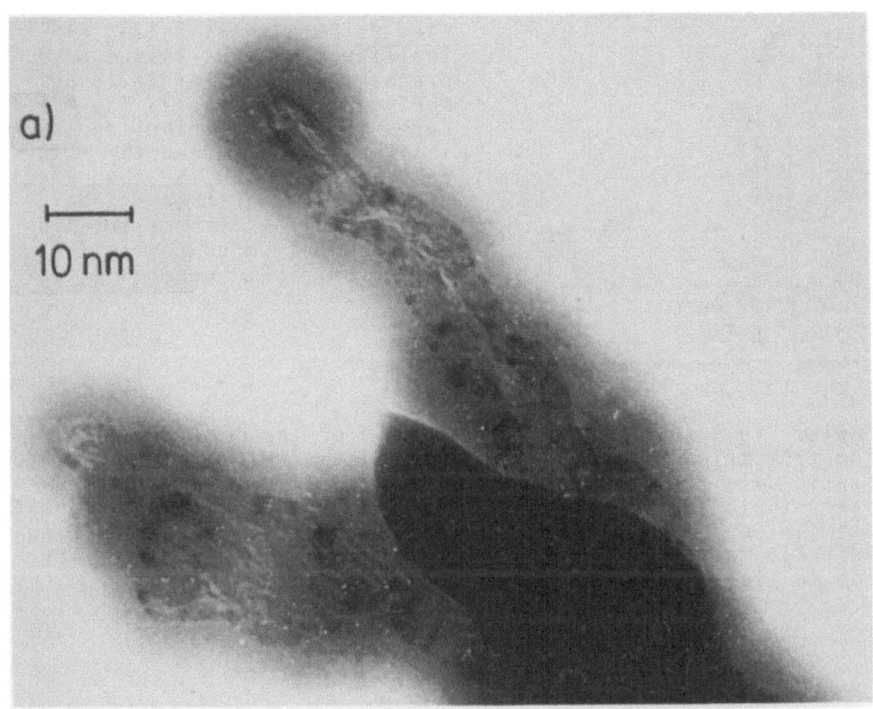

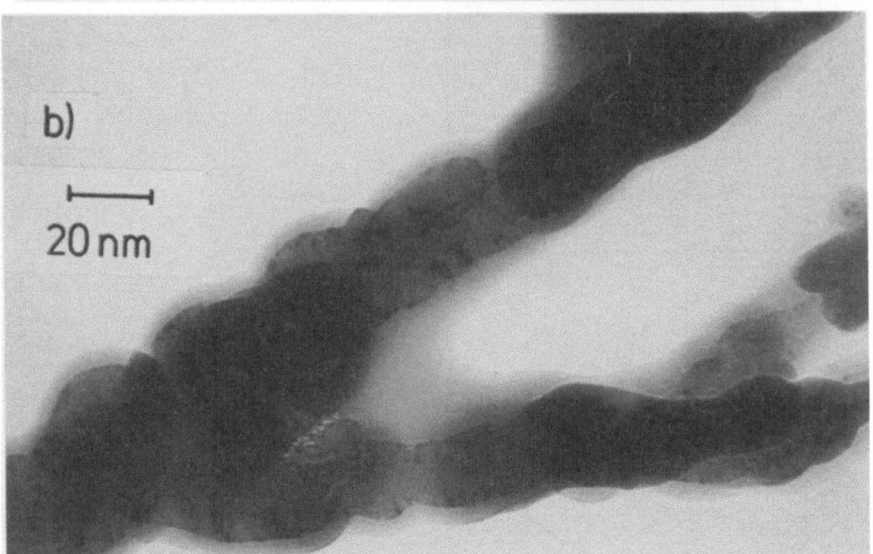

Fig. 8. Transmission-electron-micrograph of an FD emitter after adsorption of an aqueous solution of glucose (0.01 molar). The adsorbed sample forms droplets at the tips of the microneedles (a) and thin layers along the surface of the branched structures (b) (beamvoltage: 100 kV)

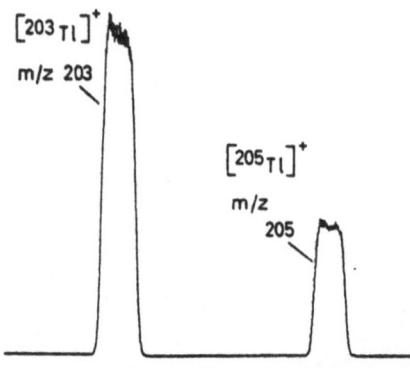

Fig. 9. FD quantification of thallium in uterus tissue of a pregnant mouse after acute poisoning with 8 mg kg⁻¹. The measured isotope distribution is intermediate betweeen the natural abundance and the abundances of the ²⁰³Tl-enriched standard and thus allows the calculation of the thallium concentration in the original specimen [14]

sample. Fig.9 shows such an estimation of the heavy metal thallium from animal tissue. It has been demonstrated that pharmacokinetic data from test animals (Fig.10) can be obtained for the highly toxic Tl⁺ ion [20] and the placental transfer in pregnant mice can be monitored [21]. The first examples for determining the trace metal pattern in healthy humans and patients suffering from multiple sclerosis have been established [22].

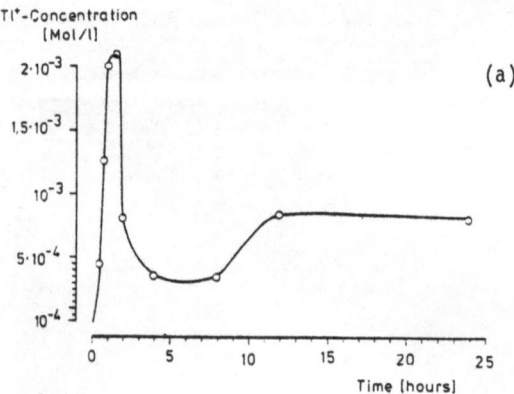

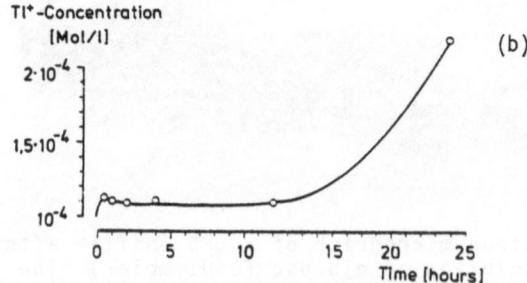

Fig. 10. Time course of the thallium concentration in mice after acute poisoning a) in kidney and b) in brain [20]

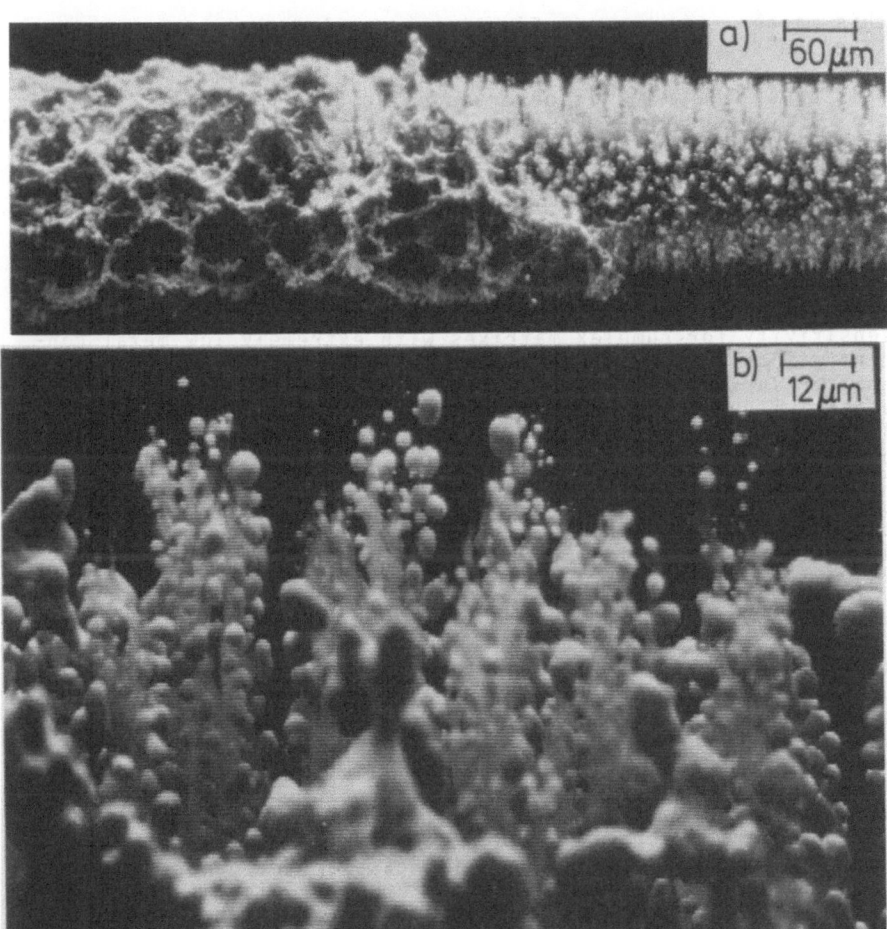

Fig. 11. Scanning-electron-micrographs of an FD emitter on which gold was deposited by electrolysis and desorbed, subsequently, by laser-assisted FD-MS. a) The left part shows the gold layer after electrolysis, to the right the target area of the laser beam is seen. b) At larger magnification the formation of metal droplets is observed after recording of an intense Au$^+$ ion current

With respect to the mechanism of FD-MS, however, the investigation of adsorbed powders of metals and alloys by laser-assisted desorption [23] is interesting. Even more so as the FD emitter can serve as electrode (cathode) in an untreated physiological fluid or homogenized tissue material and enables the electrolytical precipitation of metals on its surface [22]. Practically all nonradioactive metals(> 60) could be ionized in this manner. As shown in Fig.11 the metal deposit forms a thick layer on top of the carbonaceous microneedles (see left part of Fig.11a) from relatively concentrated solutions ($\sim 10^{-4}$ molar). Melting of the metal is observed coincident with high intensities of ion currents of singly charged metal cations [23]. The target area of the laser beam (Fig.11a, right part and Fig.11b) shows gold droplets along the emitter needles when desorption is stopped just after onset of intense [Au]$^+$ ions. For organic compounds mixed melting points and cationization make observation of a defined temperature during FD-MS difficult. In contrast, for the high purity metals it seems correct to state that melting takes place before or together with ionization. The most dramatic effect, however, is the enhancement of FD sensitivity for complex matrixes with extremely low metal concentrations ($\leq$ ppt) since an enrichment of the metal on the emitter surface is achieved by electrolytical depositing. The sensitivity for the quantification of a metal at ultratrace level from a physiological fluid has been reported to improve by one to two orders of magnitude [22]. On the other hand gold coating of the FD emitter surface can be utilized to reduce the catalytic effects of the carbon surface for sensitive organic compounds and to produce FD spectra with strongly reduced fragmentation [30].

Summary and Evaluation

An organic chemist might naively say that FD-MS appears to be more a *chemical* ionization method than chemical ionization (CI) mainly because reactions occur in the condensed (molten) sample layer rather than in the gas phase. Obviously, cationization is a general phenomenon of charge stabilization and cannot therefore be described as *the* typical feature of FD, although the conditions for cation attachment with this technique are favorable. The observation of acidic solvolysis in heteroatom-linked substances is again just a special case of the general, governing principle of *"chemistry on the emitter"*. In this respect it seems that by controlling chemical reactions in the adsorbed layer, the appearance of FD spectra can be varied and a better understanding of the field desorption mechanisms in analogy to common chemistry can be achieved.

From the viewpoint of an analytical chemist this all boils down to the fact that the unique feature of FD-MS is its *versatility*, as the method allows the detection, identification and determination of thermally labile organic compounds and of inorganic materials such as metals.

References

1. H.-R. Schulten: Int.J.Mass Spectrom.Ion Phys. 32, 97-283 (1979) and references cited.
2. H.-R. Schulten: in Soft Ionization of Biological Substrates H.R. Morris (Ed.) Heyden, London, 1981 and references cited.
3. A.B. Foster, L.J. Griggs, M. Jarman, J.M.S. van Maanen, H.-R. Schulten: Biochem. Pharmac. 29, 1977 (1980).

4. J. van der Steen, J.G. Westra, C. Benckhuysen, H.-R. Schulten: J. Am. Chem. Soc. 102, 5691 (1980).
5. O.R. Hommes, F. Aerts, U. Bahr, H.-R. Schulten: J. Neurol. Sci., in press.
6. U. Bahr, H.-R. Schulten: Biomed. Mass Spectrom., 8, 553 (1981).
7. A.B. Foster, M. Jarman, D. Manson, H.-R. Schulten: Cancer Lett. 9, 47 (1980).
8. H.-R. Schulten, F. Soldati: J. Chromatogr. 212, 37 (1981).
9. I. Stöber, H.-R. Schulten: Sci. Total Environ. 16, 249 (1980).
10. U. Bahr, H.-R. Schulten, O.R. Hommes, C. Aerts: Clin. Chim. Acta 103, 183 (1980).
11. K.E. Murray, H.-R. Schulten: Chem. Phys. Lipids, 29, 11 (1981).
12. T. Komori, I. Maetani, N. Okamura, T. Kawasaki, T. Nohara, H.-R. Schulten: Liebigs Ann., 1981, 683-695.
13. T. Komori, M. Kawamura, K. Miyahara, T. Kawasaki, O. Tanaka S. Yahara, H.-R. Schulten: Z. Naturforsch. 34c, (1979) 1094 and references cited.
14. H.-R. Schulten, W.D. Lehmann: Trends Biochem. Sci. 5, 142 (1980).
15. H.-R. Schulten, W.D. Lehmann: Biomed. Mass Spectrom. 7, 468 (1980).
16. H.-R. Schulten, H.M. Schiebel: Naturwissenschaften 67, 256 (1980).
17. L.J. Altman, R.E. O'Brien, S.K. Gupta, H.-R. Schulten: Carbohydr. Res. 87, 189 (1980).
18. H.-R. Schulten, R. Müller, R.E. O'Brien, N. Tzodikov: Z. Anal. Chem. 302, 387 (1980).
19. U. Bahr, H.-R. Schulten: In Topics in Current Chemistry, Analytical Problems, F.L. Boschke (Ed.) Springer Verlag, Heidelberg, Vol. 95 pp.1-48 (1981).
20. C. Achenbach, O. Hauswirth, C. Heindrichs, R. Ziskoven, F. Köhler, U. Bahr, A. Heindrichs, H.-R. Schulten: J. Toxicol. Environ. Health, 6, 519 (1980).
21. R. Ziskoven, C. Achenbach, U. Bahr, H.-R. Schulten: Z. Naturforschung 35c, 902 (1980).
22. H.-R. Schulten, B. Bohl, U. Bahr, R. Müller, R. Palavinskas: Int. J. Mass Spectrom. Ion Phys. 38, 281 (1981).
23. H.-R. Schulten, R. Müller, D. Haaks: Z. Anal. Chem. 304, 15 (1980).
24. H.-R. Schulten, H.-J. Düssel: J. Anal. Appl. Pyrol. 2, 293 (1980).
25. H.-R. Schulten: GIT Fachz. Labor. 24, 916 (1980).
26. U. Bahr, H.-R. Schulten, C. Achenbach, R. Ziskoven: Z. Anal. Chem. 312, 307 (1982).
27. H.-R. Schulten: GIT Fachz. Labor. 26, 533 (1982).
28. H.-R. Schulten, P.B. Monkhouse, R. Müller: Anal. Chem. 54, 654 (1982).
29. H.-R. Schulten: Z. Anal. Chem. 311, 336 (1982).
30. H.-R. Schulten, U. Bahr, R. Palavinskas: J. Trace Microprobe Techn. 1, 115 (1982).
31. H.-R. Schulten: Fuel 61, 670 (1982).
32. H.-R. Schulten: J. Chromatogr. 251, 105 (1982) and references cited.

Part 2

^{252}Cf-Plasma Desorption

2.1 High Energy Heavy-Ion Induced Desorption (Review)

Ronald D. Macfarlane

Department of Chemistry, Texas A&M University,
College Station, TX 77843, USA

1. Introduction

Since the first studies on high-energy heavy-ion induced desorption were reported in 1974 [1], there has been not only considerable progress in the
understanding of the mechanisms and utilization of the phenomenon of mass
spectrometry but also a diversification of the general principles that has
spawned "molecular" - SIMS [2], laser desorption [3] and fast atom bombardment [4]. What was for a few years a unique capability for ^{252}Cf-plasma
desorption (^{252}Cf-PD) in the analysis of complex biomolecules is now being
shared amongst these other techniques. There remain, however, some curious
fundamental questions relating to mechanism and an extra incentive that in
studying the dramatic action of a hundred megavolt heavy ion coursing through
a matrix of biomolecules, one is observing a part of nature that is on the
fringe of the fabric of accessible knowledge, that cannot be completely
simulated by a million or a billion keV particles. While the source of these
high-energy heavy ions was initially the fission fragments from ^{254}Cf decay
(^{252}Cf-PD), this has now been expanded to include ion beams from nuclear
accelerators that can give ions in the same mass and energy range as fission
fragments [heavy-ion induced desorption (HIID)]. The added advantage is that
the parameters relating to the incident ion can be controlled. Most of the
fundamentals of the primary process have been obtained from these studies
[5], [6].

2. Main Features of the Mass Spectra of Species Desorbed by Fast Heavy Ions

One feature of organic molecules is that they are unstable at high excitation,
fragmenting into a wide spectrum of smaller molecular species. The higher
the excitation, the greater is the number of fragmentation pathways. The
^{252}Cf-PD (HIID) spectra generally comprise a mix of ions that includes the
complete molecule (M) and a set of fragment ions. Ions that include the essentially intact molecule are of the type $M\pm\cdot$, $(M \pm H)^{\pm}$, $(M + Na)^{+}$, $(M + Cl)^{-}$
[7]. Which of these or which combination of these are present is related to
the chemical structure of the particular molecule, the availability of the
reagent species in the matrix and the details of the ionization process for
that particular matrix. The fragment ions seem to comprise both high and
low-energy processes. In the case of low-energy processes, the fragmentation
resembles that of chemical ionization, singling out those fragmentation channels that produce small, stable molecules (such as HCOOH), even-electron
species, and in the case of high molecular weight biopolymers such as synthetic oligonucleotides, specific cleavages at the weakest links of the molecule and in such a manner as to form the most stable ion (tertiary carbonium
ions, resonance stabilized species) [8]. High-energy processes lead to the

formation of small ions such as CN⁻, ions typically observed in high-tempera-
ture pyrolysis processes. The molecular ion/total ion intensity ratio is
very much dependent on the nature of the molecule, but is generally only a
few percent. In some cyclic molecules such as α-cyclodextrin that requires
two bonds to be cleaved to form a fragment ion, this ratio is several tens
of percent. A detailed study of this partitioning has been carried out by
FIELD and CHAIT for some selected small molecules [9].

The positive and negative ion spectra are generally complementary in terms
of obtaining structurally related information on a molecule. From relation-
ships between the masses of the ions that include the complete molecule, the
molecular weight (M) of the molecule can be deduced. The relative intensities of the positive and negative fragment ions reflect the chemical stabi-
lities of the ions that can be formed. The HIID studies have shown that the
mass spectra are essentially independent of the mass, energy, or charge of
the incident ion [10]. Only the intensity of the spectrum is affected.

The neutral component of the desorbed species is at present an unknown
quantity for biomolecular samples. Some estimates can be deduced from studies of inorganic species (5 MeV $^{19}F^{+2}$ on UF_4) which indicate that 20 percent
of the desorbed species are ions [11]. The neutral component for biomolecules is an obvious parameter to be measured; several groups are considering
how this kind of measurement could be made and it is an important quantity
relating to the ultimate sensitivity for detection.

The role of metastable species has been a relatively recent revelation
and is due to the work of CHAIT and FIELD [12]. One of the features of
^{252}Cf-PDMS spectra that has long been a puzzle has been why the mass peaks
in the time-of-flight (TOF) spectra are so often broader than the instrumental resolution, and why the transmission of atomic ions is so much more ef-

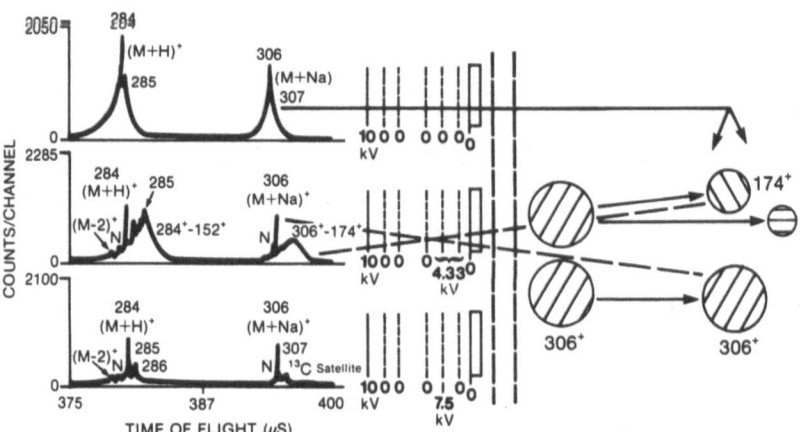

Fig. 1 Metastable analysis of guanosine. The top figure includes metastables, neutrals, and stable molecular ion. The middle figure is the spectrum
with a small retarding potential. The broad distributions on the high time
side of the peaks are the separated metastable ion component. The lower
spectrum is for a stronger retarding potential which clearly separates the
metastable from the stable ions. The time shift of the metastable peaks can
be used to calculate the m/z of the metastable ion [12].

ficient than the molecular ions of interest. These puzzles are now known to be related to the fact that a significant fraction of desorbed molecular ions are metastable. Although metastable fragmentation in flight diminishes mass resolution, this "problem" can be circumvented by the use of electrostatic filters. Conversely, the abundant metastable ion contribution, coupled with the retention of time-dependent information inherent in TOF offers the possibility of utilizing metastables effectively in studying the dynamics of desorption-ionization-fragmentation. Some results of these metastable studies are shown in Fig. 1.

The high mass range covered thus far by ^{252}Cf-PD (HIID) is the result of a combination of two factors: the surface area excited by an incident ion and the use of the unlimited mass range capability of the TOF method. Mass spectra have been reported up to m/z ∿ 6980 [13] for a monomer species and for a dimer, m/z 12,500 [14]. Molecular ions of protected oligonucleotides with a long dimension approaching 10 nm have been detected by ^{252}Cf-PDMS [13]. What the details of the dynamics of desorption process are for large molecules is not known; how many surface bonds must be broken, mechanism of the ionization process, the relationship between molecular size and the surface area excited are quantities that ultimately determine what the limit in the mass (or size) of a desorbed species will be. Figure 2 is a ^{252}Cf-PDMS spectrum of a protected oligonucleotide and its dimer in the high mass region.

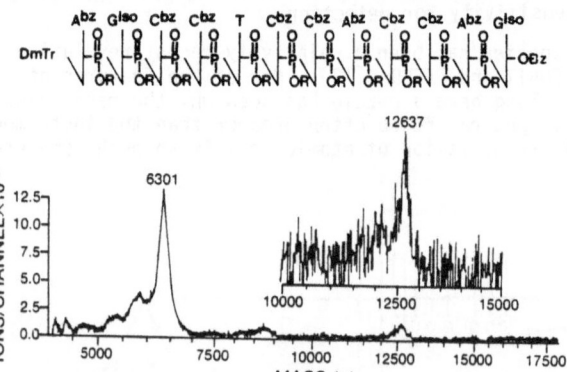

Fig. 2. ^{252}Cf-PDMS position ion spectrum of a chemically protected oligonucleotide and its dimer. An abbreviated representation of the structure is also given [14].

3. Energy and Angular Distribution of the Generated Ions

The energy and angular distribution of the desorbed ions reveal details of the desorption-ionization process. First measurements were made by FURSTENAU et al. were for atomic ions (H$^+$, Li$^+$, Cs$^+$) [15]. The spectra are statistical in shape and except for the H$^\pm$ ion peak at ∿ 2eV. (The H$^\pm$ ion spectrum peaks at ∿ 5eV.) The shapes can be represented analytically as an energy-shifted MAXWELL-BOLTZMANN distribution. The only published spectra for molecular ions (valine and a fragment ion) give a slightly less energetic distribution peaking at ∿ 1 eV [16]. The observation that very few ions are desorbed with very low energies (<0.2ev) is a clue to the desorption-ionization process since this is a characteristic feature of a surface-related phenomenon and

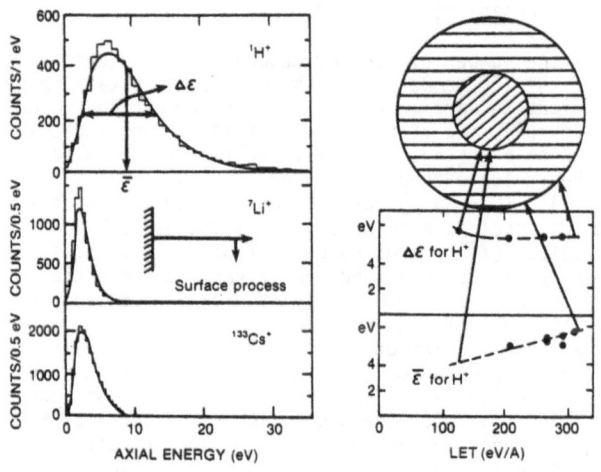

Kinetic energy spectrum independent of incident energy

Fig. 3. Kinetic energy distribution of desorbed atomic ions in ^{252}Cf-PDMS. The distribution for H$^+$ is more energetic than other ions [15]. The right-hand figure is from a study of the variation of average energy ($\bar{\varepsilon}$) and distribution width ($\Delta\varepsilon$) with incident ion energy. The weak dependence suggests that increasing the energy density does not produce a hotter excitation but rather a larger region of excitation [6].

an indication of the presence of an energy of activation for desorption. A few ions like C$_2$H$^-$ have extremely low desorption energies; this effect appears as a very narrow line in the TOF spectrum [17]. It is possible that these are pyrolysis products of hydrocarbon surface contamination that are easily desorbed in the cooler periphery of the heavy ion track. Recently, a very nice method for obtaining more precise kinetic energy spectra by TOF using ^{252}Cf-PD has been reported by BECKER [18]. There is evidence from these first studies that the kinetic energy spectrum of an ion is influenced by the matrix. For example, H$^+$ from a LiI layer has a slightly higher average energy than from a LiF layer of the same thickness. The complications of surface charging have also been addressed in these studies. The angular distribution of desorbed ions is perpendicular to the plane of the surface varying as $\sim \cos^{2.5} \theta$ for small atomic ions [15]. This also indicates that the desorption of these species is under the influence of surface forces and that the surface structure is still intact at the time of desorption. There is yet no information on the kinetic energy and angular distribution of desorbed species (m/z > 300). To measure these by TOF involves a complicated deconvolution of the isotopic satellite peaks and this has not yet been attempted. It would be informative to have this kind of information for the very large biomolecules in order to determine whether decreased yields are a consequence of insufficient energy available for desorption and whether the surface potential is having any influence on the desorption process. Some measured kinetic energy spectra are shown in Fig. 3.

35

4. Yields and Transformation Possibilities

The yield is defined here as the number of species desorbed per incident ion. Because of the simplicity of the TOF geometry, it is possible to obtain absolute yields in a rather straight-forward manner. For an isotropic source of heavy ions (as for a ^{252}Cf source), the average total yield of ions including fragment ions varies from $\sim$ 3 to 6 ' [15]. The distribution of multiplicities for a single incident heavy ion extends to beyond 50 [19]. The yield of molecular ions [M^+, $(M + H)^+$, $(M + Na)^+$] varies from $\sim$ 1 for simple biomolecules like the amino acids, nucleic acid bases, to as low as 10^{-4} for the more complex large molecules [7]. Directed beams from nuclear accelerators have been used to study yields as a function of various properties of the incident ion. This work has been done mainly at Uppsala [5] and Erlangen [6]. Some of the findings relating to molecular ion yield are given here. The yield of molecular ions (M^+ ion of ergosterol) can be correlated with the electronic energy loss $(dE/dx)_e^2$ of the incident ion [5] which, in the electronic stopping energy region (> 1 MeVu^{-1}) is dependent on mass, energy, and atomic charge state. For example, the desorption yield for a 54 MeV ^{127}I ion is 100 times higher than for an ^{16}O ion with the same velocity. For a 20 MeV ^{16}O beam, the yield of molecular ions $(M+H)^+$ of glycyl glycine is eight times higher for a +8 ion than for a +2 incident ion. The implication of the quadratic dependence of $(dE/dx)_e$ is that the desorption process is not associated with a single ionizing event but rather involves the effect of overlapping or interacting ionizations. The angle of incidence also influences the ion yield in a manner reflecting the amount of energy deposited in the surface layers: the yield of ergosterol molecular ion (M^+) is $\sim$ a factor of 10 higher at grazing angles ($\sim$80%) than when the beam is nearly perpendicular to the surface ($\sim$10°). These results are summarized in Figs. 4, 5 and 6.

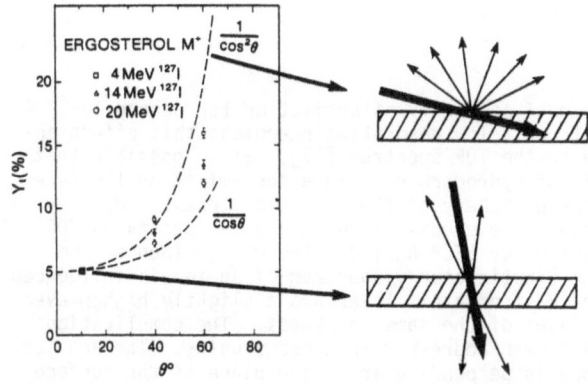

Electronic energy deposited in surface region.

Fig. 4. Effect of angle of incidence on molecular ion yield [5]

In summary, the HIID studies reveal the specific contribution of the incident ion: the greater the energy deposited into the surface region, the greater is the yield of desorbed molecular ions and because the relative yields of fragment and molecules yields are independent of the incident ion parameters, one can only conclude that the incident ion is determining how large a surface area is being excited by each ion.

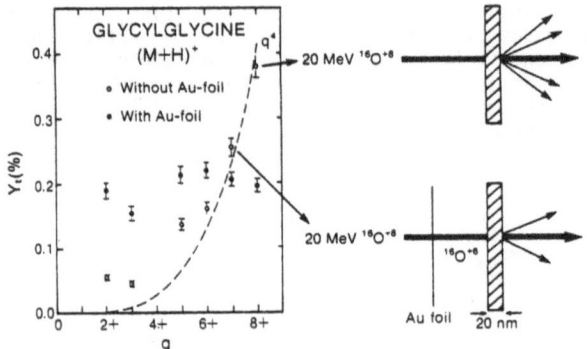

Fig. 5. Study of the effect of charge state of incident ion on desorption yield. Passage of 20MeV 6^{16} ion through the foil returns ion to its equilibrium charge state at that energy [5].

Effect of projectile charge state — energy loss at surface.

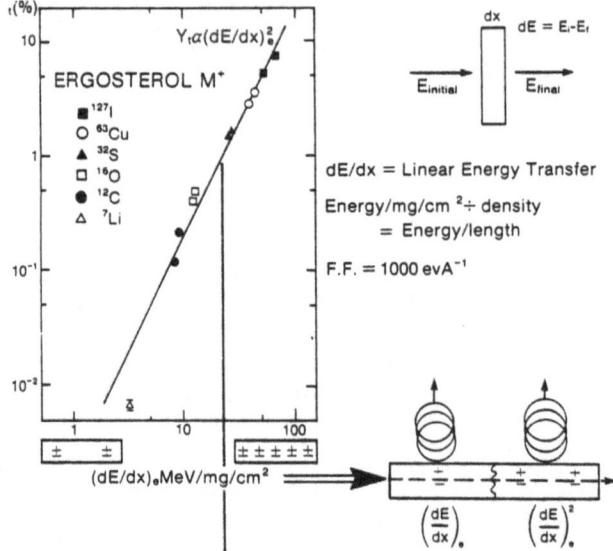

Fig. 6. Study of the yield of ergosterol molecular ion as a function of (dE/dx). The quadratic dependence implies collective excitation of the matrix [5].

One further point regarding yields remains to be discussed. An additional feature of ^{252}Cf-PD (HIID) spectra, as revealed by the use of time as a variable in the measurement, is the existence of correlations in fragment and molecular ion yields [15]. How this is manifested is that if in a particular incident ion interaction, a molecular ion (e.g., $M+Na^+$) is desorbed, there is a very high probability that the correlated fragment ion is also formed in the same event. This phenomenon was observed in the first ^{252}Cf-PD measurements and was studied in some detail by the Darmstadt group. This implies that it is the excitation of a <u>cluster</u> of molecules at the surface that is mainly responsible for the mass spectrum with one member of the cluster forming the molecular ion and another fragmenting in the same excitation event.

Some fragment ions show no yield correlation with the molecular ion. This kind of study may be revealing some important details of the desorption-ionization process at the surface and it is anticipated that more yield correlation measurements will be carried out in the future. There is undoubtedly a connection between this and the metastable studies from Rockefeller where they find $2M \rightarrow M$ occurring as a metastable decomposition [12]. An example of these correlation results is shown in Fig. 7.

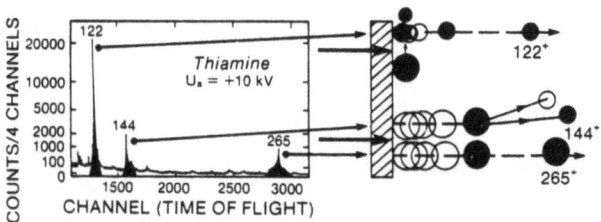

Correlation in Molecular and Fragment Ion Yields

Fig. 7. ^{252}Cf-PDMS spectrum of thiamine. If a particular fission fragment produces a molecular ion (m/z 265), there is a very high probability that the m/z 144 fragment ion is desorbed in the same event. This is not the case for the more intense m/z 122 ion implying that there are at least two sites where fragmentation occurs [15].

5. Influence of the Chemical Environment on Yields

This is a topic which has been explored extensively experimentally but is the least understood. The chemical environment does have an effect on the nature of the molecular ions formed. The presence of alkali metal ions as impurities or as added salts to the matrix is manifested in the formation of species such as $(M+Rb)^+$ or $(M+Cs)^+$ as well as $(M+Na)^+$ where the Na^+ is presented as a trace impurity. From studies of multilayers, the alkali metal ions can diffuse through as many as a hundred monolayers in the heavy-ion track and participate in a surface reaction [20]. Multiple $H^{\pm}$ alkali metal ion exchange can occur at high alkali metal ion concentrations in the matrix. Halogen ion attachment [e.g.,$(M+Cl)^-$] occurs only if the Cl is in the form of an "organic" chlorine (e.g., $CHCl_3$). Inorganic chlorides (LiCl) do not participate in these reactions [21]. No systematic studies have been carried out on the addition of protonating acids or bases to the matrix to "preform" ions in the matrix.

6. Influence of Physical Environment on Yields

It is in this aspect of the problem that significant differences in high and low-energy ion-induced desorption may become apparent. It now appears that the mechanism for energy loss by a high-energy ion via electronic excitation that stimulates the desorption of molecules necessarily involves a "coherent" insulating matrix of molecules at the molecular level, molecules that form an ordered aggregate [22]. What is the minimum size of the aggregate is not known but it may be less than 10 molecules. Recent studies at CalTech on molecular desorption from SiO_2 excited by 20 MeV ^{35}Cl ions show that no desorption occurs when the surface coverage of a carbon film is so low that only isolated SiO_2 molecules exist on the surface [23]. Approximately three atomic layers are required to affect desorption. These studies have been directed toward the general problem of desorption of molecules (neutrals and ions) which is part of another field, the electronic erosion of insulating materials by high-energy ions whose origins go back to the work of FLEISCHER et al. in 1965, and has been actively pursued and developing since then [24]. It is mainly through the efforts of SUNDQUIST that an association was made with these studies and the ^{252}Cf-PD (HIID) work. This was the central theme of the Uppsala conference in 1981 [25]. The role of the "coherent" insulating matrix in the desorption process appears to be intimately connected with the mechanism for conversion of electronic excitation (hole-electron formation) to the intermolecular vibrations that ultimately lead to desorption, involving a collective or delocalized set of electronic states with a valence band, conduction band and a significant band gap that inhibits energy dissipation by electron conduction. More will be said about this in the section on energy and momentum transfer. How this requirement for the presence of coherent insulating molecular aggregates affects ^{252}Cf-PDMS (HIID) is as follows. If the matrix is a conducting medium with a clean surface, the desorption yield is extremely low ($\sim 9 \times 10^{-3}$ for Ni^+ from Ni), a yield that can be essentially accounted for by the nuclear excitation component of the energy loss for a 100 MeV heavy ion [26]. The electronic energy loss although dominating by orders of magnitude at this energy is ineffectual for desorption because the electronic energy is dissipated by electron conduction. If now the Ni is covered by a few monolayers of a molecule that forms an insulating layer (hydrocarbon contamination from vacuum or a deposited layer of biomolecules), the desorption from that layer is orders of magnitude higher. While the electronic erosion studies indicate that the depth for erosion of neutral atoms (solid Xe excited by 1 MeV He^4) [27] by a single incident ion extends to ~ 2000 monolayers, ^{252}Cf-PDMS studies of molecular ion yields ($M+Na^+$) from thin layers of bleomycin indicate that saturation of the yield occurs with only a few monolayers (less than 10) of material [2]. This indicates that the ionization part of the desorption-ionization only involves the surface molecules and that molecules desorbed from lower layers are all neutrals. Returning to the suggestion that ordered aggregates of molecular species having an appreciable band gap must be present in the matrix to convert electronic excitation into molecular translational motion, we can perhaps now understand some of the observations (published and unpublished) that have been perplexing for the ^{252}Cf-PD (HIID) work. Small polar molecules such as the amino acids readily form ordered aggregates, and there is no problem in establishing a coherent, insulating layer with these molecules. Even mixtures of amino acids form mixed crystals and it is possible to obtain interpretable ^{252}Cf-PD spectra from these. However, if we mix chemically dissimilar molecules together (cholesterol and tetrodotoxin), this produces an amorphous film which gives an uninterpretable ^{252}Cf-PD spectrum with a low desorption-ionization yield [21]. Chlorophyll a isolated in a matrix of paraffin is another example of an amorphous medium which when excited in a ^{252}Cf-PD experiment produced only high excitation fragment ions [28]. Ultra-violet irradiations can melt coherent ordered molecular aggregates of molecules which then reform as randomized aggregates. In the process, molecular ion yields by ^{252}Cf-PD decrease more than an order of magnitude [29].

As polar molecules become larger, the ease with which molecules form ordered aggregates diminishes. The process can be enhanced by preparing films from concentrated solutions where the formation of molecular aggregates has already taken place to some extent. Films of protected oligonucleotides prepared by electrospraying from relatively concentrated solutions (10^{-3} M) have much higher molecular ion yields than these prepared from 10^{-5} M solutions even though they were the same thickness [21].

Thus, for ^{252}Cf-PD (HIID) where the conversion of electronic excitation to molecular motion is a key process, the nature of the matrix, particularly the presence of a coherent insulating matrix, is of particular importance. This is not the case for keV ion desorption where desorption is directly coupled with the nuclear excitation of the collision cascade.

7. Energy and Momentum Transfer, Time Scale

As detailed in the previous section, the energy transfer from the incident ion in the matrix is via electronic excitation and this must be converted to molecular motion. To effect this conversion, the energy must be deposited in an insulating layer that is coherently ordered. The fact that molecular ion yields depend on $(dE/dx)^2$ also can be connected to coupling energy to a coherent, delocalized structure in contrast to a collection of isolated molecules [23]. If the desorption yield varied as dE/dx, this would imply that desorption is associated with a single ionization event. The fact that it is quadratic in electronic energy loss indicates that the overlap of ionization events is necessary for desorption, that the energy is delocalized prior to desorption [30]. The time scale of events has been estimated by RITCHIE and CLAUSSEN [31]. Formation of electron-hole pairs occurs within 10^{-17} s of the passage of the incident ion. Some electrons return to the positive ion track within 10^{-14} s if not trapped. Conversion to molecular motion is initiated at 10^{-13} s. Most of the secondary electrons (delta rays) are thermalized by 10^{-12} s and return to the positive ion core by 10^{-11} s. Some electrons remain trapped in excited valence states which radiatively decay up to 10^{-9} s. These are, of course, theoretical estimates because most of these times are too short to be amenable to experiment.

The only experimental data relating to the time dependence of the process derive from the time decay tails of ions in ^{252}Cf-PD (HIID) TOF spectra, particularly the alkali metal ions, and the Rockefeller studies on metastable ion formation [12]. Both studies show that events continue to evolve long after the initial impact: ion emission continues for several microseconds and metastable decays cover a similar time span. The angular distribution and kinetic energy data discussed earlier indicate that desorption-ionization occurs from a surface potential well so that the electronic energy ultimately partitions to the surface binding vibrations involving, probably, several quanta of energy to release the adsorbate molecules from the surface.

8. Thermal Effects

The electronic erosion studies using frozen gases show that there is no direct connection between temperatures of the matrix and erosion yield so long as the temperature is well below the melting point. For H_2O it is constant from 10K to 110K [32]. There is evidence from ^{252}Cf-PD studies that diffusion controlled migration of reactant ions to polar sites near the molecules comprising the matrix does occur and this effect enhances the formation of molecular ions of the type $(M+Cs)^+$ [16]. Annealing the sample prior to analysis accelerates this process.

9. Models Relating to ^{252}Cf-PD (HIID)

There are parts of the ^{252}Cf-PD process that are well understood and some that are very difficult to characterize. There is a component of the process, the final stages, that must be very similar to SIMS (FAB) because the mass spectra of the desorbed ions are so similar [22]. Part of the challenge in developing a complete model is the recognition of the work of researchers in other fields of research that has relevance to our studies, incorporating their findings, learning their language so that what ultimately distills from our endeavors is a model that not only "explains" ^{252}Cf-PDMS (HIID) and has predictive value but is also consonant with the rest of the body of science.

The primary interaction of a high-energy fission fragment with a matrix of molecules produces ionization in the medium, the formation of separated positive ions and electrons (excitons). The concentration of excitons is dependent on the velocity and charge of the projectile. It is the diffusion and decay of these excitons that delocalizes the energy deposited into the surrounding medium and determines the range of excitation. For an incident fission fragment, this can extend to as much as 100 nm. This part of the process is reasonably well understood. What happens after this is the subject of several models that have been recently suggested. The question of how the initial electronic excitation is transformed into intermolecular motion has been discussed in terms of a thermal spike model [31], [33], a Coulomb explosion model [11] and two models based on a fast electronic perturbation [34], [35]. This question remains very much open. Once molecular motion has been excited but the matrix is still describable as a condensed medium, energy transport again becomes a fairly well understood process involving phonon and exciton migration. The desorption process for a species leaving the surface that has a well-defined surface potential can be characterized as a multi-phonon excitation into the continuum of the surface bonds. [36]. If, however, desorption is from a chaotic, random matrix of excited molecules resembling a gas or supercritical fluid, then the dynamics of the desorption process are quite different. Angular distribution and kinetic energies give information on this aspect of the problem.

The ionization process is not well understood. How important is surface ionization where quantum-mechanical models developed by LUNDQUIST and NØRSKOV [37] and SROUBEK [38] have relevance? What processes occur on the surface, and in the gas phase? What is the role of the coherent cluster in the desorption-ionization process? Is it only needed for energy dissipation through gas phase dissociation of the cluster [39] in ^{252}Cf-PDMS (HIID) to catalyze electronic relaxation to intermolecular vibration? MICHL has recently proposed a cluster desorption model where the role of the cluster is to provide a mechanism for energy dissipation through gas phase dissociation of the cluster [39]. Models for all these aspects of the problem do seem to be within the realm of possibility. Finding the experimental measurements to give unequivocal tests of their validity is a familiar chore that has been bestowed on the experimentalist.

10. Reproducibility, Quantification

For molecules that are small and give good yields, reproducibility for a set of samples of the same material prepared in an identical manner is about 20% [21]. For larger molecules, like insulin, the same comparison can show fluctuations in molecular ion yield varying more than an order of magnitude [19].

The only significant, detailed study on this problem using ^{252}Cf-PDMS has been carried out by the Marburg group who have developed an HPLC-^{252}Cf-PDMS based procedure for analyzing serum levels of verapamil. Sensitivity, reproducibility, and quantitation have been realized at a level sufficient to give meaningful and useful analyses. These studies show that if a good chemical purification procedure is developed, and the molecule gives a strong ^{252}Cf-PDMS spectrum, it is possible to use ^{252}Cf-PDMS for quantitative analysis [40].

11. Analytical Applications (Present/Future Trends)

The most significant application of ^{252}Cf-PDMS in recent years has been in the molecular weight determination of new biomolecules. These have generally been relatively complex (m/z > 1000), highly polar, thermally unstable, and have not yielded to any other MS method [7]. Since the introduction of FAB and molecular SIMS, most of these molecules have been found to be also amenable to this method as well. As a result, new areas of application are evolving to take advantage of some of the potentially unique aspects of the ^{252}Cf-PD process, particularly the larger area excited by the incident ion. It is possible that ^{252}Cf-PD coupled with TOF will be able to significantly extend the mass range. Coupled with this is a desire that more interest in developing the TOF method will occur (utilizing the information contained in meta stables, achieving higher mass resolution, taking advantage of the ability to study the process at the individual event level). The analytical application will then involve problems with new large molecules: molecular weight determination of unknowns and the use of fragmentations to assist in structure elucidation. If this development is realized, new applications in the analysis of large biomolecules for problems in molecular biology will be possible.

12. Specific Problems in Instrumentation and Measurement

The problems discussed here relate to the TOF method not only because it has been exclusively used in ^{252}Cf-PD (HIID) measurements but also because we feel that more effort should be devoted to developing its full capabilities.

A major problem is the low mass resolution, due not only to the velocity spread of the ions but also to the contribution of metastables. The Rockefeller group has shown to what extent the metastables contribute to line broadening (essentially all molecular ions of biomolecules above m/z 500 have a very significant component) and how resolution can be improved (to ∿ 2000 presently) by electrostatically separating the metastable neutrals and ions from the metastable neutrals and ions from the stable component [12]. DUCK has compiled a useful analysis of the basic configuration of the TOF as it is used in ^{252}Cf-PD (HIID) which shows that with the experimental velocity spread including a resolution of 5000 is achievable by using higher acceleration voltages (> 20 KV) and short accelerator grid-target spacings [41]. The Mamyrin reflectron which provides second-order velocity focussing has not yet been satisfactorily employed [42]. It is now known that a large part of the problem of low transmission is connected with metastable ion formation. Systems with very short flight paths might be required to circumvent the problem of metastables.

The detection efficiency of the low-velocity ions of large molecules is in a confusing state. All of the detection methods employed are based on the ejection of secondary electrons from a surface when impacted by the ion to be detected. A velocity threshold for the process (4.5×10^6 cm s^{-1}) has been measured and discussed by FRIEDMAN [43] yet molecular ions much lower than this velocity (1.2×10^6 cm s^{-1}) have been detected by ^{252}Cf-PDMS using microchannel plates. It is clear that the average electron yield is less than unity at these velocities but it is greater than zero. There are indications that new mechanisms are being manifested in electron emission following massive ion impact, possibly involving photon-induced processes [21]. More work on this problem, including a study of different kinds of surfaces and using the massive molecular ion beams now available will be an important part of the effort to extend the mass range. The Friedman study on secondary electron yields for various incident molecular ions is shown in Fig. 7.

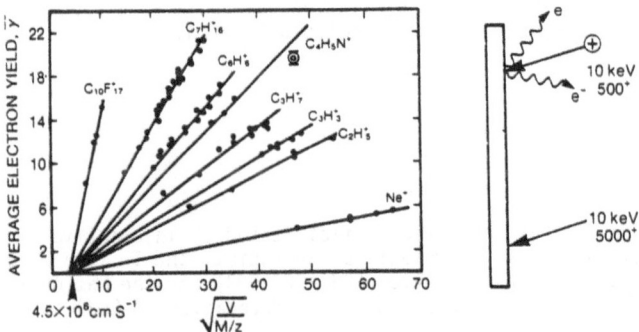

Fig. 7. Study of Secondary Electron Yield for various ions incident on Cu. A 10keV m/z 500 ion has close to unit probability for ejecting at least one $\bar{e}$ while a 10keV m/z 5000$^+$ ion has <u>close to</u> (but not zero) probability (v < 4.5 $\times$ 10^6 cm s^{-1}) [43].

The electronics for the TOF measurement are now well developed. High time resolution (< 0.1 ns), multiple-stop capability with minimal dead time are now some of the features of currently developed (but not commercially available) time interval digitizers [44], [45].

One problem that remains is the background in the TOF spectrum. This is a background that appears to be correlated with the heavy-ion excitation process because it decays with time between heavy ion impacts [21]. Although it is easily subtracted because it is smoothly varying, its presence limits the sensitivity for detecting weak intensity peaks. The present limit is $\sim$ 10 ions m^{-1}, but in the absence of this troublesome background, this could be reduced to perhaps 1 ion h^{-1}, the kind of sensitivity one would desire in searching for "super-heavy" molecular ions (m/z > 20,000). The average molecular weight of the background measured by a combination of TOF and magnetic deflection seems to correlate with the average mass of the spectrum that is being measured [21]. One possible origin of the background is that it is due to delayed ion emission, delayed by several μs. This is a specific problem that may be solvable by an instrumental approach. Figure 1 shows a typical ^{252}Cf-PDMS spectrum that contains this troublesome time-dependent background.

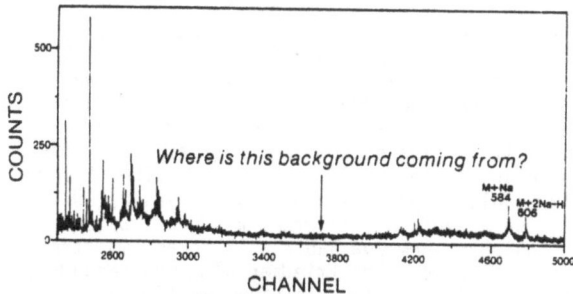

Fig. 8. ^{252}Cf-PDMS time-of-flight spectrum of an unknown molecule. The background ions come from fission fragment excitation but the time correlation has been lost due to delayed emission and metastable decay during the acceleration period.

Acknowledgements

The compilation of this review has been greatly facilitated by informative discussions with A. Benninghoven, C. J. McNeal, B. Sundquist, and T. Tombrello. The support of the National Science Foundation (CHE-8206030) and the National Institute of Health (GM25096) and the Robert A. Welch Foundation (A-258) is gratefully acknowledged.

References

1 D. F. Torgerson, R. P. Skowronski, and R. D. Macfarlane, Biochem. Biophys. Res. Comm. 60, 616 (1974).

2 A. Benninghoven, D. Jaspers, and W. Sichtermann, Appl. Phys. 11, 35 (1976).

3 M. A. Posthumus, P. G. Kistemaker, H. L. C. Meuzelaar, and M. C. Ten Noever de Brau, Anal. Chem. 50, 985 (1978).

4 M. Barber, R. S. Bordoli, R. D. Sedgwick, and A. N. Tyler, J. C. S. Chem. Comm. 325 (1981).

5 P. Håkansson, I. Kamensky, and B. Sundquist, Nucl. Instr. and Meth. 198, 43 (1982).

6 P. Duck, W. Treu, W. Galster, H. Frohlich, and H. Voit, Nucl. Instr. and Meth. 168, 601 (1980).

7 R. D. Macfarlane, Biomed. Mass Spec. 8, 449 (1981).

8 C. J. McNeal, S. A. Narang, R. D. Macfarlane, H. M. Hsiung and R. Brousseau, Proc. Nat'l. Acad. Sci. U.S.A. 77, 735 (1980).

9 B. T. Chait, W. C. Angosta, and F. H. Field, Int. J. Mass Spectrom. Ion Phys. 39, 339 (1981).

10 P. Dück, H. Fröhlich, W. Treu, and H. Voit, Nucl. Instr. and Meth. 191, 245 (1981).

11 L. E. Seiberling, J. E. Griffith, and T. A. Tombrello, Rad. Effects 52, 201 (1980).

12 B. T. Chait and F. H. Field, Int. J. Mass Spectrom. Ion Phys. 41, 17 (1981).

13 C. J. McNeal, Anal. Chem. 54, 43A (1982).

14 C. J. McNeal and R. D. Macfarlane, J. Amer. Chem. Soc. 103, 1609 (1981).

15 N. Furstenau, W. Knippenberg, F. R. Krueger, G. Weiss, and K. Wien, Z. Naturforsch 32a, 711 (1977).

16 R. D. Macfarlane, Nucl. Instr. and Meth. 198, 75 (1982).

17 R. D. Macfarlane and K. Wien (unpublished results).

18 O. Becker, Nucl. Instr. and Meth. 198, 53 (1982).

19 R. D. Macfarlane, and B. Sundquist (unpublished results).

20 R. D. Macfarlane, C. J. McNeal and J. E. Hunt, Adv. in Mass Spec. 8A, 349 (1979).

21 R. D. Macfarlane and C. J. McNeal, unpublished results.

22 R. D. Macfarlane, Acc. Chem. Res. 15, 268 (1982).

23 T. A. Tombrello, CalTech Report, BAP-28 (1982).

24 R. L. Fleischer, P. B. Price, and R. M. Walker, J. Appl. Phys. 36, 3645 (1965).

25 Proceed. Nordic Symposium on Ion Induced Desorption of Molecules from Biorganic Solids (Nucl. Instr. and Meth. 198, 1-173 (1982).

26 W. Knippelberg, O. Becker, and K. Wien, Nucl. Instr. and Meth. 198, 59 (1982).

27 R. W. Ollerhead, J. Böttinger, J. A. Davies, J. L. 'Ecuyer, H. K. Haugen and N. Matsunami, Rad. Effects 49, 203 (1980).

28 R. D. Macfarlane and J. E. Hunt, unpublished results.

29 R. D. Macfarlane and D. Jacobs, unpublished results.

30 L. Larsson and R. Katz, Nucl. Instr. and Meth. 138, 631 (1976).

31 R. H. Ritchie and C. Claussen, Nucl. Instr. and Meth. 198, 133 (1982).

32 W. L. Brown, W. M. Augustyniak, E. Simmons, K. J. Marcantonio, L. J. Lanzerotti, R. E. Johnson, J. W. Boring, C. T. Reimann, G. Fotti, and V. Pirronello, Nucl. Instr. and Meth. 198, 1 (1982).

33 R. D. Macfarlane and D. F. Torgerson, Phys. Rev. Lett. 36, 486 (1976).

34 F. R. Krueger, Surf. Science 86, 246 (1979).

35 C. C. Watson and T. A. Tombrello, Proc. Lunar Planet Sci. Conf. 135th, 845 (1982).

36 H. J. Kreuzer and D. N. Lowy, Chem. Phys. Lett. 78, 50 (1981).

37 J. K. Norskov and B. I. Lundquist, Phys. Rev. B 19, 5661 (1979).

38 Z. Sroubek, K. Zdansky and J. Zavadil, Phys. Rev. Lett. 45, 580 (1979).

39 R. G. Orth, H. T. Jonkman, and J. Michl, J. Amer. Chem. Soc. 104, 1834 (1982).

40 L. Schmidt, H. Danigel, and H. Jungelas, Nucl. Instr. and Meth. 198, 165 (1982).

41 P. Dück, Ph.D. Dissertation, Erlangen (1981).

42 B. A. Mamyrin and D. V. Shmikk, Eng. Trans. Soviet Phys., J. Exp. Theor. Phys. 49, 762 (1979).

43 R. J. Beuhler and L. Friedman, Int. J. Mass Spectrom. Ion Phys. 23, 81 (1977).

44 R. F. Bonner, D. V. Bowen, B. T. Chait, A. B. Lipton, F. H. Field and W. F. Sippach, Anal. Chem. 52, 1923 (1980).

45 E. Festa and R. Sellem, Nucl. Instr. and Meth. 188, 99 (1981).

2.2 Secondary Ion Emission from Metals Under Fission Fragment Bombardment*

Karl Wien and Otmar Becker

Institut für Kernphysik, Technische Hochschule Darmstadt,
D-6100 Darmstadt, Fed. Rep. of Germany

1. Ion Emission from Clean Metal Surfaces

Recently we reported about yields of secondary ions emitted from clean metal surfaces under the bombardment of ^{252}Cf fission fragments [1]. In a next step we tried to measure the energy distributions of these ions. In both cases comparison with current sputter theories is problematic, because ionization is usually not taken into consideration. Using empirical ionization probabilities from low-energy SIMS experiments the yields of metal ions agreed reasonably with values calculated by means of linear cascade theory [2].

As described elsewhere [3] the distribution of the axial energy E_z (= energy component perpendicular to the target surface) can be derived from the shape of a mass line measured with a TOF technique. The studies were done with 1.4 mg/cm^2 Al, o.4 mg/cm^2 Ni and Cu foils, which were cleaned by ion etching with 2.5 keV Ar$^+$ (2 mA/cm^2, 3o min). The foils were bombarded from the backside with ca. 5o fission fragments per sec. As shown in Fig.1 the Al$^+$ mass line, for instance, is considerably broadened and shape analysis proved that the energy distribution extends to at least 4oo eV. That means the energy distribution of metal ions ejected from a metal surface is quite different from that of molecular ions ejected from organic layers. The later distributions extend to ca. 1o eV [3]. On the other hand our results for metal ions give only a hint that the linear cascade model is once more applicable. Up to now we are not able to measure complete energy distributions for these ions, because the TOF spectrometer was not working at full acceptance. If the radial energy is too high, the ions miss the stop detector [3]. An improved method will be discussed later.

Compared to our earlier experiments the mass spectra were taken at lower vacuum pressure (3 x 1o^{-10} mbar) and with cleaner surfaces. In some cases even the contamination of Na and K was rather low (Fig.1). Taking advantage of the improved counting statistics, absolute ion yields per fission fragment (=ff) were determined. They are listed in Table 1. Because of the imperfect acceptance and the rather uncertain detection probability of the secondary ion detector in most cases only lower or upper limits are given. This time the semi-empirical values of the sputter theory turn out to be lower by a factor 5 or more. But because of the uncertain ionization probability the comparison is some-

---

This work was supported by the Deutsche Forschungsgemeinschaft.

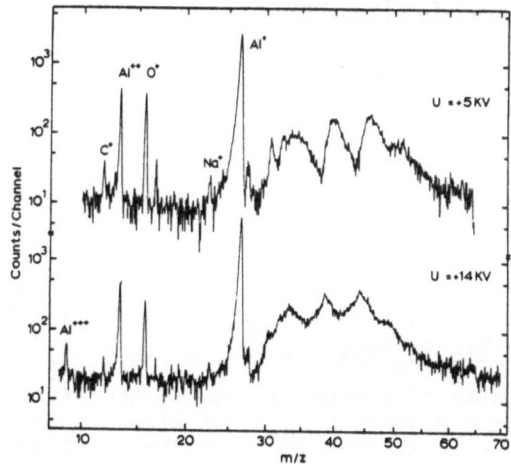

Fig.1. Mass spectra of positive ions emitted from an Al surface after an oxygen exposure of 5ooo Langmuir

what arbitrary. Note that the relative yields of the Al ions are not far from values taken at much lower bombarding energies (keV range [4]). No negative metal ions could be observed. From a Cu surface the only negative ion, which has been detected, was H^- with a yield of $1o^{-3}$/ff.

Table 1. Yields of metal ions [10^{-4}/ff]. New values have been obtained by improved methods. See Nucl.Instrum. Methods 189,59 (1982)

Ions	Experiment	Sputter Theory*	Relat. SIMS values**
Al^+	> 5o	8.5	14o
Al^{++}	2o		2o
Al^{+++}	2		2
Al^-	< o.5		
Ni^+	> 23	4	
Ni^-	< o.5		
Cu^+	> 17	2	
Cu^-	< o.5		

*evaluated using experimental ionization probabilities
**refer to Ref. [4]

When the ion emission from Al was studied, the oxygen pressure in the spectrometer chamber was lower than 5×10^{-11}mbar. The Al foil was then exposed to oxygen increasing its pressure step by step from 1×10^{-9} to 8×10^{-7}mbar. As illustrated in Fig.2 the Al^+ yield increased by a factor 5 until an oxygen exposure of 200 Langmuir, whereas the Al^{++} yield remained at the same value as before. No change of the line shape of the Al ions was observed. That indicates that the sputter process remains the same if aluminum Oxyde is present at the surface or not.

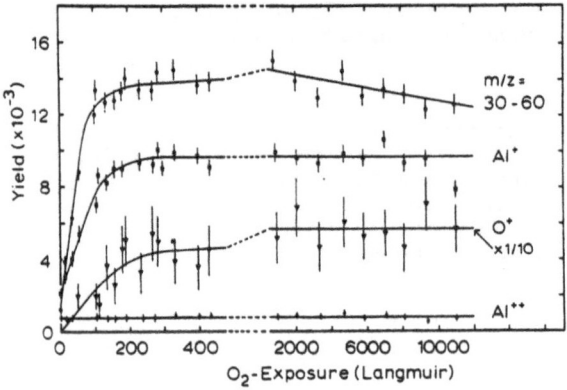

Fig.2. Yields of ions emitted from an Al surface under oxygen-exposure

In all positive ion spectra obtained for clean metals a relative intensive distribution appeared divided up in several broad peaks. For Al the distribution extends from mass number 29 to 60, for Ni and Cu from 33 to 75 (see Fig.1 and 3). These ions come from the target foil, because they are accelerated by the electrical potential in front of the target in the same way as the other ions. The peaks change their shape if the acceleration potential is increased, which is obviously a consequence of the energy- and angular-distribution of the ions. In the case of Al their yield is enhanced by a factor of more than 10, when the target is exposed to oxygen of 200 Langmuir. Up to now we have no explanation for this phenomenon.

2. Improved Method for the Measurement of Energy Distributions

As discussed in Ref [3] two problems arise if the energy distribution $\Delta N/\Delta E_z$ is evaluated from a TOF-mass line:

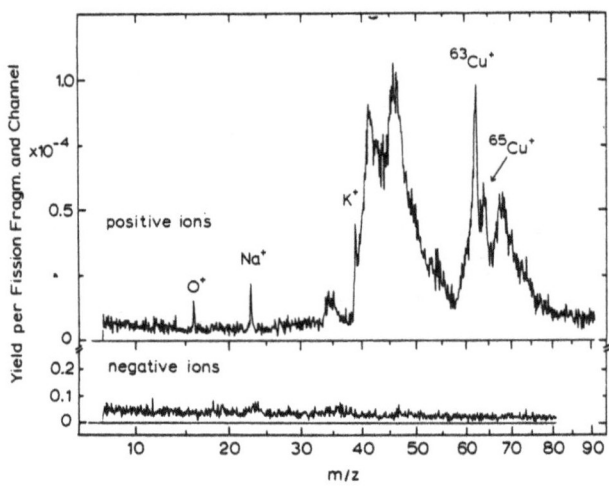

Fig.3. Mass spectra of negative and positive ions emitted from a cleaned Cu surface at U = -8 KV and +10 KV

a) The TOF spectrometer has to work at full acceptance, i.e. the radial energy of an ion leaving the surface has to be

$$E_r \le eU \cdot R^2 \cdot (2d_1 + d_2)^{-2} \qquad . \tag{1}$$

(Here is R = radius of the stopdetector, U = acceleration voltage, d_1 = distance between target and acceleration grid, d_2 = drift distance). The values of U and d_2 cannot be chosen arbitrarily, because they determine the extension of a mass line in time scale. For a proper shape analysis the width of a mass line should be well above the time resolution of the spectrometer.

b) In order to appoint the time, which corresponds to E_z = O under a mass line, a calculation has to be carried out including precise values of U, d_1 and d_2. Especially the measurement of the drift distance d_2 causes uncertainties.

To overcome both problems a second grid was mounted inbetween the usual acceleration grid and the target as illustrated in Fig.4. Using (1) two times the condition for E_r is now

$$E_r \le R^2 (2d'/\sqrt{eU'} + (2d_1 + d_2)/\sqrt{eU})^{-2} \qquad . \tag{2}$$

With d' = 0.3 cm and U' = 500 volts, for instance, (2) is still hard to satisfy for E_r = 400 eV. It needs a short drift distance d_2 (~ 10 cm) and a high acceleration voltage U (~ 15 KV). But the mass lines are spread over a reasonable time intervall (for Al^+ 60 nsec instead of 13 nsec) due to the low acceleration by the intermediate grid.

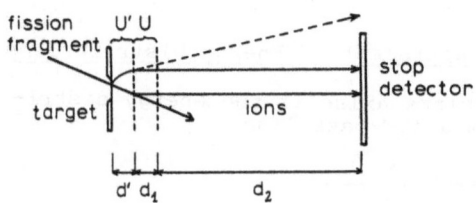

fission fragment
U' U
target
ions
stop detector
d' d₁
d₂

Fig. 4. New arrangement of the acceleration grids in the TOF spectrometer

Fission fragments, which strike the grid bars or - if the grid material is sufficiently thin - penetrate the bars, produce secondary ions also in the plane of the intermediate grid. These ions cause additional peaks in the TOF spectrum as shown in Fig. 5. Their width is given mainly by the time resolution of the apparatus and they can be used to determine the E_z = O point of regular mass lines.

The corresponding time interval Δt is almost independent from d_1 and d_2 for U'/U << 1, if the two considered lines are generated by ions of the same mass. The quantities which have to be known precisely are now d' and U'.

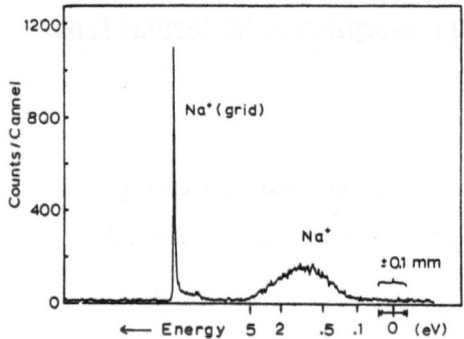

<u>Fig. 5.</u> TOF spectrum of Na$^+$ ions emitted from the target and the intermediate grid. The time of flight has been transformed into initial ion energies. The E_z = 0 point depends on the error of d'

Usually we observe a gap between the E_z = 0 point and the on-set of the energy distribution (refer to Fig.5). This gap could be interpreted as a potential barrier, which an ion moving in-side the solid has to overcome at the surface.

References

1 F.R. Krueger and K. Wien, Z. Naturforsch. <u>33a</u>, 638 (1978)
2 P. Sigmund, Phys. Rev. <u>184</u>, 2, 383 (1969)
3 O. Becker, N. Fürstenau, F.R. Krueger, G. Weiß and K. Wien, Nucl. Instr. & Methods <u>139</u>, 195 (1976)
4 A. Benninghoven, private communication

2.3 Fast Heavy Ion Induced Desorption of Molecular Ions from Small Proteins

B. Sundqvist, P. Håkansson, I. Kamensky, and J. Kjellberg

Tandem Accelerator Laboratory, University of Uppsala, Box 533
S-751 21 Uppsala, Sweden

1. Introduction

A number of new mass spectrometric methods for biomolecules have been introduced in recent years by which it has been possible to considerably extend the mass region towards higher masses. Among many new techniques introduced, those involving neutral or charged particle bombardment of a sample seem most promising. Fast (MeV) ions, like fission fragments, are used in 252Cf-PDMS introduced by R.D. MACFARLANE and collaborators [1-3] in 1974. The use of low energy (keV) ions as in SIMS was pioneered by BENNINGHOVEN and co-workers [4]. CHAIT and STANDING [5] have used a pulsed ion source and time-of-flight technique in connection with low-energy ions. Fast (MeV) ions interact primarily with the electrons in a medium while slow (keV) ions interact with the atoms and cause atomic recoil-cascades. In spite of this difference the mass spectra of desorbed ions look very much the same. However recent measurements by our own group indicate that there are differences [6]. Fast ions are more efficient, i.e. give larger yields of molecular ions than low-energy primary ions and this seems to be even more so the larger the molecule is. Recently BARBER et al. [7] have introduced the so called FAB- (Fast Atom Bombardment) technique. This method has proven to be very efficient in most cases studied so far. Although a beam of keV neutral particles is used the basic interaction with the sample is the same as with keV ions. A maybe more significant difference from the other techniques mentioned is the use of a liquid sample "holder", i.e. glycerol.

In collaboration with C.J. MCNEAL and R.D. MACFARLANE we were able to observe molecular ions of bovine insulin about a year ago [8] using a slightly modified PDMS-setup. A beam of 90 MeV 127I(20+) from the Uppsala tandem accelerator was used instead of fission fragments. Since then we have also detected molecular ions from a neurotoxin with a molecular weight of 7820 [9]. Quite recently DELL and MORRIS [10] have also reported the detection of molecular ions of insulin using the FAB-technique.

In this paper we report some preliminary results from a more detailed study of insulin mass spectra and summarize the experiences gained concerning production and detection of molecular ions of small proteins.

2. Material and sample preparation

Most of the results in this report concern bovine insulin (Sigma), a protein which consists of two peptide chains linked by two disulfide bridges. The number of amino-acid residues is 51. The molecular weight of the molecule is 5733. Results are also given for a neurotoxin. The neurotoxin was isolated from Thai cobra venom by an Uppsala group. It is a protein with a well-defined tertiary structure [11] and the molecular weight has been determined to 7820. The molecule is a single polypeptide chain with 71 aminoacid residues cross linked by five disulfide bridges. The molecules were dissolved in trifluoracetic acid and then electrosprayed [12] on 500 µg/cm^2 aluminum foils. About 20 µg of sample molecules was sprayed over an area of 80 mm^2.

3. Experimental techniques

All the data discussed in this report were collected using time-of-flight techniques. In one set of data, fission fragments from a ^{252}Cf-source were used as primary ions while in the rest of the data various fast heavy ion beams from the Uppsala tandem accelerator were exploited. In the former data set a ^{252}Cf-PDMS spectrometer very similar to the one described by R.D. MACFARLANA et al. [13] was used and in the accelerator runs data were taken with a slightly modified spectrometer earlier described [14]. Throughout the experiments described a time-to-digital converter (TDC) built at the laboratory was used. The TDC has a time resolution of ±1 ns, 200 ns dead time and a multistop function with possibility to accept 32 stop signals for each start signal. The TDC can also be run in an event-by-event mode making it possible to look for correlations between ions in the spectra. In the latter case data were analyzed off-line with the VAX 11/780 computer of the laboratory. In all experiments reported here the intensity of the primary ions was ~ 2000 particles per second. Time-of-flight spectra were mass calibrated by using a sample of CsI and exploiting the cluster ions $(CsI)_nCs^+$ and $(CsI)_nI^-$.

4. ^{252}Cf-PDMS spectra of bovine insulin

In Fig. 1 the positive ion spectrum of bovine insulin is shown. These data were taken with a ^{252}Cf-spectrometer while our published spectra on this molecule were collected using an accelerator beam of 90 MeV ^{127}I(20+). The

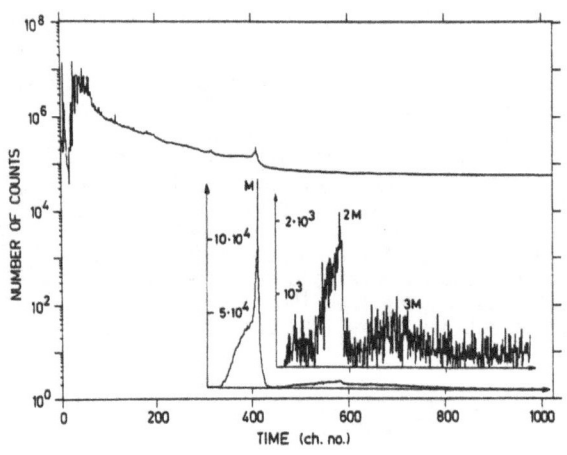

Fig. 1. Positive ion spectrum of bovine insulin. The insets show background subtracted spectra

high mass region is characterized by an almost exponentially decreasing background. In the inset in Fig. 1 the background is subtracted. The m/z region shown contains peaks which can be associated with the molecular ion at m/z 5742±5 the dimer at m/z 11447±15 and trimer ions. The molecular ion is most likely a protonated molecule. It is interesting to note that the molecular ion peak has three components of which the two broadest distributions may correspond to metastable decay. In the negative ion spectrum a molecular ion peak was observed at m/z 5743±5 indicating a deprotonated molecule. The yields, i.e. the number of secondary ions per primary ion, were 0.5% and 0.05% for positive and negative molecular ions respectively.

5. The influence of primary ion parameters on the yield of molecular ions of insulin

Earlier measurements by us have shown how the yield of M^+ ions of ergosterol depends on primary ion parameters such as stopping power, velocity, atomic number [14], angle of incidence [15] and charge state [16] of the primary ions.

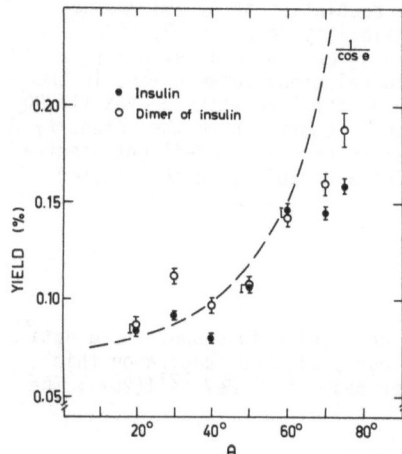

Fig. 2. The yield of monomer and dimer ions of bovine insulin as a function of angle of incidence of 90 MeV $^{127}I(18+)$ primary ions

In Fig. 2 the influence of the angle of incidence for 90 MeV $^{127}I(18+)$ ions on the yield of molecular ions of insulin is shown. An angle θ of 90° corresponds to oblique incidence. Also the distribution for the dimer ion is given. The distributions are in fair agreement with distributions measured earlier with 20 MeV ^{127}I and for lighter molecules [15]. We have also studied how the yield of molecular ions from insulin varies with the amount of energy deposited in a surface layer, i.e. the stopping power of the primary ions. A primary ion velocity of 1.2 cm/ns was chosen, i.e. close to the maximum of the electronic stopping power and then the stopping power was changed by using different ions ranging from ^{16}O to ^{127}I. Also the yields of two protected oligonucleotides of MW 1884 and 3609 were studied in this way.

The results are plotted in Fig. 3 as the yield divided by the stopping power as a function of stopping power. Also shown are old data on ergoste-

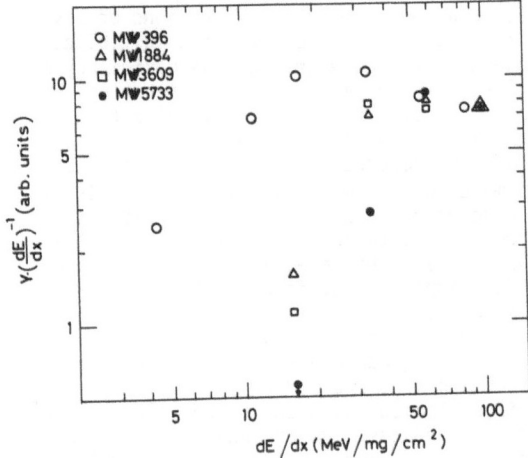

Fig. 3 The yields divided by stopping power for molecular ions of ergosterol (MW 397), two protected oligonucleotides (MW 1884 and MW 3609) and bovine insulin (MW 5733) as a function of stopping power of the primary ion

rol [14] but displayed as the yield divided by the square of the stopping power. The interesting feature of the data which we like to illustrate in this way is that there is a threshold stopping power increasing with the size of the molecule, before a slower dependence on stopping power sets in. In fact below this threshold in the distributions the molecular ion peaks are very hard to observe at all in the spectra.

6. Multiplicity and correlations of ions desorbed from bovine insulin

The multistop TDC produces also a spectrum of the multiplicity of ions per primary ion, i.e. the number of molecular ions desorbed for each fast heavy ion.

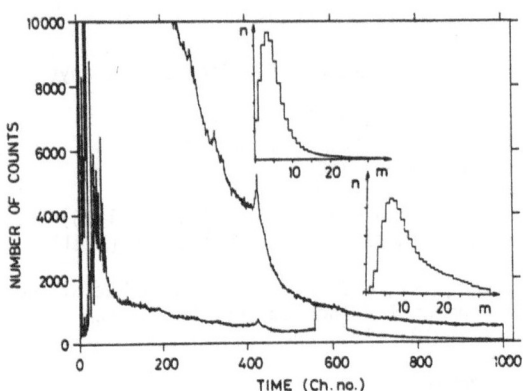

Fig. 4. A complete spectrum of positive ions of bovine insulin as in Fig. 1 and a spectrum of ions coming in the same events as ions in the dimer window. The insets show multiplicity distributions for the complete spectrum (left) and molecular ions of insulin (right). n means number of counts and m multiplicity

The left inset in Fig. 4 shows such a spectrum for positive ions of bovine insulin. The average multiplicity is about 4 but events with multiplicity up to 32 are detected. It is important to remember that the efficiency of the whole system (transmission losses in grids and the efficiency of the channel plates) is assumed to be ~ 50%, i.e. the actual multiplicities are considerably higher. The TDC can also be run in an event-by-event mode registering the whole sequence of ions desorbed per incoming fast ion. A first run on bovine insulin was made in this mode. The final analysis of the data is not ready but some conclusions can be made already at this stage. If only flight times corresponding to an insulin molecular ion are selected the multiplicity (shown in the right inset in Fig. 4) is higher than the average multiplicity. This implies that insulin molecular ions are mainly desorbed in high multiplicity events. The same is true for dimers and trimers. If a gate is put on the dimer distribution (Fig. 4) and all ions in events containing ions in that window are selected one can see that the molecular ions and other high mass ions are strongly correlated to such events as compared to low mass ions. The correlation to ions close to the gate may be due to metastable decay products. One explanation of these "super"-events may be that there are preformed clusters in the sample.

7. Observation of molecular ions from a neurotoxin protein of MW 7820

A sample of the neurotoxin was prepared in the same way as described for insulin. An accelerator beam of 90 MeV $^{127}I(20+)$ was used and both positive and negative ion spectra with distinct molecular ion peaks were obtained. Molecular ion peaks were observed at m/z=7842±14 and m/z=7861±15 in the positive and negative spectrum respectively. This is to our knowledge the largest protein studied by mass spectrometric techniques [9].

This work was supported by the Swedish Natural Science Research Council and the Swedish Board for Technical Development.

References

1. D.F. Torgerson, R.P. Skowronski and R.D. Macfarlane, Biochem. Biophys. Res. Commun. 60 (1974) 616
2. R.D. Macfarlane, Chapter 38 in Biochemical Applications of Mass Spectrometry, First Supplementary Volume, Ed. G.R. Waller and O.C. Dermer, John Wiley & Sons Inc. (1980) 1209
3. C.J. McNeal and R.D. Macfarlane, JACS. 103 (1981) 1609
4. A. Benninghoven, D. Jaspers and W. Sichterman, Appl. Phys. 11 (1976) 35
5. B.T. Chait and K.G. Standing, Int. J. Mass. Spec. Ion Phys. 40 (1981) 185
6. I. Kamensky, P. Håkansson, B. Sundqvist, C.J. McNeal and R.D. Macfarlane, Proceedings of the Nordic Symposium on Ion Induced Desorption from Bioorganic Solids, Uppsala, June 15-18, 1981, Nucl. Instr. Meth. 198 (1982) 65
7. M. Barber, R.S. Bordali, R.G. Sedgwick and A.N. Tyler, J. Chem. Soc. Chem. Commun. (1981) 325
8. P. Håkansson, I. Kamensky, B. Sundqvist, J. Fohlman, P. Peterson, C.J. McNeal and R.D. Macfarlane, JACS. 104 (1983) 2948
9. P. Håkansson, I. Kamensky, J. Kjellberg, B. Sundqvist, J. Fohlman and P. Peterson, (submitted for publication in Biochem. Biophys. Res. Commun.)

10. A. Dell and H. Morris, Biochem. Biophys. Res. Commun. <u>106</u> (1982) 1456
11. E. Karlsson, H. Arnberg and D. Eaker, Eur. J. Biochem. <u>21</u> (1971) 1
12. C.J. McNeal, R.D. Macfarlane and E.L. Thurston, Anal. Chem. <u>51</u> (1979) 2036
13. R.D. Macfarlane and D.F. Torgerson, Int. J. Mass. Spectrom ion Phys. <u>21</u> (1976) 81
14. P. Håkansson and B. Sundqvist, Rad. Eff. <u>61</u> (1982) 179
15. P. Håkansson, I. Kamensky and B. Sundqvist, Surface Sci. <u>116</u> (1982) 302
16. P. Håkansson, E. Jayasinghe, A. Johansson, I. Kamensky and B. Sundqvist, Phys. Rev. Lett. <u>47</u> (1981) 1227

2.4 Problems in Standardization of ^{252}Cf Fission Fragment Induced Desorption Mass Spectrometry

Hartmut Jungclas, Lothar Schmidt, and Harald Danigel

Kernchemie, Fachbereich Physikalische Chemie, Philipps-Universität
D-3550 Marburg, Fed. Rep. of Germany

1. Introduction

The ^{252}Cf fission fragment induced desorption mass spectrometry, spreading
from Texas A&M University [1], is now utilized world-wide in more than
ten laboratories. Before applying a new analytical technique for quantitative measurements one should investigate the sensitivity and reproducibility of the given equipment [2,3]. Besides standardization of this
time-of-flight mass spectrometry is discussed.

2. Experimental

Our mass spectrometer has been described in detail [2,3], thereby we
focus on the differences between this set-up and most others. Samples of
nonvolatile compounds are prepared by: desolving in a suitable solvent,
cleaning on a HPLC column if needed, injecting the solution in a rough
vacuum stage at constant flow (≈ 0.5 ml min^{-1}), collecting the nonvolatile compounds on a thin sample foil in a vacuum-drying process. A step-
ping disk carries twelve samples, which are consecutively collected and
analysed, and thus serves as an interface for monitoring a HPLC separation.
Accelerated ions are detected via conversion electrons, which are deflec-
ted by a magnetic field to a micro channel plate detector arrangement
(Fig. 1, stop-MCP).

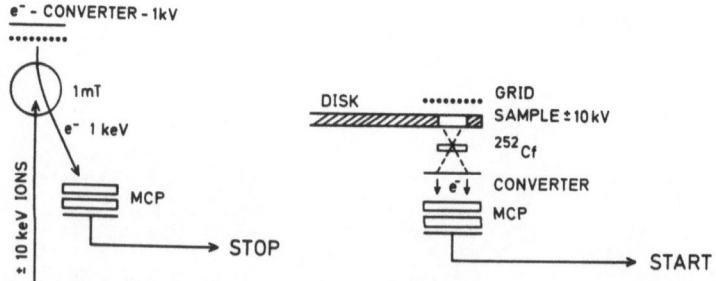

Fig. 1. Schematic drawing of time-of-flight mass spectrometer utilizing
fission fragment induced desorption of nonvolatile compounds

3. Yield Parameters

This section is based on our observations, but it points to parameters which affect the yield of Cf-252 fission fragment induced desorption time-of-flight mass spectrometry in general.

3.1 Fission Fragments

The intensity of fission fragments transmitting the sample foil has been investigated as a function of Cf-252 source - sample distance [4] questioning for the effective sample area A_{eff}. As fission fragments suffer an energy loss while passing the source cover foil (2 μm Ti) and the sample backing (here ⩾ 2 μm Ti), the velocity of primary ions and A_{eff} (i.e., the solid angle of emission used for ionization effectively) depends on the thickness of these foils. Besides the yield of secondary ions varies with the velocity of primary ions [5]. We studied the influence of the foil thickness using different Ti-backings and the sample: aduct of the Eu-chelate (hexafluoracetylacetonate) with tri-n-butylphosphineoxide (mol.wt. 1211). As the intensity distributions for H^+ ions and the fragments m/z 786, 1004 resemble the slope of A_{eff}/A_{total} (Fig. 2, A_{total} = 64 mm^2), the foil thickness does not influence relative but absolute intensities.

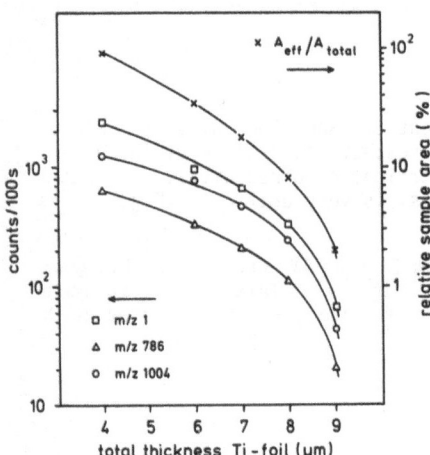

Fig. 2. Intensities of three selected ions (see text) as a function of Ti backing thickness (including 2 μm Ti cover foil of Cf-252 source)

3.2 Ionization

In order to demonstrate the influence of sample impurities on the ionization yield, we list the following observations obtained with the antitumor drug Etoposide [2]:
(i) When the extract from serum samples was analysed directly, the spectrum did not show quasimolecular ions (QMI) of Etoposide. When using HPLC for additional cleaning, the QMI became visible. (ii) The relative intensities of QMI, i.e., $[M+H]^+$, $[M+Na]^+$ and $[M-H+2Na]^+$, depended on the Na concentration. (iii) Etoposide samples were prepared using different solvents and preparation procedures. In some cases no QMI were obtained.

Problems as described here may differ from compound to compound, but obviously the formation and desorption of QMI is related to a fast surface chemistry [6] which is not understood completely (preformed ions?).

3.3 Ion Acceleration

When a metastable ion survives the acceleration period (< 1 μs, depending on the spacing sample-grid, the acceleration potential and m/z), it contributes to the parent ion intensity. But if its lifetime is < 1 ns, the decomposed ion appears at the fragment m/z. However, metastable ions with an intermediate lifetime (1 ns - 1 μs) form a background which extends from the fragment m/z up to the parent m/z. This undesired background can be shifted by retarding potentials [7] or suppressed by an ion mirror in order to improve the line shape. In publications the acceleration potential and the sample-grid spacing should be listed or standardized.

3.4 Ion Beam Divergence

Experimental beam profile measurements were performed with various settings of Cf-252 source - sample distance, sample - acceleration grid spacing, acceleration potential and length of drift path [4]. As a result the divergence is independent of $m/z \neq 1$, thus these parameters do not influence relative intensities. Just H^+ ions showed a larger divergence consistent to the line shape analysis of H^+, Li^+ and Cs^+ ions [8].

3.5 Ion Detection

Various settings of the acceleration potential and the bias on microchannel plates (MCP) have been tested in respect to absolute yields of m/z 39, 165 (fragment) and 455 (quasimolecular ion) obtained with a Verapamil sample [9]. The influence of the MCP-bias is very critical (Fig. 3) as even relative intensities vary.

Increasing the acceleration potential from 2 to 12 kV absolute yields grow until they reach a saturation level (Fig. 4). This potential must be chosen high enough, thus all ions of interest are detected with saturated yields.

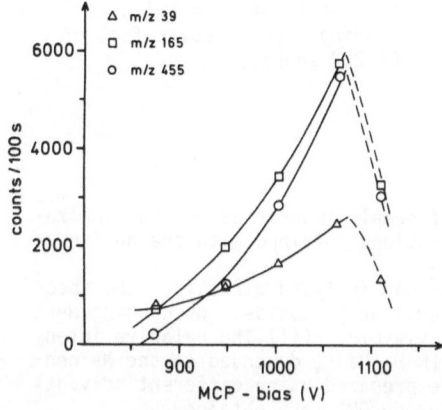

Fig. 3. Intensities of three selected ions (10 keV) in Verapamil (mol. wt. 454) spectrum as a function of bias on microchannel plates (stop-MCP)

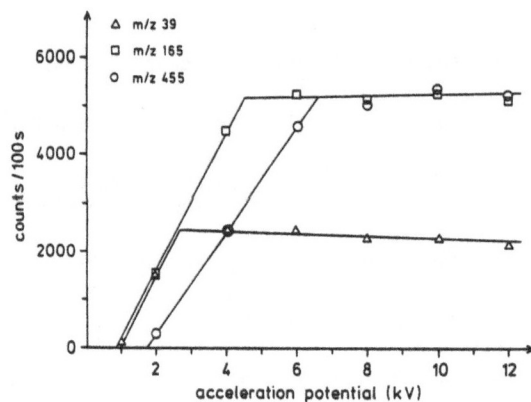

Fig. 4. Intensities of three selected ions in Verapamil (mol. wt. 454) spectrum as a function of ion energy, i.e., acceleration potential

Measurements were performed to compare two detection systems: ions directly impinging on MCP and detection via conversion electrons. The latter method improves relative yields of very heavy masses (m/z > 500), but systematic data as shown here for the direct detection are published elsewhere [3].

3.6 Time-of-Flight Registration

Not all ions generating a logic "stop" signal lead to a proper increment in the time-of-flight spectrum. First, there is the dead time of the time counting devise (attention with time-to-amplitude converters). Second, there is a chance to correlate a "stop" with the false "start" event contributing to a smooth background in the spectrum. Both phenomena gain importance with growing flight time and increasing decay rate of the Cf-252 source. For quantitative analysis a time-to-digital converter with short dead time and buffer logic [10,11] is very useful. A dead-time correction for absolute intensities is recommended.

4. Results and Discussion

We analysed pure samples of alanine and guanosine to compare the fragmentation pattern and relative intensities with published data [12]. Variations concerning the cationized ions due to the Na concentration appeared to be the evident difference. When summing up intensities of protonated and cationized ions, the agreement was fair. We found 20% quasimolecular ions for alanine (21% in ref. [12]) normalized to the total ion current but excluding H^+ and Na^+ ions. For guanosine the ratio quasimolecular to quasibase ions varied between 3 and 5 depending on the sample preparation (5.6 in ref. [12]).

As a result, standardization of the Cf-252 method is strongly related to sample preparation procedures and the efficiency determination of ion detectors. Probably both problems are present in all desorption mass spectroscopies.

The financial support of the Bundesministerium für Forschung und Technologie, Bonn, FRG, is gratefully acknowledged.

References

1 D.F. Torgerson, R.P. Skowronski and R.D. Macfarlane, Biochem. Biophys. Res. Commun. 60, 616 (1974)

2 H. Jungclas, H. Danigel, L. Schmidt and J. Dellbrügge, Org. Mass Spectrom. (in press)

3 H. Danigel, H. Jungclas and L. Schmidt, "A ^{252}Cf fission fragment induced desorption mass spectrometer: design, operation and performance (to be published)

4 H. Danigel and R.D. Macfarlane, Int.J.Mass Spectrom. Ion Phys. 39, 157 (1981)

5 A. Albers, K. Wien, P. Dück, W. Treu and H. Voit, Nucl.Instrum.Methods 198, 69 (1982)

6 R.D. Macfarlane, Nucl. Instrum. Methods 198, 75 (1982)

7 B.T. Chait and F.H. Field, Int.J.Mass Spectrom.Ion Phys. 41, 17 (1981)

8 N. Fürstenau, W. Knippelberg, F.R. Krueger, G. Weiß and K. Wien, Z.Naturforsch. 32a, 711 (1977)

9 H. Jungclas, H. Danigel and L. Schmidt, Org.Mass Spectrom. 17, 86 (1982)

10 R.F. Bonner, D.V. Bowen, B.T. Chait, A.B. Lipton and F.H. Field, Anal. Chem. 52, 1923 (1980)

11 E. Festa and R. Sellem, Nucl.Instrum. Methods 188, 99 (1981)

12 B.T. Chait, W.C. Agosta and F.H. Field, Int.J.Mass Spectrom. Ion Phys. 39, 339 (1981)

Part 3

Secondary Ion Mass Spectrometry (SIMS) Including FAB

3.1 Secondary Ion Mass Spectrometry of Organic Compounds (Review)

Alfred Benninghoven

Physikalisches Institut der Universität Münster, Domagkstraße 75,
D-4400 Münster, Fed. Rep. of Germany

1. Historical Development

In the last decade secondary ion mass spectrometry (SIMS) became a well-established technique for bulk, thin film and surface analysis [1,2]. It was mainly applied to the determination
of element concentrations, and only in surface analysis molecular
cluster ions became important [3,4]. Experience with organic
material was very limited. There exists an early investigation
of catalyst surfaces covered by organic and inorganic anion complexes [5]. These catalysts show a strong emission of the complete
anion complexes in the negative secondary ion spectrum (Fig.1).
Another well-known fact was the emission of $C_m H_n^{\pm}$ ions during
sputtering of surface contamination layers [6]. The intensities of these ions decrease rapidly with increasing number of
carbon atoms. They were considered as fragments of larger
hydrocarbon compounds.

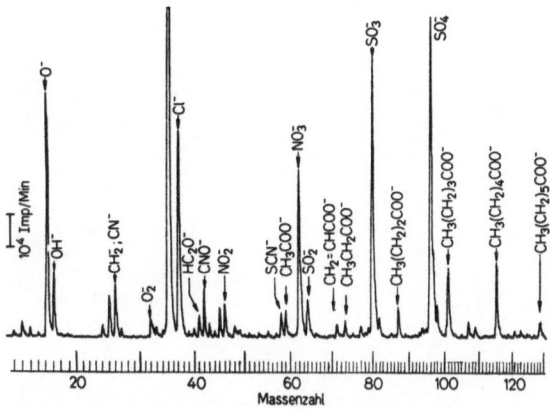

Fig.1 Negative secondary ion spectrum of the uppermost monolayer of a silver catalyst. Primary ions: Ar^+, 3 keV;
10^{-9} A cm^{-2} (static SIMS) [5]

This work was partially supported by the "Bundesminister für
Forschung und Technologie"

From this experience one did not expect a considerable molecular or parent ion emission from larger organic molecules. It was a completely unexpected result at this time, when we found by a systematic investigation of amino acids (1976) [7] , and of a large number of other groups of compounds (1977) [8,9] , that even involatile and thermally labile organic molecules showed a very strong emission of $(M\pm H)\pm$ secondary ions during their bombardment with primary ions in the keV range. These results were very similar to those obtained by MACFARLANE for ^{252}Cf-fission fragment bombarded metal foils. He called this ion formation "plasma desorption" (PDMS) and considered it as a special result of the interaction between the high-energy (MeV-range) fission fragments and the solid [10,11] , and not as a simple sputtering process. At this time we already assumed the same process of ion formation for the ^{252}Cf-fission fragment induced ion emission (PDMS) and the keV sputtering process (SIMS) [7-9] .

Following our first conference reports and publications on this parent ion formation during sputtering of large organic molecules, a rapidly increasing number of laboratories worldwide entered this new field of mass spectrometry [12-16] . The main reason was the obviously wide capacities of organic SIMS as an analytical tool. This analytical application of organic SIMS was extraordinarily pushed by the introduction of the glycerol matrix by BARBER in 1981 [16] . This regenerative target reduces the radiation damage effects considerably and enables high absolute secondary ion currents, so that the large number of existing double-focussing high-resolution mass spectrometers became available for organic SIMS, simply by adding an appropriate excitation source.

Today an extended knowledge of the secondary ion emission behaviour of many groups of compounds and the influence of many parameters on this emission is available. The analytical application in structure elucidation, in trace detection, and in combination with separation techniques as HPLC and TLC is still increasing rapidly.

2. Fundamentals of Secondary Ion Emission

2.1 Ion Surface Interaction

The interaction of a fast atomic or molecular particle with a solid or liquid surface results in two groups of processes occurring in an area near the point of the primary ion impact: primary particle implantation and radiation damages in the excited area on one hand, and photon, electron, and ion emission from the surface on the other hand (Fig.2).

In a simple model it is assumed that the incoming bombarding particle transfers its momentum and energy to a small area around the path of the primary particle in the target. The deposited energy is further dissipated by collision processes. A certain fraction of this energy may return to the surface and result in the ejection of a surface particle (sputtering) [17] .

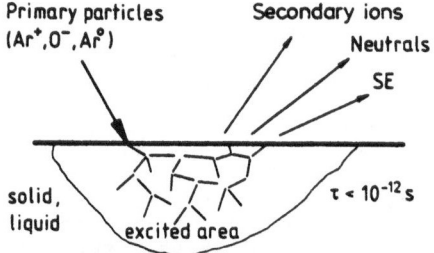

Fig.2 Interaction of a primary particle (atomic or molecular, neutral or charged) with a solid or a liquid surface. The kinetic energy of the primary particle is transferred to an excited area near the surface. As a result of this energy transfer, secondary ions and neutral particles are removed from the surface, in addition to secondary electrons (SE) and photons

Today this sputtering process is widely applied as a thin film deposition technique and for surface cleaning. The fact that a fraction of the emitted atomic or molecular particles is positively or negatively charged is the base of secondary ion mass spectrometry.

Understanding the secondary ion emission process means answering two main questions: What is the reason for the different charge states of the emitted surface species, and secondly, how the formation of the emitted molecular clusters occurs. The question about the origin of the emitted clusters is discussed contradictory in the literature. There exist two main proposals: The cluster emission model, assuming the direct emission of a cluster already preformed on the surface before any bombardment [3,4] , and the recombination model, based on the assumption of single atom emission followed by a recombination of these atomic particles near the surface [18,19] . It is evident that for large organic molecules the recombination model can be excluded.

2.2 Quantitative Description

A more detailed investigation of the sputtering and secondary ion formation process needs its quantitative description. For this purpose we consider one monolayer of an organic molecule M deposited on a metal substrate, e.g. During ion bombardment this monolayer is removed and various positive, negative or neutral secondary species are emitted. X_i may be one of these species (Fig.3).

The emission of the secondary ion X_i from a closed monolayer M can be characterized by the ion yield $Y(X_i)$, indicating the average number of emitted secondary ion species X_i per incident primary ion. The total sputtering yield for the adsorbed molecules M, that is the number of molecules M disappearing from

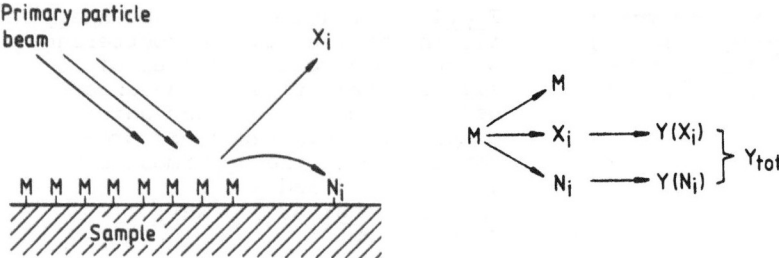

Fig.3 During heavy particle bombardment molecules M disappear
from the surface. They can be transformed into another surface
species N_i or they are emitted as secondary particles X_i
(sputtering)

the surface per primary ion impact, is described by the sputtering yield Y(M). In addition, a transformation probability
$P(M \rightarrow X_i)$ can be defined, indicating the probability that a surface molecule M disappearing from the surface during the ion
bombardment results in the emission of a secondary particle X_i.

If we start with a surface coverage $\Theta(o)$ at the time t = o,
we obtain in a first approximation the following time dependence
of the actual relative surface coverage $\Theta(t)$ during ion bombardment:

$$\Theta(t) = \Theta(o) \exp - \frac{Y(M)}{N_o} \int_0^t \gamma(t)dt. \qquad (1)$$

γ is the primary ion flux density, and N_o the number of molecules M in a closed monolayer on the substrate.

$$\frac{Y(M)}{N_o} = \sigma(M) \qquad (2)$$

may be considered as a damage or disappearance cross section.

For the rate $\dot{N}$ of secondary ions X_i detected in the mass
spectrometer after their mass separation, we obtain the following expression

$$\dot{N}(X_i,t) = \Theta(t)Y(X_i) \cdot f \cdot A \cdot \gamma \qquad (3)$$

for a bombarded target area A and if we assume that the transformation probability $P(M \rightarrow X_i)$ (eq.4) does not depend on Θ. f is
the overall transmission of the secondary ion mass spectrometer
for the particles X_i, i.e. the number of detected ions X_i divided
by the number of generated ions X_i at the target surface. As an
example Fig.4 presents the decrease of some secondary ions during sputtering of a monolayer.

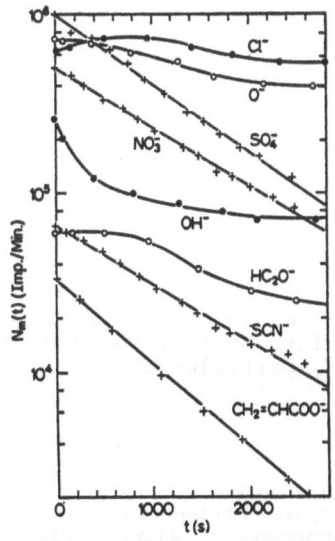

Fig.4 Decrease of some secondary ion intensities during sputtering of the uppermost monolayer of a silver catalyst [5] . The exponential decrease of some anion complexes indicates their presence in only the uppermost monolayer (eqs.1 and 3).
Primary ions: Ar$^+$, 3 keV; $2 \cdot 10^{-8}$ A·cm^{-2}

Equations (1) and (2) allow the calculation of σ(M) from experimental values, equation (3) gives the secondary ion yield $Y(X_i)$ for given X_i, t and f. From the damage cross section σ(M) and the secondary ion yield $Y(X_i)$ we directly find the transformation probability:

$$P(M \rightarrow X_i) = \frac{Y(X_i)}{N_0 \cdot \sigma(M)} = \frac{Y(X_i)}{Y(M)} \quad . \tag{4}$$

In a similar way as we did it here for a monolayer of individual molecules M on a substrate surface, we can define and determine secondary ion yields, damage cross sections, transformation probabilities and the total sputtering yield for a homogeneous target as a metal or a metal oxide, e.g.

2.3 Secondary Ion Emission from Various Surface Structures

The secondary ion emission from a large number of relatively well-defined surfaces has been investigated systematically in the last decade by various laboratories. Some of the main results, as far as they are important for an understanding of the secondary ion emission behaviour of organic molecules, will be summarized in the following.

From an atomically clean metal (Me) surface, mainly positively charged secondary ions Me_n^+ are emitted. The ion yields $Y(Me_n^+)$ are low (typical 10^{-3}), and, in general, rapidly decrease with increasing n [20] . From a solid consisting of two different elements A and B, secondary ions of the general composition $A_m B_n^{\pm}$, with m,n = 0,1,2,3 ...., are emitted. The yields $Y(A_m B_n^{\pm})$ rapidly decrease for high values of m and n [21] .

Very important is the tendency of charge sign conservation. An atomic or molecular secondary ion is preferentially emitted with a charge sign corresponding to that of the species at the surface before the bombardment. This means, e.g. that from a metal oxide preferentially positively charged metal ions Me^+ and negatively charged oxygen ions appear in the secondary ion spectrum [3,4] . The probability for a cluster ion Me_mO_n to be negatively charged increases with decreasing valence state of the metal atoms at the surface and with an increasing number (n) of oxygen atoms in a cluster containing m metal atoms [21] . This charge sign conservation tendency is confirmed by many other examples as the very high yields of negative anions (Cl^-, SO_4^-, CO_3^-, NO_3^-, etc. [22]) from the corresponding salts or the emission of secondary ions $A(AHa)^+$ and $Ha(AHa)^-$ from the corresponding alkali (A) halides (Ha) [23,24] .

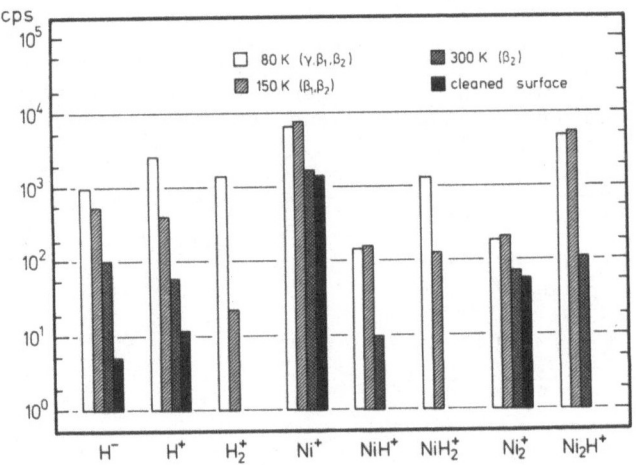

Fig.5 Secondary ion emission from clean and hydrogen-saturated nickel surfaces at various temperatures [28] . The ß states correspond to dissociative adsorption of hydrogen atoms, the γ state (80 K) corresponds to adsorbed H_2 molecules. The γ state is stable only at low temperatures. Primary ions: Ar^+, 3 keV; $5 \cdot 10^{-9}$ A on 0.1 cm^2

The secondary ion emission from metal surfaces Me covered by adsorbed atoms, radicals or small molecules B shows another characteristic regularity: secondary ions of the general composition MeB^+ or Me_2B^+ are emitted. Their yields increase proportional to the relative coverage of the surface. These ions may be considered as cationized species B. They are observed for large organic molecules, too [25,26] . A detailed investigation of this behaviour has been carried out for Ni surfaces covered by H, H_2, C, OH, C_2H_2 and C_2H_4 [27] . Figure 5 presents the emission of some characteristic secondary ions from a Ni surface covered by hydrogen atoms (ß states) and H_2 molecules (γ state).

3. The Secondary Ion Emission of Organic Molecules

3.1 Main Results

A large body of experimental results concerning the secondary ion emission behaviour of a wide variety of organic molecules is available today. It includes important groups of compounds as amino acids [7] , peptides [29] , nucleosides, nucleotides [30] , vitamins [8,31] , drugs, pharmaceutical products [32,33], etc., up to the mass range of several 1000 amu. Although very different sample preparation techniques are applied, the main features of the secondary ion emission behaviour of organic molecules are independent of the special preparation technique.

The most important secondary ions emitted from organic molecules are the protonated and deprotonated species $(M+H)^+$ and $(M-H)^-$. Occasionally dimers as $(2M\pm H)^\pm$ are emitted with relatively low intensities [31] . Other important parent-like ions are molecules attached to a positive metal (Me^+) [25,26] , or a negative halide (Ha^-) ion [33] $(M+Me)^+$ or $(M+Ha)^-$. In some cases M^- or M^+ ions are emitted. This depends strongly on the chemical nature of the sample molecules and its environment. In

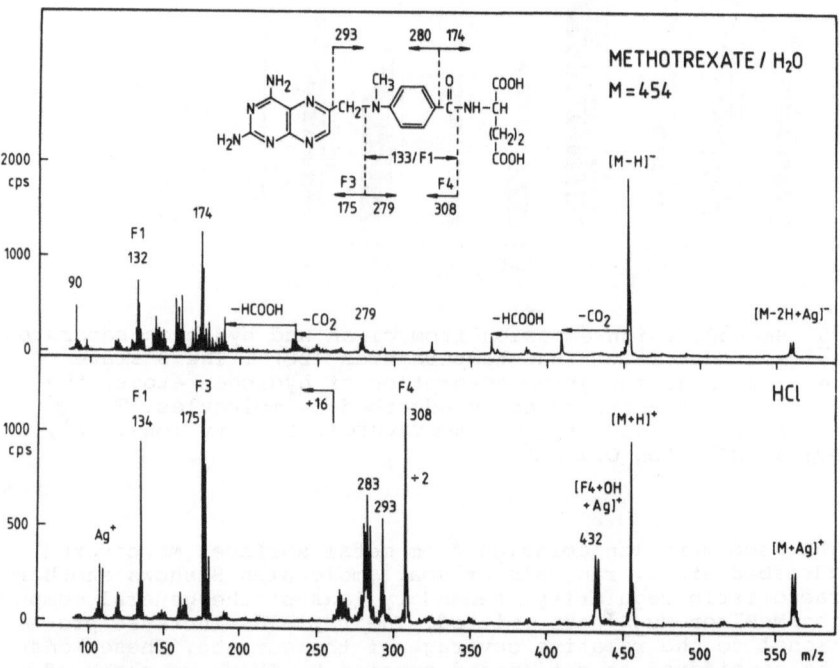

Fig.6 Positive and negative secondary ion spectrum of methotrexate on Ag. 2 nmol, dissolved in 1 μl of distilled water (upper spectrum), respectively in 0.1 n HCl (lower spectrum) were deposited on 1 cm². Primary ions: Ar^+, 3 keV; $2.5 \cdot 10^{-10}$ A on 0.1 cm² [34]

addition to these so-called parent-like ions, characteristic fragments in the high mass range and unspecific fragment ions in the low mass range are emitted.

A most striking fact is the high intensity of the parent-like secondary ion emission - provided that the sample molecules are in an appropriate chemical environment. The yield of parent-like secondary ions as $(M-H)^-$ or $(M+H)^+$ may vary by orders of magnitude depending on this chemical environment. Yields of more than 0.1 have been observed [9] . This emission behaviour, especially for involatile and thermally labile molecules, was not expected from the previous results on the secondary ion emission from solids. From this experience one mainly expected the emission of small fragments and a decrease of their intensities with increasing number of atoms in a fragment.

As an example, Fig.6 presents the positive and negative secondary ion spectrum of methotrexate on a silver surface. In addition to characteristic fragments, parent-like secondary ions $(M+H)^+$, $(M-H)^-$ and $(M+Ag)^+$ appear in the spectra with high yields. The fragments in the low mass range originate from the sample molecules as well as from surface contaminations.

The secondary ion yields depend very strongly on the chemical environment of the sample molecules. The properties of the bombarding particles have only a minor influence on the yields. This can be explained on the base of our present knowledge of the sputtering process. The primary particle bombardment, however, has a severe effect on the organic molecules. They are destroyed in an area of some 10^{-7} cm around the point of primary particle impact. This corresponds to a typical damage or disappearance cross section σ of some 10^{-14} cm^2 for these molecules.

3.2 Parameters Governing the Secondary Ion Formation Process

For a more detailed analysis of the secondary ion formation process, it is useful to distinguish three different steps of this process: the energy and momentum deposition in a surface near area of the target, the transfer and distribution of this energy into an excited volume, and the separation and ejection of the secondary ion from the surface, provided that a sufficient amount of kinetic energy is transferred to the corresponding surface species (Fig.7).

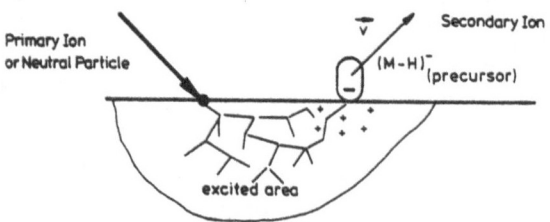

Fig.7 In the precursor model the emission of a parent-like secondary ion as $(M-H)^-$,e.g., is considered as the result of a fast $(t < 10^{-12}$s) transfer of kinetic energy to a preformed ion (precursor)

It follows that the parameters determining the secondary ion formation can be separated into two groups: the bombardment conditions, and the chemical environment of the sample molecules.

The bombardment conditions are mainly determined by the mass, composition, energy and angle of incidence of the bombarding particles. These parameters determine the properties of the excited volume. The chemical environment of the sample molecules is mainly determined by the chemical nature of the substrate, the relative coverage of the substrate by the sample molecules M, and the concentration of coadsorbed atoms, radicals or molecules, originating from the solution, its additives or the ambient atmosphere. The chemical environment is only indirectly determined by the temperature of the sample and its ion or electron prebombardment.

3.3 Sample Preparation Techniques

Due to the extraordinary influence of the chemical environment on the secondary ion formation process, the sample preparation is a crucial step in secondary ion mass spectrometry of organic compounds. It should be carefully described in all publications. The sample may be prepared by depositing and drying an appropriate solution of a solid surface, a metal substrate,e.g. by depositing the molecules under investigation on an atomically clean surface, by a molecular beam technique, or a liquid matrix may be applied (Fig.8).

Sample Preparation

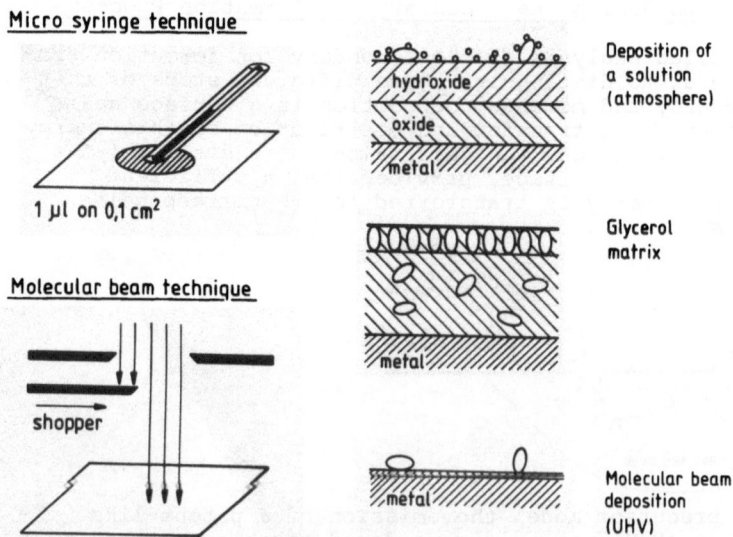

Micro syringe technique

1 μl on 0,1 cm²

Molecular beam technique

shopper

Deposition of a solution (atmosphere)

hydroxide
oxide
metal

Glycerol matrix

metal

Molecular beam deposition (UHV)

metal

Fig.8 Different sample preparation techniques result in different sample compositions

Table 1 Surface concentrations resulting from the homogeneous
deposition of different solutions (1 µl) of the sample molecules
(M = 300) on 1 cm^2 of a substrate surface

Concentration in the solution (mol/l)	10^{-2}	10^{-3}	10^{-4}	10^{-5}	10^{-6}
Surface concentration (mol/cm^2)	10^{-8}	10^{-9}	10^{-10}	10^{-11}	10^{-12}
Relative surface coverage Θ	10	1	0.1	0.01	0.001
Number of deposited molecules in a bombarded area of 0.1 cm^2	$6 \cdot 10^{14}$	$6 \cdot 10^{13}$	$6 \cdot 10^{12}$	$6 \cdot 10^{11}$	$6 \cdot 10^{10}$
Amount of deposited material (ng) in a bombarded area of 0.1 cm^2	300	30	3	0.3	0.03

The most frequently applied preparation technique for analytical purposes is the solution of the sample molecules in
a liquid. A small amount of the solution is deposited on the
target surface, and the solvent is evaporated. Depending on
the amount and concentration of the deposited solution, this
preparation results in submonolayers, monolayers or multilayers
of the sample molecules on the substrate surface. Table 1 gives
the relation between sample concentrations in the solution and
the corresponding concentrations of the sample molecules on the
surface.

For a more detailed investigation of the influence of the
chemical environment on the ion formation probability, more defined surface compositions are required. They can be achieved
by applying a molecular beam technique for the deposition of
controlled amounts of sample molecules on atomically clean metal surfaces under UHV-conditions (Fig.8).

A very important technique for practical analysis is the deposition of the sample molecules in a liquid matrix of very low
vapour pressure, in general glycerol. In such a liquid the target molecules are mobile and, in general, will suffer a certain
surface segregation, resulting in a considerable surface enrichment. During enhanced ion bombardment of such a liquid matrix,
sample molecules will be destroyed on or removed from the surface. Due to the high mobility of the sample molecules in the
liquid, however, the disappearing molecules are rapidly replaced
so that by this regeneration effect a stable secondary ion signal may be obtained over a long period of time despite the permanent destruction and sputtering of sample molecules by the
primary particle bombardment.

3.4 Secondary Ion Emission from Samples Prepared by Depositing
and Drying of a Solution

In secondary ion mass spectrometry of organic compounds most of
the solid samples are prepared by the deposition of an appropriate

solution on a metal substrate. In our standard preparation technique, 1 μl of a 10^{-3} solution, containing about $6 \cdot 10^{14}$ molecules, is deposited on 1 cm^2 of an etched silver surface. Due to the strong interaction between the sample molecules and the metal surface, this virtually results in the formation of a single monolayer (Table 1). Only for sample concentrations exceeding one monolayer equivalent, a multilayer and crystallite formation on the substrate surface starts. Lower coverages can be obtained by applying lower concentrations of the sample molecules in the deposited solution [35] .

The chemical nature of the substrate has a very strong influence on the emission of the different parent-like secondary ions. Their yields are high for noble metals as Ag, Au, and Pt , they are low for reactive metals as Ni or Fe, e.g. The reason seems to be a decomposition of the organic molecules on the more reactive metals.

Up to a monolayer equivalent we found a linear increase of the parent-like ion yields with the surface concentration of the sample molecules. The yields are very sensitive to the chemical composition of the solution. For leucine on Ag, e.g., the yield for $(M\pm H)^\pm$ can be changed by more than one order of magnitude for a constant surface concentration of the leucine molecules, only by changing the acidity of the solution. A similar behaviour is observed for the emission of the Li-cationized molecules $(M+Li)^+$ and $(M-H+2Li)^+$, if the lithium concentration is changed in the solution (Fig.9).

In contrast to the extreme influence of the chemical environment on the secondary ion yields, the properties of the bombar-

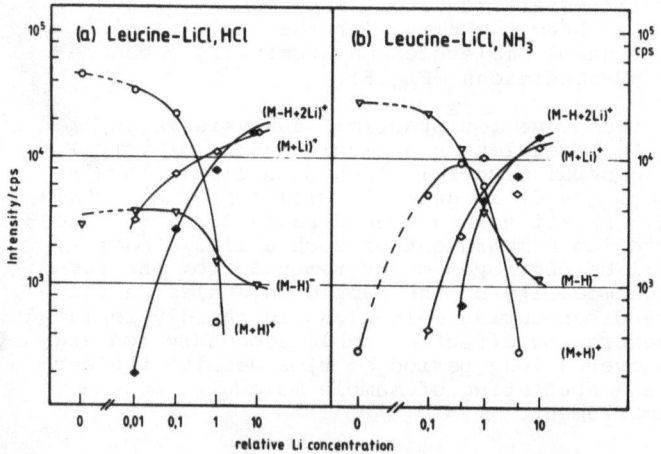

Fig.9 Cationization of leucine at different lithium concentrations [26,35] . Sample solutions: 10^{-3} mol/l leucine in 0.1 mol/l HCl (a) and 0.01 mol/l NH_3 (b) respectively. Constant surface concentration of leucine: 100 ng·cm^{-2}, Primary ions: Ar^+, 3 keV; $5 \cdot 10^{-10}$ A on 0.1 cm^2

ding particles have only a minor influence on the secondary ion formation. It can be explained by changes in the total sputtering yield. So the ion emission increases with increasing mass and energy of the primary ions, but the damage cross section does this in a similar way (Fig.10). As a result the transformation probability does not change too much - in contrast to the very strong influence of the chemical environment on this probability. The charge state of the bombarding particle has no detectable influence on the secondary ion emission [35] .

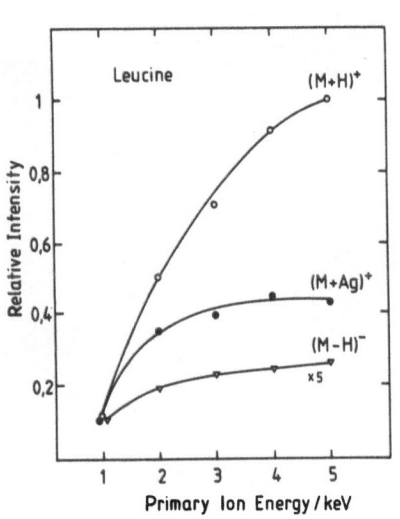 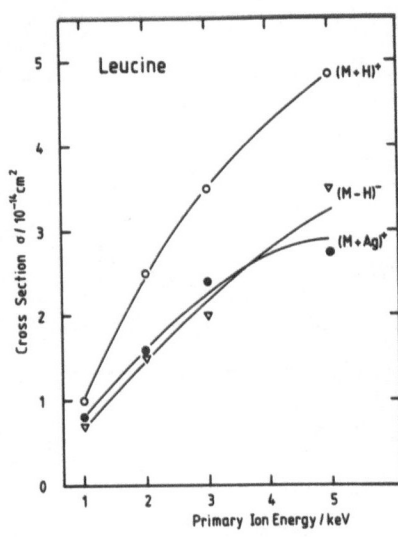

Fig.10   Energy dependence of some secondary ion yields and damage cross sections for an Ar^+-bombarded leucine monolayer on Ag. The transformation probabilities $P(M \rightarrow X_i)$ (eq.4) are relatively stable [35]

3.5   Molecular Beam Deposited Amino Acids on Clean Metal Surfaces

In order to get a more detailed insight into the secondary ion formation process, we have extensively studied the secondary ion emission from metal-supported amino acids prepared under ultra-high vacuum conditions by a molecular beam technique [36, 37]. The coverage determination was based on static AES and microbalance measurements. The metal substrate surfaces were cleaned by heating and ion bombardment and controlled by dynamic AES and SIMS. From deposition and reevaporation experiments at various substrate temperatures, we found two different bonding states of the amino acid on the surface of reactive metals, including Ag. The first monolayer is strongly bound to the metal surface (no reevaporation at room temperature), whereas the second and the following layers are only weakly bound. On the noble metals Pt and Au, even the first monolayer is only weakly bound.

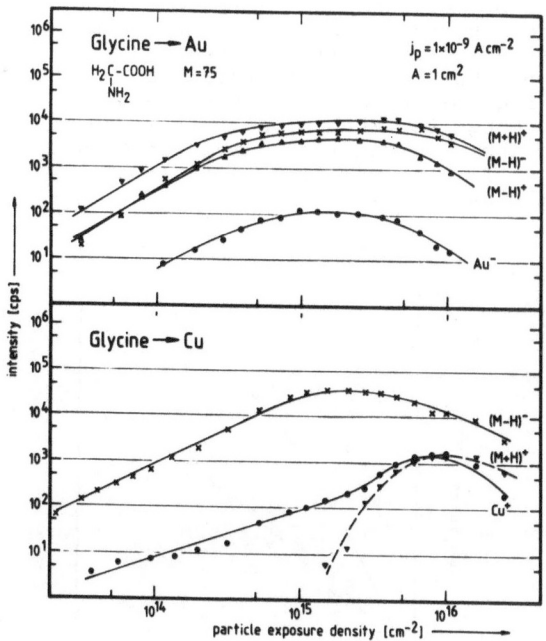

Fig.11. Changes in the secondary ion emission from an Au and a Cu surface during molecular beam deposition of glycine [36]. A closed monolayer is formed at an exposure of about 10^{15} glycine molecules on 1 cm². Primary ions: Ar^+, 3 keV; $1 \cdot 10^{-9}$ A cm^{-2}

By changes of the relative surface coverage, the chemical nature of the substrate, its temperature, and the ion or electron prebombardment, we found a very different secondary ion emission behaviour for these different bonding states. This may be illustrated by the coverage dependence of the parent-like secondary ion emission from Au and Cu during an increase of Θ (Fig.11). For Cu, a reactive metal, the $(M+H)^+$ emission starts only after the completion of the first strongly bound monolayer, whereas the $(M-H)^-$ emission increases proportional to the surface coverage in the submonolayer range. On a gold surface, however, $(M+H)^+$ and $(M-H)^-$ show the same increase in the submonolayer range.

This behaviour together with the results of the evaporation behaviour of amino acids from metal substrates support the assumption that the first monolayer on reactive metals is bound as an anion $(M-H)^-$. It explains the strong emission of this complex as a negative secondary ion and the absence of a $(M+H)^+$ emission, as well as the behaviour of the secondary ion emission during heating and enhanced ion or electron bombardment. More details of these investigations are summarized in another contribution to these proceedings [37] .

By isotope exchange experiments, we found that a proton transfer at the amino acid group is responsible for the formation of the $(M+H)^+$- and $(M-H)^-$-ions, whereas the $(M-H)^+$ formation on an Au surface is related to the loss of a directly carbon-bound hydrogen atom.

4. The Precursor Model of Secondary Ion Formation

From the experimental facts summarized in the preceding paragraph, some general features concerning the secondary ion emission from organic compounds result. These general features allow the formulation of a qualitative model of the ion formation process in SIMS. The model meets two important requirements: it describes the experimental facts and allows predictions. It seems to be applicable to ^{252}Cf-plasma desorption and laser desorption, too.

The model is based on the generally accepted assumption that an excited area is generated around the point of primary ion impact [17] . The momentum and energy transfer from the bombarding particle to the lattice and the dissipation of this energy in a volume of some 10^{-7} cm diameter is responsible for the creation of this excited area. Parent-like secondary ions are generated by the fast transfer of relatively small amounts of kinetic energy to a precursor (preformed ion) on the surface of the excited area (Figs 2 and 7). The extremely short period of time ($< 10^{-12}$ s) in which the energy transfer occurs seems to be responsible for the fact that the vibrational excitation of the emitted molecules M and their subsequent fragmentation is drastically reduced, compared with the relatively slow thermal evaporation.

The finally emitted parent-like secondary ion species may leave the surface as a neutral particle or as an ion. This strongly depends on its bonding state at the surface. All experimental results confirm that the charge sign of the emitted secondary ion is determined by the charge state of its precursor. During its separation from the surface the charge sign has the tendency to be conserved, i.e. a negatively charged surface precusor, as an anion complex, e.g., will be emitted with a high probability as a negatively charged ion.

Some considerations on the energy transfer to surface species during sputtering are described in another contribution to these proceedings [38] . Figure 12 presents in a schematic way the average energy $\bar{E}(r)$ transferred to a surface species as a function of r, indicating its distance to the point of primary ion impact. In the direct vicinity of this point ($r < R'$) a relatively large amount of energy is transferred to a surface species. Hence the probability of its destruction and the emission of fragment or atomic ions is very high, too. If only very small amounts of energy are transferred to a precursor (large distance $r > R$), no desorption can occur. It is only in the intermediate area $R' < r < R$, where the energy transfer can result in the emission of an unfragmented parent-like secondary ion, provided the molecule M was in the appropriate precursor state.

The assumption that only by a very limited energy transfer the formation of the parent-like ions can be achieved is supported by their relatively low kinetic energy, compared with the high kinetic energy of the atomic ions. Figure 13 presents as an example retarding field measurements for Ag^+ and $(M+Ag)^+$, both emitted from the same surface. Similar energy distributions as for $(M+Ag)^+$ are found for other parent-like secondary ions as $(M+H)^+$ and $(M-H)^-$,e.g.

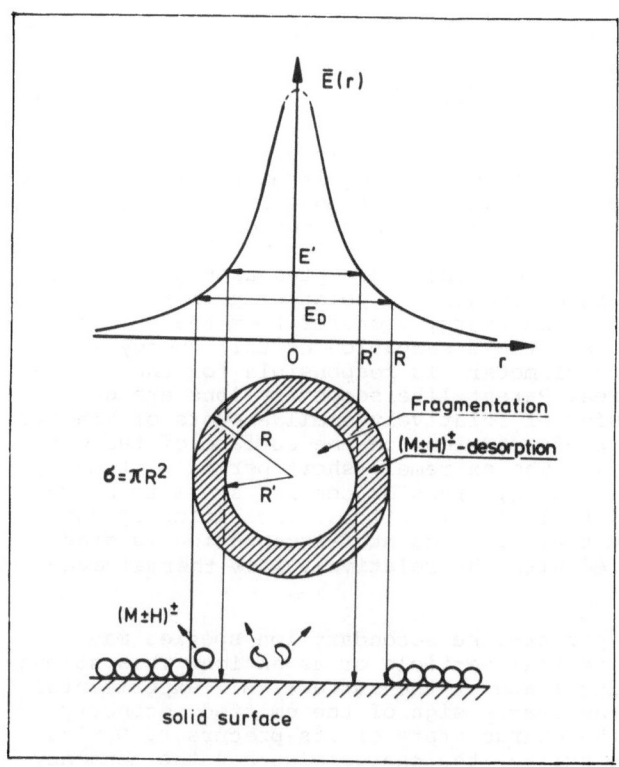

<u>Fig.12</u> Energy transfer to parent-like secondary ion precursors
during their sputtering from a surface. The energy distribution
$\bar{E}(r)$ represents the average energy transferred by an impact cascade to a surface molecule M, located in a distance r from the
point of primary ion impact [2] . This energy transfer results
in the disappearance of M for r < R. Only for R'< r < R (energy
transfer between E_D and E') does this result in the emission of an
intact parent-like secondary ion as $(M-H)^-$,e.g. Note that E (r)
does not describe one single impact process, but averages a
large number of these events

The precursor model can be summarized in four main points
[2,35,36] :

1. <u>Precursor formation:</u> Before any particle bombardment, a
    precursor of the finally emitted secondary ion exists on
    the target surface. This concerns the atomic composition
    as well as the charge sign of the finally emitted secondary
    ion.

2. <u>Fast transfer of kinetic energy:</u> The very fast (< 10^{-12} s)
    transfer of a small amount of kinetic energy to the pre-
    cursor results in a high probability of being emitted as an
    unfragmented parent-like secondary ion.

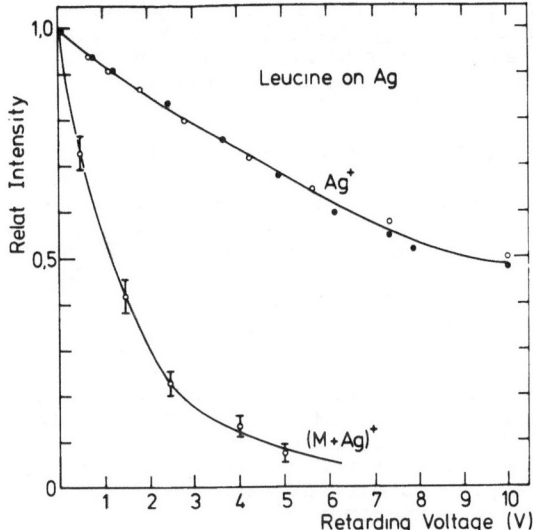

Fig.13 Results of retarding field measurements for Ag$^+$ and (M+Ag)$^+$ ions, emitted from a submonolayer leucine on silver. Primary ions: Ar$^+$, 3 keV. All parent-like secondary ions are emitted with an energy distribution similar to that of (M+Ag)$^+$

3. Charge sign conservation: There exists a tendency to maintain the average charge sign of the precursor during its separation from the surface and its transformation into the corresponding secondary ion.

4. Fragmentation: A high energy transfer to a surface molecule results in a high probability for this particle to be fragmented. This may cause the emission of characteristic fragments as well as atomic ions. In addition, fragmentation may occur by the decomposition of excited molecular ions.

There exists an excellent agreement between the experimental facts and this model. For a given precursor, the secondary ion emission should be determined by the extension of, and the energy and momentum distribution in the excited area. Therefore, in a first approximation, one expects a similar change in the secondary ion yields $Y(X_i)$ as we know it for the total sputtering yield of a solid (a metal,e.g.), and relatively constant transformation probabilities $P(M \rightarrow X_i)$, if the mass, the energy, or the composition of the bombarding primary particles is changed. This is in excellent agreement with the experimental results. On the base of these considerations we can predict and understand the increase of $Y(M)$, $Y(X_i)$ and $\sigma(M)$ with increasing mass and energy of the bombarding particle, as well as the fact that the highest secondary ion emission from a solid or a liquid matrix is obtained if it is bombarded by heavy [39] and molecular [40] ions. The model also explains the complete independence of the secondary ion emission behaviour of a given target on the charge state of the bombarding particle [36,37] . If, however, we consider the dependence of the energy distribution $\bar{E}(r)$ (Fig.12) on the primary ions' properties as their mass, energy and composition, we expect changes in the transformation probabilities $Y(M \rightarrow X_i)$ if we change these parameters.

The chemical state of a molecule M on the target surface is determined by its chemical nature, by the composition of the deposited solution, and by the chemical nature of the substrate. The assumption of a precursor formation and a conservation of its average charge sign explains the strong emission of Me^+ and O^- from an oxide [2-5] as well as the emission of SO_4^-, NO_3^-, CO_3^-, etc. from the corresponding surface complexes [22] .

The secondary ion emission behaviour of metal-supported amino acids [37] can easily be explained within the frame of the model. Obviously amino acid molecules can be in different bonding states on a metal surface: as Me-anion complexes $(Me^{\pm}(M-H)^-)$ or as amino acid dimers $((M-H)^- - (M+H)^+)$. The first state results in the emission of $(M-H)^-$, and virtually no $(M+H)^+$. From the second state both of these ions are expected with approximately the same intensity. This is perfectly confirmed by the experimental results (Fig.11) [36,37] .

The precursor model is an empirical model. Up to now it is not possible to deduce it from simple physical laws. It explains the influence of the chemical environment on the secondary ion formation by the assumption of a precursor. The other main assumptions of the model are the sputtering of the precursor by fast energy transfer and the charge sign conservation, which is a general observation in secondary ion mass spectrometry of organic as well as inorganic material. Whereas it is relatively easy to understand the precursor formation, it is difficult to explain the effect of fast energy transfer on the vibrational excitation and the charge sign conservation during the separation of the precursor from the surface.

Also in ^{252}Cf-plasma desorption and laser desorption the precursor formation seems to play an important part in the ion formation process, and the fast transfer of kinetic energy to these precursors seems to be the main reason for the appearance of the parent-like ions in the corresponding spectra. This explains the striking similarity in the SIMS , PDMS , and laser desorption spectra. In field desorption, the energy transfer is different, but from recent results [41] it is evident that the preformation of ions plays an important role in this ionization process, too.

5. Analytical Applications

5.1 General Considerations

The outstanding analytical power of SIMS is based on the fact that by sputtering even large and thermally labile organic molecules can be transformed with a high probability into parent-like secondary ions. Thermally labile large organic molecules play a most important role in biochemical, medical and environmental sciences. This is the reason for a rapidly increasing application of SIMS in biomedical research, in toxicology, in environmental control, for the detection of drugs and pharmaceuticals, in clinical analysis, etc., and its importance as an on-line or off-line detector in HPLC [41] and TLC [43] .

Regarding the wide variety of analytical questions in the various fields mentioned above, it is helpful to distinguish three main groups of analytical problems:

A. The identification and structural investigation of a uniform compound, available in a large quantity.

B. The identification and possibly structural analysis of a very small quantity of an isolated uniform compound.

C. The identification and possibly structural analysis of a small quantity of a certain compound, in a mixture together with many other compounds.

For an assessment of the analytical power of SIMS for these different problems, four main steps in a SIMS analysis should be distinguished:

-   The sample preparation
-   The transformation of the precursors in secondary ions
-   The e/m analysis of these ions
-   The detection of the separated ions.

These different steps must be optimized with regard to the special problem to be solved. This optimization will be different for the different groups of analytical problems mentioned above.

The sample preparation is a crucial step in every SIMS analysis. It determines the chemical state of the sample molecule on the bombarded surface. Therefore it must be the intention of the sample preparation to transfer the sample molecules in appropriate precursors of analytically useful parent-like ions. A wide variety of measures is available in order to optimize this precursor formation: the choice of the substrate material, the solvent, and the surface concentration, the addition of promotors as alkali ions or acids, etc.

The transformation of the precursor into a free secondary ion occurs through the energy transfer from the bombarding particle to the excited area. As the primary particle properties as mass, energy and composition have only a limited influence on the transformation probabilities $P(M \rightarrow X_i)$, these parameters are not very critical, provided that the excitation source allows a sufficient variation of the primary particle flux density $\sqrt{}$. If this flux density is limited, the secondary ion current can be increased by applying high energy, high mass and molecular bombarding particles. But it should be emphasized that this does not considerably influence the transformation probabilities.

Depending on the kind of information one needs, the e/m analysis of the emitted secondary ions should be carried out by different types of instruments. In most of the early SIMS investigations quadrupole mass spectrometers with a relatively high transmission but limited mass range have been applied [3,4] . By the recent introduction of the liquid matrix, SIMS became available for low transmission instruments as high resolution double-focussing magnetic mass spectrometers. A most promising instrument for the analysis of secondary ions, especially if only a very limited number of sample molecules is available, is the time-of-flight instrument with its unlimited mass range, its very high transmission and its capacity of simultaneous detection of all masses. A serious limitation of these instruments, however, is their relatively low mass resolution [31,44] .

The detection of the mass-separated secondary ions is the last step in a SIMS analysis. The mass-separated secondary ion currents are in general very low, so that single ion detection should be applied. For a high efficient and stable detection of large organic molecules, a high post acceleration is mandatory [31] . By a post acceleration of 20 keV a detection probability of about 50 % for ions with the mass number 5000 can be achieved.

In the following, special solutions for the different groups of analytical problems - identification of sample molecules available in a large quantity, identification of separated traces, and analysis of complex mixtures - will be discussed in some more detail.

5.2 Liquid Matrices

Until recently the application of secondary ion emission to low transmission instruments was limited by the fact that the secondary ion current from a monolayer of sample molecules on a solid substrate rapidly decreases during enhanced ion bombardment (eq.3). The deposition of the sample molecules as multilayers on the substrate surface did not solve the problem: during sputtering of the uppermost monolayer the next layers are considerably destroyed by the primary beam. The problem can be solved, however, by depositing the sample molecules in a liquid that has to meet some important requirements:

-   It should guarantee a high mobility of the sample molecules, so that the loss of particles during sputtering from the surface can be compensated.

-   The liquid should promote a favourable precursor formation, so that with a high probability a sample molecule will be transformed into a parent-like secondary ion.

-   The vapor pressure of the liquid should be sufficiently low ($< 10^{-4}$ Torr), so that the mean free paths of the bombardding particles and the secondary ions remain sufficiently large.

Glycerol was found to have all these properties [45] . It is the most successfully applied liquid matrix in SIMS [39] .

The processes in a liquid matrix may be illustrated by the results of Fig.14, where the emission of some representative

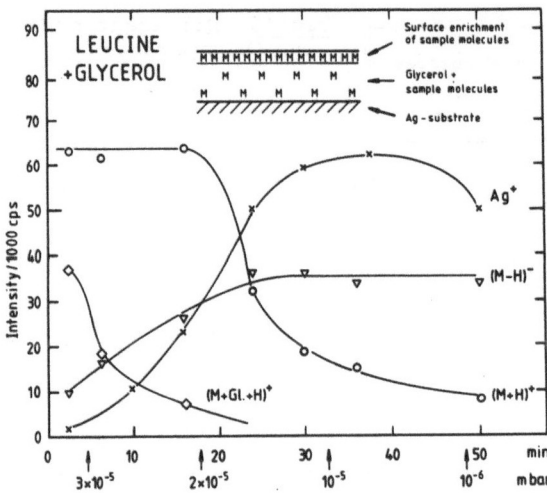

Fig.14 Secondary ion emission from leucine in a glycerol
matrix, deposited on an Ag substrate, during evaporation
of the glycerol.
Sample preparation: 4 µl glycerol mixed with 1 µl of a 10^{-2} mol/l
leucine solution in 0.2 mol/l HCl, deposited on 1 cm^2.
Primary ions: Ar^+, 3 keV; $5 \cdot 10^{-10}$ A on 0.1 cm^2

secondary ions from a silver surface covered by leucine in a
glycerol matrix is presented. Parent-like secondary ions are
emitted from the liquid matrix with high intensities for about
20 minutes. The secondary ion yields are stable, even if the
primary ion flux density ∨ is equivalent to the removal of several monolayers per second. Obviously sample molecules disappearing from the surface during the sputtering process are
replaced by sample molecules diffusing from the bulk of the
glycerol matrix to the surface. After the evaporation of the
glycerol, which is indicated by a corresponding decrease of the
pressure in front of the target and the appearance of Ag-con-
taining secondary ions, the emission of glycerol-related secondary ions decreases. By a more detailed investigation [35] we
found that the transformation probabilities $P(M \rightarrow (M^{\pm}H)^{\pm})$ are
of the same order of magnitude for sputtering the molecules M
from the glycerol matrix or from the silver surface.

The application of a glycerol matrix was the crucial step
that made low transmission double-focussing instruments available
for secondary ion mass spectrometry of organic compounds. A minor
problem to be solved was the primary beam transfer into the
ion source which is, in general, on a high potential. In the
first applications of liquid matrices, all ion optical problems
were avoided by using a beam of neutral atoms as bombarding particles. Unfortunately this modification of the bombardment
conditions was overestimated by calling the secondary ion mass
spectroscopy with a liquid matrix "fast atom bombardment (FAB)".
The confusion was increased by the fact that the crucial step

| | Excitation | Fast Heavy Particles (atoms,molecules) keV-MeV range | |
Target		IONS	NEUTRALS
SOLID (Mono-,Multilayer)			
LIQUID (Glycerol)			

Fig.15 Different modes of SIMS operation

in the application of SIMS to high resolution double-focussing mass spectrometers - the liquid matrix - was not mentioned in the first publications on "FAB".

5.3 Time-of-Flight Instruments

If only a very limited number of sample molecules is available for an analysis, a high efficient and stable precursor formation is important. A maximum use of the secondary ions generated from this precursor state in the following e/m analysis can be achieved by the application of a time-of-flight (TOF) instrument. Its typical overall transmission of some 10 % exceeds that of a quadrupole by more than 3 orders of magnitude. The simultaneous detection of all masses in such an instrument gives an additional factor of at least 100 in practical sensitivity, so that compared with the quadrupole mass spectrometer used up to now as the most sensitive static SIMS instrument, we obtain a theoretical increase of sensitivity by a factor of at least some 10^5 !

A time-of-flight instrument for secondary ion mass spectrometry may be of a straight forward construction or it may be more sophisticated and adapted to the special needs of secondary ion analysis. In our laboratory we have developed and operate now a TOF instrument that meets some important requirements. It is equipped with a pulsed, mass separated primary ion source that enables the removal of a monolayer on 0.1 cm^2 in some minutes, an energy and angle focussing time-of-flight path, and a post-acceleration combined with an ion- electron-photon converter, providing a stable and efficient detection of large molecular ions up to several 1000 amu.The instrument can be operated for positive and negative ions. It is described in detail in another contribution to these proceedings [31] .

If we consider the application of a TOF instrument with 10 % overall transmission to sample molecules which are transformed into useful secondary ions X_i with a probability of 10 %, this means that 1 % of the sample molecules is transformed into detected parent-like secondary ion species X_i. Under these assumptions 10^{-15} mol result in 10^6 detected parent-like ions.

The successful use of this very high sensitivity, however, remains a question of sample handling, sample contamination and background spectra.

5.4 Planar Separation Techniques

The identification and quantitative determination of a certain molecular species in a sample containing a large number of compounds presents a most difficult but most realistic and exciting analytical problem. It plays a central role in environmental control, in the analysis of biological samples, in clinical diagnostic and in many other fields.

In general, the direct SIMS analysis of the original mixture is not possible due to the interferences of parent-like and fragment ions in the spectra. The normal way to solve the problem is the preseparation of one or a limited number of components by liquid chromatography (HPLC) or thin layer chromatography (TLC), and the subsequent SIMS identification of the separated species. Successful coupling of SIMS with HPLC and TLC, both in the offline and in the on-line mode, are reported in literature [47] .

There is, however, a general problem in the combination of a monolayer-sensitive detection technique as SIMS with a preseparation technique as HPLC or TLC, which are based on surface separation processes. In a combined HPLC-SIMS analysis, e.g., the different species in a solution are separated by their different residence times in the adsorbed state at the solid-liquid interface. These different residence times finally result in their separation. For the SIMS analysis the eluate of the HPLC is then again deposited on a solid surface. Here the solvent is evaporated and the sample molecules are transferred in an appropriate precursor state. It is an exciting idea to combine these processes - the preseparation and the precursor and secondary ion formation - on the same surface [46] .

The stationary phase materials in HPLC or TLC are highly porous. Only a very small and negligible fraction of their total surface is an outer surface and accessible to the probing ion beam. This is the reason why this stationary phase material is inappropriate for a combination of the separation and secondary ion formation steps on the same solid surface. The combination needs a flat and planar surface.

The central idea of a planar separation is the isolation of one or a limited number of components of a more or less complex mixture in a solution on a plane surface, from which a direct secondary ion emission can occur. The isolation can be achieved by selective precipitation on a specially prepared surface, or the resolution of all disturbing compounds from a surface on which the complete mixture has been deposited.

We have carried out some preliminary investigations on the separation of mixtures on solid surfaces [46] . Figure 16 presents as an example of this planar separation technique the isolation of adenine on a silver surface from a mixture of alanine, adenine, phenylalanine and sulphanilamide.

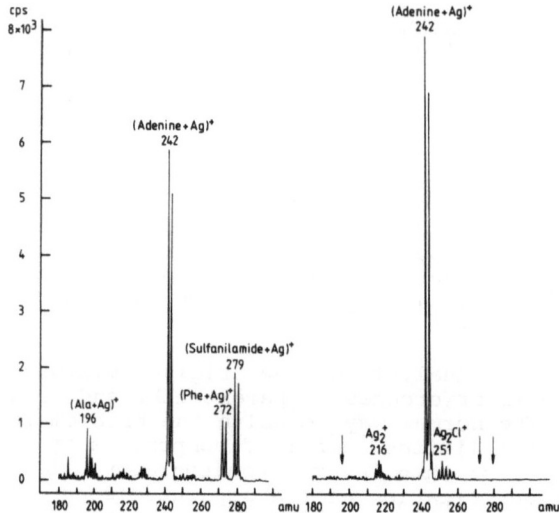

Fig.16 Planar separation technique. Example of a planar separation on a silver surface. Spectrum (A) was obtained from an
etched silver surface, after dipping it in an aqueous solution
of alanine, adenine, phenylalanine and sulphanilamide
(10^{-3} mol/l each). Spectrum (B) is from the same surface
after rinsing it with an appropriate solvent. Of the four
compounds of the initial solution only the adenine molecules
are now isolated on the surface

Regarding the large number of parameters that determines the
residence time of a molecule at the solid liquid interface, it
should be possible to develop specific isolation techniques by
special choices of substrate compositions, solvents and additives that allow the isolation of any compound out of any mixture by such a planar separation technique.

6. Future Developments

Secondary ion mass spectroscopy of organic molecules is still in
a very formative stage. Its development in the near future will
follow two main directions: the study of the ion formation and
emission event and the related processes in the bulk and on the
surface of the bombarded target on one hand, and the extension
and improvement of the analytical applications in various fields
on the other hand.

Up to now, we do not know too much about details of the processes following the impact of a primary particle on a surface
that finally results in the emission of molecular secondary ions.
Therefore the investigation of the ion emission process will include the question of energy transfer from the bombarding particle to the solid or liquid and finally to the emitted surface
species, the question of vibrational excitation during the trans-

fer of kinetic energy, and the charge exchange between the emitted particle and the surface during the emission process. A very important question is how changes in the bombardment conditions change the energy distribution in the excited surface area around the point of primary particle impact and how these changes influence the transformation probabilities $P(M \rightarrow X_i)$.

Another question concerning the ion formation and emission process that should be investigated more extensively in the future is the surface chemistry and the precursor formation. The strong dependence of the precursor formation and the secondary ion emission on the chemical environment of the surface molecules is important for an understanding and for the prediction of secondary ion formation from samples prepared by different techniques. On the other hand, SIMS opens the possibility to study details of this surface chemistry, indirectly even at the solid-liquid interface.

Besides the study of the ion formation and emission process, the improvement and extension of its analytical application will be the second important direction in future development of secondary ion mass spectrometry. Also in this field some main problems can be isolated. One is the optimization of the ion formation probabilities by applying the results of an extensive study of the ion formation and emission process. The intention must be to produce a maximum of analytically useful ions from a given number of sample molecules by optimizing mainly the precursor formation but also the energy transfer to these precursors. This optimization will be focussed mainly on the sample preparation, but also on the bombardment conditions.

Another important problem that will be treated in the near future is the analytical application of SIMS to the analysis of mixtures of organic molecules and the adaption of known pre-separation techniques, considering that secondary ion mass spectrometry is primarily a monolayer technique. These developments will include planar separation and electrical field assisted separation techniques. The real use of the extreme sensitivity of SIMS with time-of-flight instruments, the problem of handling small numbers of molecules, the control and limitation of contaminations during sample preparation, surface deposition and sample introduction into the secondary ion source are further problems.

References

1. SIMS II, Proceedings of the Second International Conference on Secondary Ion Mass Spectrometry, Stanford 1979, A.Benninghoven et al., eds., Springer-Verlag, Berlin-Heidelberg-New York, 1980
2. SIMS III, Proceedings of the Third International Conference on Secondary Ion Mass Spectrometry, Budapest 1981, A.Benninghoven et al., eds., Springer-Verlag, Berlin-Heidelberg-New York, 1982
3. A.Benninghoven: Surf.Sci. 35, 427 (1973)
4. A.Benninghoven: Surf.Sci. 53, 596 (1975)
5. A.Benninghoven: Z.Physik 230, 403 (1970)

6.   A.Benninghoven, A.Müller: Surf.Sci. $\underline{39}$, 416 (1973)
7.   A.Benninghoven, D.Jaspers, W.Sichtermann: Appl.Phys. $\underline{11}$, 35 (1976)
8.   A.Benninghoven, W.Sichtermann: Org.Mass Spectrom. $\underline{12}$, 595 (1977)
9.   A.Benninghoven, W.Sichtermann: Anal.Chem. $\underline{50}$, 1180 (1978)
10. D.F.Torgerson, R.P.Showronski, R.D.Macfarlane: Biophys. Res.Comm. $\underline{60}$, 616 (1974)
11. R.D.Macfarlane, these proceedings
12. R.J.Colton, J.S.Murday, J.R.Wyatt, J.J.DeCorpo: Surf.Sci. $\underline{84}$, 235 (1979)
13. R.J.Day, S.E.Unger, R.G.Cooks: Anal.Chem. $\underline{52}$, 557 (1980)
14. J.A.Gardella, D.M.Hercules: Anal.Chem. $\underline{52}$, $\overline{2}$26 (1980)
15. H.Kambara, S.Hishida: Org.Mass Spectrom. $\underline{16}$, 167 (1981)
16. M.Barber, R.S.Bordoli, R.D.Sedgwick, A.N.Tyler: J.C.S.Chem.Comm. $\underline{325}$ (1981)
17. P.Sigmund in Sputtering by Particle Bombardment, R.Behrisch ed., Springer-Verlag, Berlin-Heidelberg-New York, 1981
18. H.Oechsner, W.Gerhard: Surf.Sci. $\underline{44}$, 480 (1974)
19. B.J.Garrison, N.Winograd, D.E.Harrison, Jr.: Phys.Rev. $\underline{B\ 18}$, 6000 (1978)
20. A.Benninghoven, L.Wiedmann: Forschungsber. des Landes NW, Westdeutscher Verlag, Opladen, 1978
21. C.Plog, L.Wiedmann, A.Benninghoven: Surf.Sci. $\underline{67}$, 9 (1977)
22. A.Benninghoven, A.Müller: Phys.Lett. $\underline{40\ A}$, 169 (1972)
23. H.Franke, H.Hoinkes, H.Wilsch: Surf.Sci. $\underline{63}$, 121 (1977)
24. R.J.Colton, J.E.Campana, T.M.Barlak, J.J.DeCorpo, J.R.Wyatt: Rev.Sci.Instrum. $\underline{52}$, 1685 (1980)
25. H.Grade, N.Winograd, R.G.Cooks: J.Am.Chem.Soc. $\underline{99}$, 7725 (1977)
26. W.Sichtermann, A.Benninghoven: Intern.J.Mass Spectrom. Ion Phys. $\underline{38}$, 351 (1981)
27. A.Benninghoven, P.Beckmann, D.Greifendorf, M.Schemmer: Surf.Sci. $\underline{114}$, L62 (1982
28. A.Benninghoven, P.Beckmann, D.Greifendorf, K.H.Müller, M.Schemmer: Surf.Sci. $\underline{107}$, 148 (1981)
29. C.E.Costello, A.M.Van Langenhove, S.A.Martin, K.Biemann: these proceedings
30. A.Eicke, W.Sichtermann, A.Benninghoven: Org.Mass Spectrom. $\underline{15}$, 289 (1980)
31. P.Steffens, E.Niehuis, T.Friese, A.Benninghoven, these proceedings
32. A.Benninghoven, W.Sichtermann: Int.J.Mass Spectrom.Ion Phys. $\underline{38}$, 351 (1981)
33. W.Sichtermann, M.Junack, A.Eicke, A.Benninghoven: Fresenius Z.Anal.Chem. $\underline{301}$, 115 (1980), and $\underline{311}$, 410 and 411 (1982)
34. A.Eicke, V.Anders, M.Junack, W.Sichtermann, A.Benninghoven: Int.J.Mass Spectrom.Ion Phys. $\underline{46}$, 479 (1983)
35. M.Junack, A.Eicke, W.Sichtermann, A.Benninghoven: submitted to Anal.Chem.
36. W.Lange, M.Jirikowsky, A.Benninghoven: submitted to Surf.Sci.
37. W.Lange, D.Holtkamp, M.Jirikowsky, A.Benninghoven, these proceedings
38. D.Greifendorf, P.Beckmann, M.Schemmer, A.Benninghoven, these proceedings

39. C.Fenselau, these proceedings
40. F.W.Röllgen, these proceedings
41. H.R.Schulten, these proceedings
42. A.Benninghoven, A.Eicke, M.Junack, W.Sichtermann, J.Krizek, H.Peters: Org.Mass Spectrom. $\underline{15}$, 459 (1980)
43. S.E.Unger, A.Vincze, R.G.Cooks, R.Christman, L.D.Rothman: Anal.Chem. $\underline{53}$, 976 (1981)
44. K.G.Standing, W.Ens, R.Beavis, these proceedings
45. M.Barber, R.S.Bordoli, G.J.Elliot, R.D.Sedgwick, A.N.Tyler: Anal.Chem. $\underline{54}$, 645 A (1982)
46. A.Benninghoven: Trends in Anal.Chem. 1, 311 (1982)
47. Proceedings of the 2nd Workshop on LC/MS and MS/MS, Montreux 1982; J.Chromatography, to be published

3.2 Fast Atom Bombardment (Review)

Catherine Fenselau

Department of Pharmacology, Johns Hopkins University,
Baltimore, MD 21205, USA

The combination of techniques which has been marketed as fast atom bombardment -- presentation of the sample in a liquid matrix, and bombardment by a neutral primary beam -- has proven so successful for analytical work that it has rapidly come into use in scores of analytical labs around the world without the publication of much physico-chemical data addressing parameter optimization or mechanisms. Indeed, the first commercial FAB sources were delivered within a month of the appearance of the first announcements of the technique in the scientific literature [1,2].

Experimental Setup

The experimental setup is both simpler and more flexible than those described for SIMS work. A major impetus in its development both at the University of Manchester Institute for Science and Technology [3] and by various instrument manufacturers is the compatibility of the technique with high voltage sources on magnetic instruments of Nier Johnson and Mattauch Herzog geometry. With some attention to the line-of-sight transmission of x rays and energetic neutrals, FAB sources are also used on quadrupole analyzers. On any instrument a pumping system comparable to the kind used to pump chemical ionization sources is required to handle the gas flow through the atom gun. An open source is preferable to a closed one to minimize fouling.

Atoms guns are available in several configurations, e.g., the saddle field [4], capillaritron [5] and Wahlin [6] guns which produce neutral atoms with energies as high as 10 Kv. The actual velocities and fluxes of neutral primary beams are difficult to measure. Velocities are assumed to be similar to those of ions formed as precursors or concomitantly in the same gun [3]. The neutral atom velocities can be varied to produce optimal spectra. It is difficult to provide a good interpretation of these studies, since generally the voltage controls on FAB guns also produce changes in the discharges and the resulting atom flux. However the velocity and flux of the primary neutral beam are clearly less critical than in SIMS work. Glancing angles have been favored, analogous to those used in the parent SIMS technique, e.g., 60-80° from the norm. However the question must be asked, what is the angle when the target is a convex droplet?

A variety of neutral species have been used in the primary beam, including argon, krypton, xenon, nitrogen, methane and mercury. Xenon appears to be generally favored because it produces a spectrum with good signal to noise ratio. Figure 1 compares the molecular ion intensity of a decasaccharide measured with argon and xenon primary atom beams.

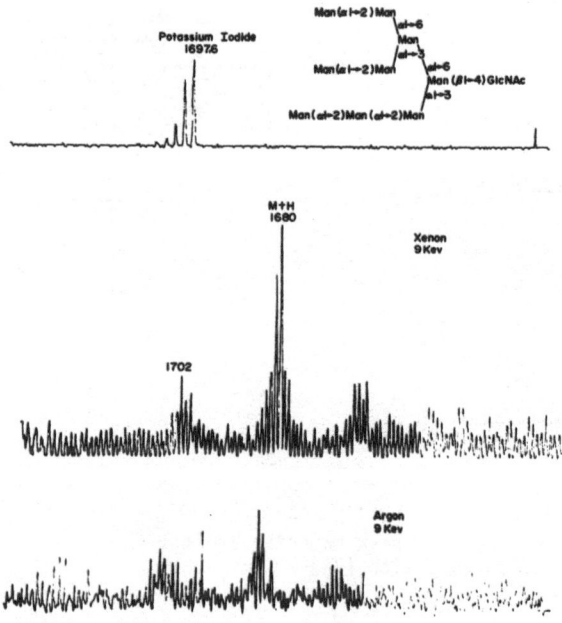

Fig. 1. Partial positive ion spectra obtained with xenon and with argon keeping other instrumental conditions constant. The masses were assigned by manual comparison with the KI reference cluster. Spectra were recorded on a multichannel analyzer [34]

Target and Matrix

Targets have generally been constructed of conducting metals, copper, silver, stainless steel or gold plate. A polyimide surface has been used with no apparent change in spectra [7]. The target can be heated to increase viscosity of the sample matrix.

Presentation of the sample in a liquid matrix has provided a major advance in secondary ion mass spectrometry when either a neutral or a charged primary beam is used. A variety of liquids have been employed, including glycerol, thioglycerol, sulfolane, diamylphenol, carbowaxes, dimethylsulfoxide and mixtures. Glycerol is the most commonly used. The matrix should be involatile, viscous, and capable of dissolving the sample. Acidic matrix, e.g., thioglycerol, enhance protonation of the sample and production of positive ion spectra, and nonacidic matrix are best for negative ion spectra. The pH of the matrix relative to the pK_A of the sample also influences the quality of the spectrum [8]. Basic samples are readily protonated to provide positive ion spectra, and negative ion spectra are readily obtained from samples with acidic groups such as carboxylates, phosphates and sulfates.

Additives are often used to improve the quality of a spectrum, e.g., lithium chloride promotes the formation of M+Li ions from relatively nonpolar species such as polypropylene glycol and ammonium acetate or ammonium thiocyanide promotes formation of $M+NH_4$ ions from

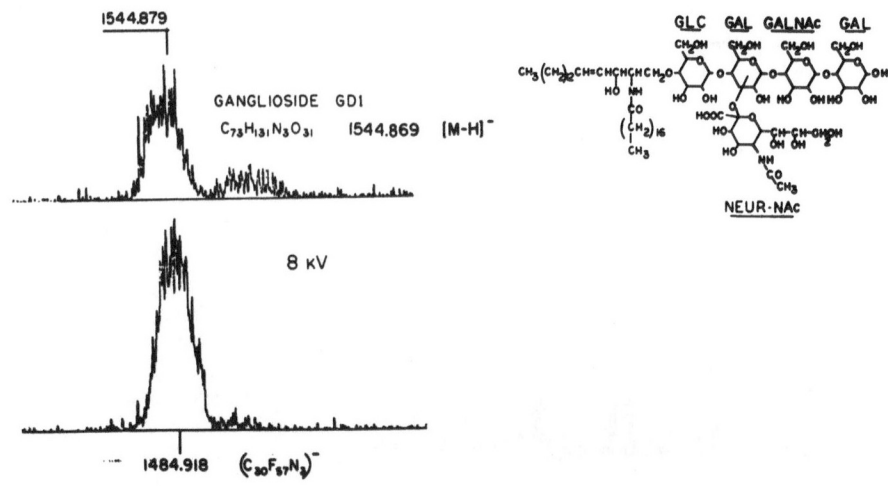

Fig. 2. Accurate mass assignment made by peak matching against the triazine reference anion at 11000 resolution [34]

carbohydrates. The addition of acids to glycerol has been found to improve formation of M+H ions in some samples. Derivatization reactions have also been carried out in the matrix [30].

Sample concentration affects the quality of the spectrum with both a minimum and a maximum bracketing the optimum range for sensitivity and spectral quality. The optimum concentration differs for different samples, however most reports are in the range 0.5 - 10 µg sample per µl of matrix. The size of the target and the amount of matrix required to cover it directly influence the amount of sample required.

Preparation of the matrix on the target varies. Some investigators dissolve the sample in the matrix prior to depositing it on the target, others deposit the sample onto the target (e.g., in methanol) and then cover it with the matrix.

Performance Characteristics

The technique is easy to use.

Positive ions, negative ions and neutral species are desorbed.

Mass range has been demonstrated to exceed 6000 amu [9,10].

Current sample requirements are 50 ng - 50 µg (e.g., in 0.5 µl glycerol) varying with different classes of compounds.

A strong secondary ion flux is produced in which currents of individual (M+H)$^+$ ions can exceed 10^{-10} A.

The secondary ion flux is fairly even and can be prolonged for more than 30 min in some cases.

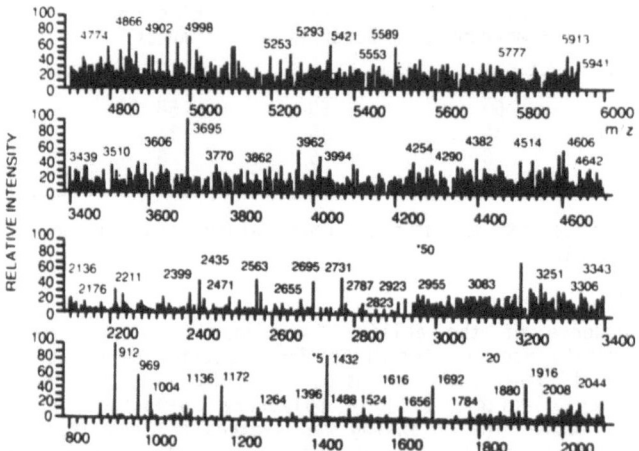

Fig. 3. Partial positive ion spectrum of CeI glycerol clusters acquired and mass assigned by the DS-55 computer system [10]

This strong, even, prolonged ion flux facilitates accurate mass measurements at low and high resolution (illustrated in Fig. 2), computer calibration and acquisition of spectra in excess of 5900 amu (illustrated in Fig. 3), collisonally induced dissociation (illustrated in Fig. 4), MSMS work [11] and reliable measurements of isotope ratios and isotope enrichment. A comparison is shown in Table I of measurements of ^{18}O enrichment made on the same sample by FAB, FD and EI (derivatized for the latter) [37].

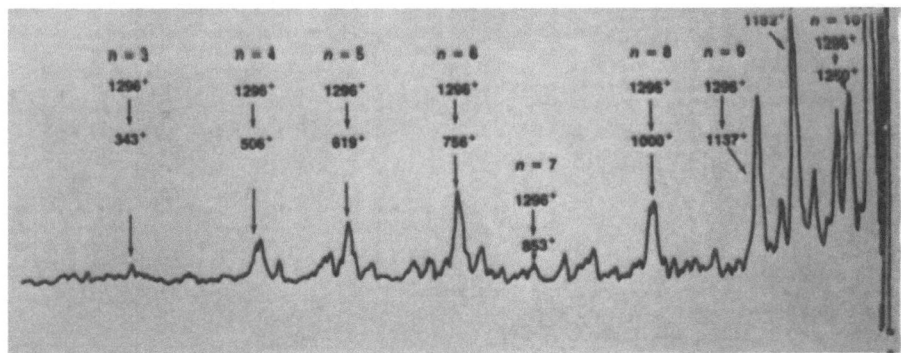

Fig. 4. MIKE spectrum of protonated angiotensin 1 (H-Asp-Arg-Val-Tyr-Ile-His-Pro-Phe-His-Leu-OH) produced by FABMS. n=1 for Asp, n=2 for Arg, etc.,[35]. Used with permission of the American Chemical Society

Table 1. Isotope analysis of ^{18}O-labeled p-nitrophenol glucuronide by fast atom bombardment, field desorption and electron impact [37]

p-Nitrophenol Glucuronide	Fast atom Bombardment	Field Desorption	Electron Impact
Unlabeled	60	59	58
Mono-isotopic labeled	34	35	34
Di-isotopic labeled	6	6	8

All values are determined with an estimated uncertainty less than or equal to 3 , e.g., 60 $\pm$3.

An important consequence of the ability to measure isotope ratios reliably is the ability to use the technique for quantitation using an isotope labeled internal standard. An assay of this kind has been developed for dipalmitoyl lecithin [12], a pulmonary surfactant which some premature infants lack. The d_9-analog was synthesized and used as internal standard for assays in samples of amniotic fluid. A standard curve is shown in Fig. 5 in which the ratios of averaged molecular ion peaks are shown to have a linear relationship to the ratios of d_0 and d_9-dipalmitoyl lecithin in standard solutions.

Accurate mass measurements and computerized acquisition of spectra require calibration with reference compounds. A number of satisfactory reference compounds have been reported for both positive and negative ion

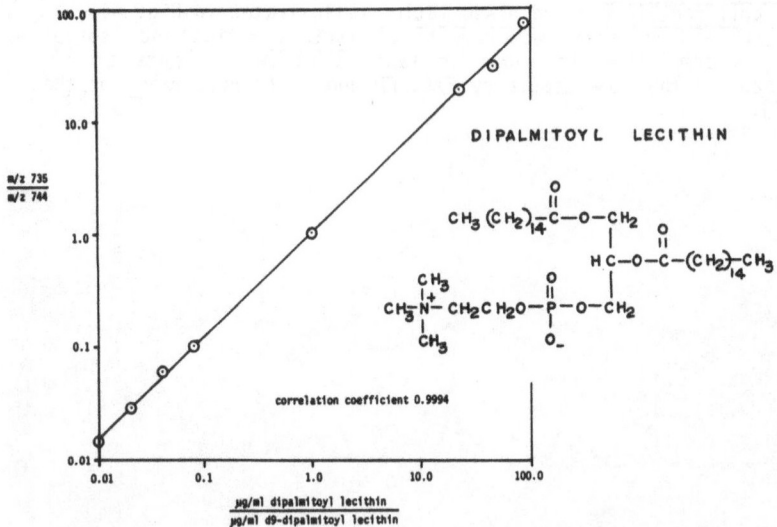

Fig. 5. The ratios of intensities of the $(M+H)^+$ peaks for dipalmitoyl lecithin and d_9-dipalmitoyl lecithin plotted (log log scale) against the ratios of those compounds in standard solutions [12]

modes. Cluster ions of potassium iodide (Fig. 1) and cesium iodide are useful. The latter provide monoisotopic cluster ions into the middle and high mass range. Because the mass increments are large between cesium iodide clusters, this salt can be mixed with other alkali halides, or as is the case in Fig. 3, with glycerol to provide reproducible series of mixed clusters. Triazines and phosphazenes are also useful, as seen in Fig. 2. Although these and other reference compounds do themselves provide good FAB spectra, they cannot usually be mixed directly with the sample. Spectra of such mixtures are not merely additive, but feature mixed cluster ions and even suppression of sample ions in some cases. Consequently, either calibration compound and sample are determined sequentially, or they are presented simultaneously but physically separated. Fig. 6 shows the spectra of 7-methylguanosine monophosphate and potassium iodide mixed together in glycerol and also presented simultaneously on a split probe. A split probe was used to make the negative ion accurate mass measurement shown in Fig. 2.

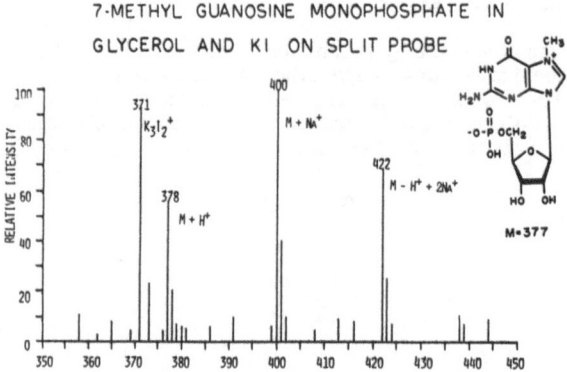

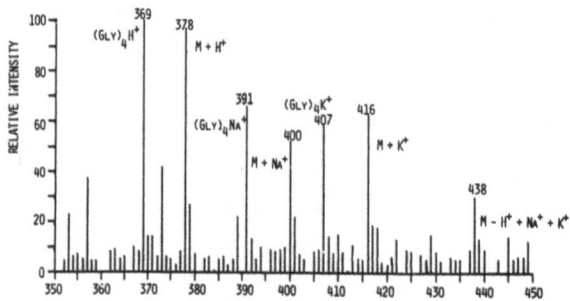

Fig. 6. Partial positive ion spectra (top) of 7-methylguanosine monophosphate in glycerol and KI presented simultaneously but separately on a split probe, and (bottom) of 7-methylguanosine monophosphate and KI mixed in glycerol. Other conditions were held constant [34]

Characteristics of the Spectra

It is important to note the characteristics of the spectra because any mechanisms advanced will have to account for these.

Positive ion spectra may contain a variety of molecular ion species, $(M+H)^+$, $(M+Na)^+$, $(M+K)^+$, etc., and occasionally M^+, $(M-H)^+$ and $(M-2H)^+$. Negative ion spectra are simpler. Usually $(M-H)^-$ is the only molecular ion species observed.

In common with other desorption techniques species are often encountered in which acidic protons have been replaced by alkali metal cations. Ions of this type can be seen in Fig. 6. Alkali metal cations which have replaced protons are carried in fragment ions as well as molecular ion species.

Alkali metal salts appear to be ubiquitous in biochemical and environmental samples, and in some stocks of glycerol. (Thioglycerol usually contains ammonium salts.) These can usually be removed from the sample by use of an ion exchange column.

Clusters are often observed from the glycerol or other matrix. Their occurrence and relative intensities vary, depending on the sample and its concentration. Occasionally cluster ions are observed composed of a sample molecule, a glycerol molecule and a proton. These mixed clusters can also contribute fragment ions to the spectrum.

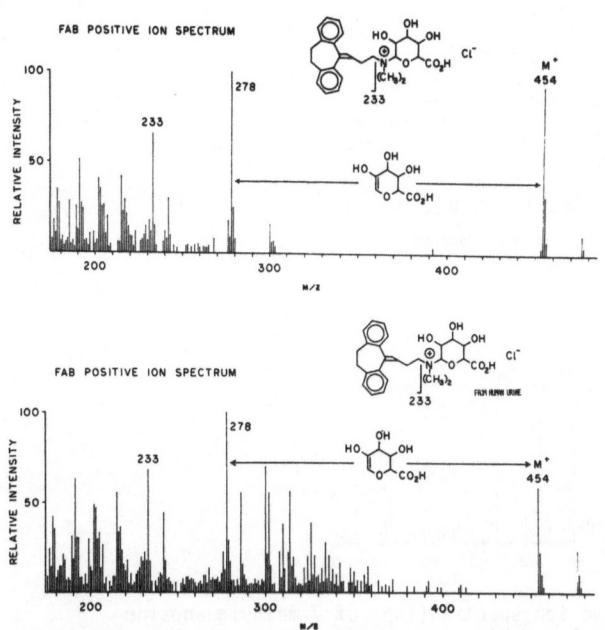

Fig. 7. Partial positive ion spectra of synthetic (top) and isolated (bottom) samples of the quaternary ammonium-linked glucuronide of amitriptyline [36]

One of the oft cited features of FAB spectra is the peak at every mass (Fig. 1) or "incoherent noise." The effects of this background on the signal to noise ratio and on the dynamic range of the spectrum vary with the sample, the solvent and the concentration. Studies of these background signals at high resolution have revealed as many as 20 isobaric contributions at a single mass [13]. These ions are thought to arise from high energy damage to the sample and matrix at the point of atom impact.

Most of the ions observed are even electron ions, anions or cations rather than radical ions. This is true for molecular ion species and for fragment ions, in both positive ion spectra and negative ion spectra. This means that most fragmentation occurs by the loss of neutral molecules and not radicals (see for example Fig. 7) analogous to chemical ionization.

Unimolecular gas phase decomposition has been confirmed by the observation of metastable decompositions [14] (Fig. 4). It also seems likely that fragments are produced in the condensed phase by processes currently under study.

Sample preparation is important to the success of the analysis. The effects of alkali metal salts have already been discussed. Organic contaminants are reported to obscure or repress desorption of sample ions. Fast atom bombardment shares with other condensed phase ionization techniques a sensitivity to the history of the sample. Thus, while a recent interlaboratory comparison [6] indicates that comparable spectra can be obtained on different instruments using aliquots of the same sample, comparisons of spectra of the same compound from different sources often do not provide matches which would be suitable for computerized library searches. Such a comparison is shown in Fig. 7 between spectra of a synthetic and human metabolite of amitriptyline.

Role of the Matrix

The arguments for using a neutral primary beam in place of ions have been clearly made [3,15,38]. Charging of the sample is minimized and the SIMS technique is readily compatible with the high accelerating voltages used on magnetic instruments with the capabilities for high resolution and high mass range.

The second innovation introduced as part of the FAB technique is the use of a liquid matrix for sample presentation. This may have the more profound effect on analytical mass spectrometry.

The major and profound effect of dissolving or suspending the sample in a liquid matrix, rather than drying it directly onto the metal target, is the remarkable prolongation of secondary ion currents from a few seconds duration with no matrix to thirty, even sixty minutes. The explanation was offered after an early comparison study that "the steady evaporation of the liquid and the resulting molecular motion are sufficient constantly to regenerate a fresh sample surface for sputtering" [16]. In addition to replacing sample sputtered or damaged by the bombardment, the more abundant matrix molecules absorb high energy at the point of impact, sustaining considerable damage themselves [17].

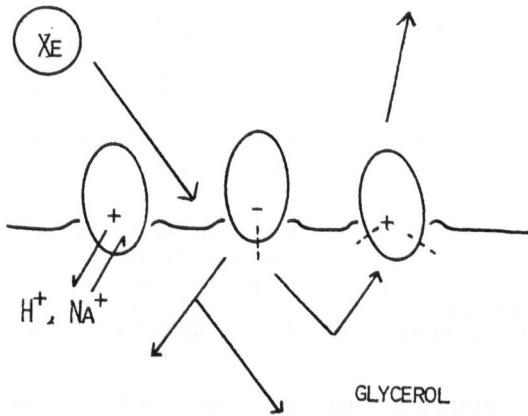

GLYCEROL

Scheme 1. The glycerol supplies sample ions and neutrals to
the surface. Solvation of the ions(-----) produces charge
separation, i.e., the bond distances between cations and
anions are undoubtably longer than in the crystalline state

Secondary ion currents are analogously prolonged by the use of liquid
matrix under bombardment with primary ion beams (liquid SIMS).

In addition to its important interaction with the primary beam, the
liquid matrix also has significant influence on the secondary beam. It
appears to facilitate formation of ions prior to desorption. Thus the
effect of pH on the abundance of $(M+H)^+$ ions and the effect of added
alkali or ammonium salts on both replacement of acidic protons and
formation of cationated molecular ion species are readily rationalized as
solution chemistry.

The liquid matrix also makes an important energetic contribution. When
the sample is presented in solution, the heat of solution contributes to
lowering the energy required for desorption. This is most readily
envisioned (see Scheme) for preformed ions, such as quaternary ammonium
halide salts. In this case solvation contributes significantly to charge
separation. Only desolvation is required to desorb these ions.
(Quaternary ammonium cations have low probabilities for recombination
with electrons.) Thus the liquid matrix may be envisioned as influencing
both the chemical nature of the molecular ion species and fragments
observed, and the energy required for desorption.

Mechanistic Models for FAB

At least three different models have been proposed to explain some or
all of the characteristics of FAB spectra:

Desorption of ions preformed in solution by a localized nonequilibrium
vibrational process analogous to that considered operational in SIMS.

Evaporation of preformed ions from splash droplets [19] analogous to
proposals by Iribarne [18] and by Vestal for aerosols.

Gas phase ion molecule reactions analogous to the thermolized processes in chemical ionization. In fact, the contribution of these reactions may vary from instrument to instrument, depending on how open the source is, and how efficiently source pressure is controlled.

Regardless of which models are eventually accepted for the physical mechanism of energy transfer, it does seem clear that the kinds of ions observed in a spectrum strongly reflect the chemistry of the matrix, and that the formation of ions and fragments can be influenced according to chemical principles.

Analytical Potential

Returning to the claim for analytical success with which this overview opened, a partial list of real samples for which FAB has provided structural information includes peptides [20,21], drug-alkylated guanosine monophosphates [22], ecdysone phosphates [23], novel porphyrins [24], a leukotriene [25] and conjugated mammalian drug metabolites [e.g., 26-28]. Feasibility studies indicate potential for identification of antibiotics [29,30], nitrogen-containing compounds in fossil fuels [31], surfactant long chain quaternary amines [32] and oligodeoxyribonucleotides of up to ten base units [33].

References

1. D. J. Surman and J. C. Vickerman, J.C.S. Chem. Commun. 1981, 324-325.
2. M. Barber, R. S. Bordoli, R. D. Sedgwick and A. N. Tyler, J.C.S. Chem. Commun. 1981, 325-327.
3. M. Barber, R. S. Bordoli, G. J. Elliott, R. D. Sedgwick and A. N. Tyler, Anal. Chem. 54, 645A-657A, 1982.
4. J. Franks and A. M. Ghander, Vacuum 24, 489-491, 1974.
5. J. F. Mahoney, D. M. Goebel, J. Perel and A. T. Forrester, Biomed. Mass Spectrom., in press.
6. K. L. Clay, L. Wahlin and R. C. Murphy, Biomed. Mass Spectrom., in press.
7. S. A. Martin, C. E. Costello and K. Biemann, Anal. Chem., in press.
8. D. H. Williams, C. Bradley, G. Bojesen, S. Santikorn and L. C. E. Taylor, J. Amer. Chem. Soc. 103, 5700-5704, 1981.
9. A. Dell and H. R. Morris, Biochem. Biophys. Res. Commun. 106, 1456-1462, 1982.
10. A. M. Buko, L. R. Phillips and B. A. Fraser, Biomed. Mass Spectrom., in press.
11. D. F. Hunt, W. M. Bone, J. Shabanowitz, J. Rhodes and J. M. Ballard, Anal. Chem. 53, 1704-1706, 1981.
12. B. C. Ho, C. Fenselau, G. Hansen, J. Larsen and A. Daniel, Clin. Chem., in press.
13. C. Costello, private communication.
14. M. Barber, R. S. Bordoli, R. D. Sedgwick and T. W. Tetler, Organic Mass Spectrom. 16, 256-260, 1981.
15. F. M. Devienne and J. C. Roustan, Organic Mass Spectrom. 17, 173-181, 1982.

16. D. J. Surman and J. C. Vickerman, J. Chem. Res. Synopsis 1981, p. 170-171.
17. F. H. Field, J. Phys. Chem., in press, 1982.
18. J. V. Iribarne and B. A. Thomson, J. Chem. Phys. 64, 2287-2294, 1976.
19. P. J. Arpino and G. Guiochon, J. Chromatog. 251, 153-164, 1982..
20. K. L. Rinehart, L. A. Gaudioso, M. L. Moore, R. C. Pandey, J. C. Cook, M. Barber, R. D. Sedgwick, R. S. Bordoli A. N. Tyler and B. N. Green, J. Amer. Chem. Soc. 103, 6517-6520, 1981.
21. D. H. Williams, G. Bojesen, A. D. Auffret and L. C. E. Taylor, FEBS Lett. 128, 37-39, 1981.
22. V. T. Vu, C. C. Fenselau and O. M. Colvin, J. Amer. Chem. Soc. 103, 7362-7364, 1981.
23. R. E. Isaac, M. E. Rose, H. H. Rees and T. W. Goodwin, J. Chem. Soc. Chem. Commun. 1982, 249-251.
24. J. P. Collman, C. S. Bencosme, R. A. Durand, R. P. Kreh and F. C. Anson, J. Amer. Chem. Soc., in press.
25. R. C. Murphy, W. R. Matthews, J. Rokach and C. Fenselau, Prostaglandins 23, 201-206, 1982.
26. P. C. C. Feng, C. Fenselau and R. Jacobson, Drug. Metab. Disp. 10, 286-288, 1982.
27. J. E. Bakke, A. L. Bergman and G. L. Larsen, Science 217, 645-647, 1982.
28. M. Stogniew and C. Fenselau, Drug Metab. Disp. 10 Nov/Dec, 1982.
29. M. Barber, R. S. Bordoli, R. D. Sedgwick and A. N. Tyler, Biochem. Biophys. Res. Commun. 101, 632-638, 1981.
30. A. Dell, H. R. Morris, M. D. Levin and S. M. Hecht, Biochem. Biophys. Res. Commun. 102, 730-738, 1981.
31. R. D. Grigsby, S. E. Scheppele, Q. G. Grindstaff, G. P. Sturm, L. C. E. Taylor, H. Tudge, C. Wakefield and S. Evans, Anal. Chem. 54, 1108-1113, 1982.
32. R. J. Cotter, G. Hansen and T. R. Jones, Analytica Chimica Acta 136, 135-142, 1982.
33. L. Grotjahn, R. Frank and H. Blocker, Nucleic Acids Research 10, 4671-4678, 1982.
34. Spectrum courtesy of D. N. Heller, Johns Hopkins University, measured on a Kratos MS-50 mass spectrometer using the Kratos FAB gun.
35. R. D. Sedgwick, reported in C. J. McNeal, Anal. Chem. 54, 43A-50A, 1982.
36. J. P. Lehman and C. Fenselau, Drug Metab. Disp., in press.
37. C. Fenselau, P. C. C. Feng, T. Chen and L. P. Johnson, Drug Metab. Disp. 10, 316-318, 1982.
38. R. S. Lehrle, J. C. Robb and D. W. Thomas, J. Sci. Instrum. 39, 458-463, 1962.

3.3 Changes in Secondary Ion and Metastable Ion Mass Spectral Patterns with Experimental Conditions

Hideki Kambara

Central Research Laboratory, Hitachi Ltd., Kokubunji,
Tokyo 185, Japan

Molecular secondary ion mass spectrometry(SIMS) has been proven to be a powerful analytical method for investigating molecular weights of thermolabile nonvolatile bio-organic compounds[1-10]. Secondary ion mass (SIM) spectra show a lot of fragment peaks which provide important information on molecular structure. One of the favourable characteristics of SIMS is that spectral patterns are quite reproducible in most cases. This enables precise discussion on molecular structure using SIM spectra, such as differenciation of isomers, etc. Although SIM spectral patterns are reproducible under the same experimental conditions, they sometimes change with conditions.

This paper reports that SIM spectral patterns are affected by primary ion current, amount of matrix substances and the presence of salts. It is also demonstrated that, being different from FD, many kinds of intense daughter ions from molecular species can be observed in a metastable spectrum.

A Hitachi M-80 double focusing mass spectrometer equipped with an extra EI ion source for SIMS was used. The primary ion(Xe^+) intensity was $0.5 \sim 1 \times 10^{-8}$A for a dry surface condition(without glycerol) and $1 \sim 3 \times 10^{-7}$A for a glycerol matrix condition. The secondary ion current monitored after passing through the source slit(0.1mm width) was around 1% of the incident beam current. The bombarding energy was 5 keV and the secondary ion acceleration was 3kV. Each sample was loaded on a silver plate(3mmX6mm) which was introduced into the ionization chamber using a side entry direct inlet probe. These are also used for EI and FD operation. The primary beam hit the target at a 20°angle. Only a part of the target (0.5mmX6mm) was bombarded with primary ions. The target was placed at the position where ions are produced in the EI or FD mode. Consequently, exactly the same ion focusing property of the instrument could be used for SIMS operation as for EI operation. The target was parallel to the source chamber slit, which enabled pulling out most of the secondary ions from the bombardment area. Each spectrum was recorded on a U.V. chart. The sample size was 1ug and glycerol quantity was $1 \sim 2$ μl.

Primary ion Current Dependence of SIM Spectra

The SIM spectra of Bradykinin were obtained at two different primary ion currents, as shown in Figure 1. The absolute ion intensity was far stronger in case(b) than in case(a), although the two spectra were displayed with their MH^+ peaks at the same height. Some of the fragment peaks grew remarkably in case(b) compared to those in case(a). The same spectral pattern as case(b) was obtained even with a low primary current when the sample was bombarded with a high primary ion beam.

Although the observed species in both cases were about the same for Bradykinin, new species appeared for fragile compounds. The SIM and daughter ion spectra for chymostatin(an inhibitor of chymotripsin) is shown in Figure 2.

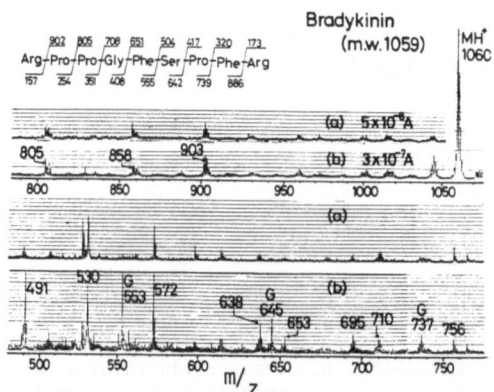

Figure 1. SIM spectra of Bradykinin
obtained at two different
ion currents

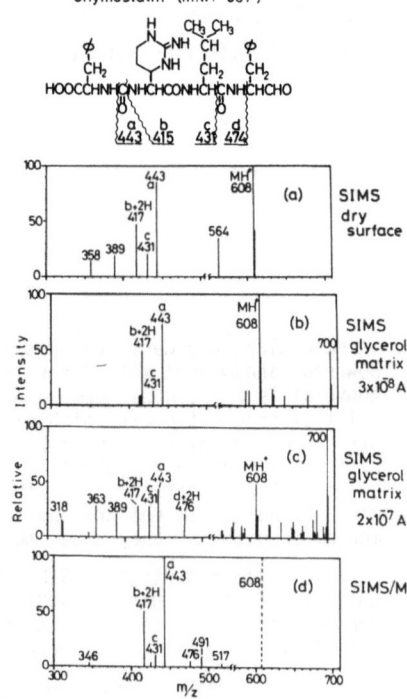

Figure 2. SIM and metastable ion mass ▷
spectra of chymostatin

The spectrum obtained under the dry surface condition(a) resembles spectrum(b) obtained using a glycerol matrix with a low primary ion current. The ions at m/z 417, 431, and 443 provide important structural information. The relative abundances of these in both cases resemble those in case(d)(daughter ion spectrum from MH$^+$). However, the spectral pattern changed when the primary ion current increased(case(c)). New ion species appeared at m/z 363, 389, and 476. The fragment at m/z 431 increased as well. The reaction products appearing in the molecular ion region of spectrum(c) also increased in contrast with those in spectrum(b). It is interesting to note that the ion intensity ratio of MH$^+$ to (MH$^+$+glycerol) is almost 2:1 in spectrum(b), while it changed to 1:2 by bombarding the same sample with a higher ion current. These changes in spectral pattern are irreversible. Once the sample was bombarded with a high ion current, the ion intensities of the fragments, glycerol adducts and reaction products in the molecular ion region stay intense even if the priamry ion current is increased.

These facts indicated that ion bombardment dissociates samples. The products stay in a glycerol layer and are observed in a spectrum as fragments. Generally, a primary ion current lower than 1×10^{-7}A and large fluidity of the matrix are favourable for keeping relative intensities of dissociated products low in a SIM spectrum. A low primary ion current does not always give fewer fragments. The spectra of a cord factor obtained under two different conditions are shown in Figure 3. Spectrum(b) was obtained with a priamry ion current of 1×10^{-8}A. This was very noisy and intense fagments were observed. On the other hand, a fine spectrum with low noise and fewer fragments was obtained with an ion current as high as 7×10^{-8}A. The base peak was the MH$^+$. This spectral change can not be elucidated by sample dissociation due to ion bombardment. A surface condition change seems to influence fragmentation.

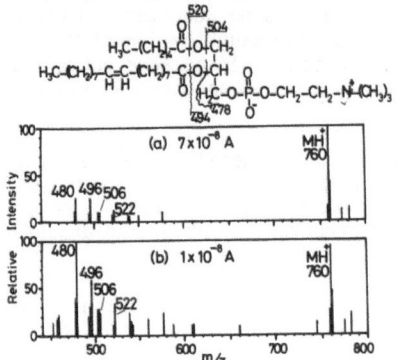

β-Oleoyl-γ-palmitoylphosphatidylcholine (m.w.759)

Figure 3.
SIM spectra of a cord factor
obtained at two different
ion currents

Figure 4. ▷
Changes in SIM spectral pattern
of aculeacin by the addition of NaCl

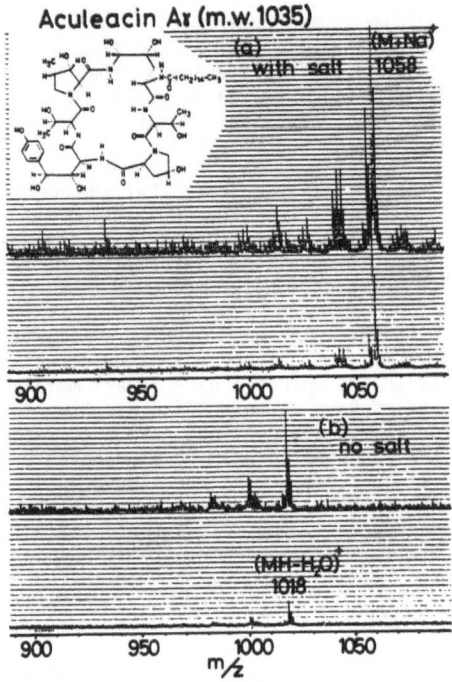

Aculeacin Aᵧ (m.w.1035)

Effects of Alkali Metal Salt on Fragmentaion

Although there are many kinds of compounds the SIM spectra of which are
not much affected by the presence of alkali metal salt, there are some compounds in which SIM spectra are much influenced. The SIM spectra of aculeacin
are shown in Figure 4. The only species appearing in the molecular ion region
is $(MH-H_2O)^+$ in spectrum(b)(pure glycerol matrix). However, an intense cationized molecule appears in spectrum(a)(glycerol matrix containing a small amount
of NaCl). An analogous spectral pattern change occured for some glyco-lipids.
The addition of NaCl to these samples is favourable to observe the molecular
species.

The addition of alkali metal salt to samples frequently produces confusing
SIM spectra. Although a protonated molecule is clearly observed in a matrix
assisted SIM spectrum for some compounds, such as lividomycin A, the spectral
pattern completely changed due to the addition of alkali metal salt and no
molecular species could be observed. The reason for this change could not be
clarified but it seemed that many kinds of dissociation products appeared in
the spectrum. An analogous study was done for saccharides previously[6].

The SIM spectra of minosaminomycin were obtained with a pure glycerol
matrix and with a glycerol matrix containing LiCl. In the latter case, lithiated molecules$(M+Li)^+$ and $(M+2Li-H)^+$ were observed together with several
lithiated fragments. It had been considered that cationization produces very
stable molecular ions, which results in less fragmentation. However, it is
likely that more fragments are produced from the lithiated molecule for
minosaminomycin. To investigate whether the lithiated molecule dissociates
unimolecularly into fragments, metastable ion spectra of $(M+Li)^+$ as well as
MH^+ were obtained, as shown in Figures 5(b) and 5(c). Intense daughter ions
are observed even from $(M+Li)^+$. All the fragments could be interpreted from
the molecular structure. Each daughter ion from $(M+Li)^+$ seems to keep a

lithium atom. Spectrum(c) is quite different from spectrum(b). The ion at m/z 491 is the base peak in spectrum(b); however, there is no daughter ion corresponding to this species in spectrum(c). Instead, the daughter ion at m/z 494(487+Li) appeared in spectrum(c), which is rationalized by bond rupture at "b". This species at m/z 494 is also observed in the SIM spectrum obtained in the presence of LiCl.

These facts indicate that the position of an alkali metal in a molecule can be different from the position of the proton in MH$^+$, resulting in the observation of different daughter ion species. This means there is some possibility to obtain new information on molecular structure by observing daughter ions of a cationized molecule.

Daughter ion mass spectra of amastatin are shown in Figure 6. Daughter ions from MH$^+$ reflect the molecular structure well. All the daughter ions from the cationized molecule have alkali metals. Although most of the cationized species have corresponding daughter ions from MH$^+$(spectrum(c)), the daughter ion at m/z 396 does not have a corresponding daughter species in spectrum(c). This daughter ion can be elucidated by the bond cleavage at "c" accompanied by a Li attachment. The corresponding species appeared in the daughter ion spectra of (M+Na)$^+$ and (M+K)$^+$. The relative daughter ion intensities are different in each case. The most intense cationized daughter ions were obtained from (M+Li)$^+$ and the weakest from (M+K)$^+$.

Changes in SIM Spectral Pattern due to Amount of Glycerol

SIM spectral patterns for many compounds obtained using a fluid matrix, such as glycerol, differ from those obtained under a dry surface condition. A matrix assisted SIM spectral pattern approaches a dry surface SIM spectral pattern when the fluid matrix almost vanished from the surface or when only a small amount of matrix substance was put on the surface.

Cationized molecules but no protonated molecule are observed in a SIM spectrum of stachyose obtained from a dry surface. Di-, and tri-saccharide moieties are all from cationized molecules. Fragments not being cationized are all smaller than the monosaccharide moiety. However, the protonated molecule is observed in the matrix assisted SIM spectrum. The sodiated molecule and its fragments also appear in the spectrum. It is noteworthy that the (M+Na)$^+$ intensity is almost the same as that of (M+Na-162)$^+$ under the matrix condition. But it became far stronger than (M+Na-162)$^+$ under the matrix condition. This indicates that less excitation during ionization occurs in a glycerol matrix.

A more dramatic spectral pattern change was observed for kemptide, as shown in Figure 7. The intense protonated molecule, together with structurally important fragments, is observed in the matrix assisted spectrum. On the other hand, the major peak in a dry or almost dry surface spectrum appears at m/z 354, which is not observed in the matrix assisted spectrum. It was confirmed by observing daughter ions from MH$^+$ using a B/E linked scan that the ion at m/z 354 can be a unimolecular reaction product of MH$^+$ produced under the dry surface condition. However, this cannot be observed in a daughter ion spectrum of MH$^+$ produced under matrix assisted conditions.

In most cases, SIM spectra are quite reproducible. However, as demonstrated here, SIM spectra depend on the target surface condition as well as on the primary ion current(or dose) for some molecules even if the same primary ion

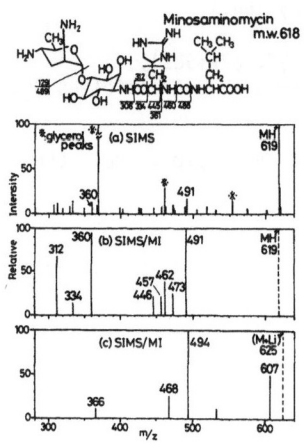

Minosaminomycin
m.w.618

Figure 5.
SIM and metastable ion mass spectra
from the MH^+ and $(M+Li)^+$ of
minosaminomycin.

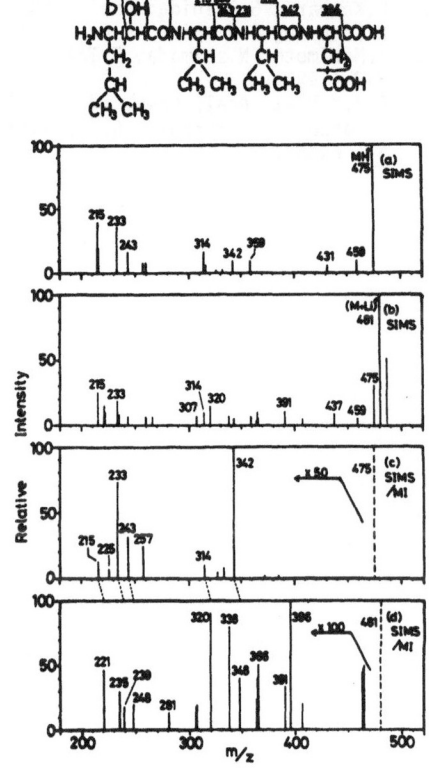

Amastatin
m.w. 474

Figure 6.
SIM and metastable ion mass
spectra of the MH^+ and $(M+Li)^+$
for amastatin

species, bombardment energy, incident beam angle, and exactly the same sample
are used. The most favourable primary ion current was around 5×10^{-8}A in this
experiment. That was the current at which most reproducible spectra and
abundant molecular ion species were obtained.

Cationization by adding salts is not a good method for producing intense
molecular species in SIMS, except for a few compounds. However, new structural
information can be obtained from metastable spectra of cationized molecules.

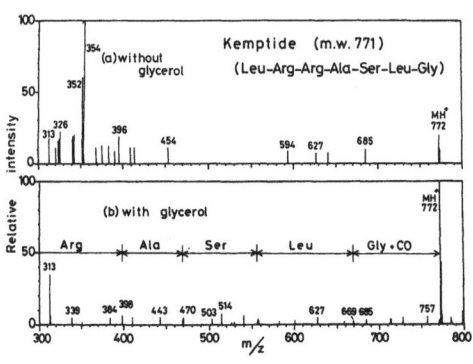

Kemptide (m.w. 771)
(Leu-Arg-Arg-Ala-Ser-Leu-Gly)

Figure 7.
SIM spectra of kemptide
obtained under a dry
surface and aglycerol
matrix conditions

105

References

1. A.Benninghoven, D.Jaspers, W.Sichtermann: Appl. Phys. <u>11</u>, 35(1976).
2. A.Benninghoven, W.K.Sichtermann: Anal. Chem. <u>50</u>, 1180(1978).
3. H.Grade, R.G.Cooks: J.Am.Chem.Soc. <u>100</u>, 5615(1978).
4. R.J.Day, S.E.Unger, R.G.Cooks: Anal. Chem. <u>52</u>, 557A(1980).
5. H.Kambara, S.Hishida: Org. Mass Spectrom. <u>16</u>, 167(1981).
6. H.Kambara, S.Hishida: Anal. Chem. <u>53</u>, 2340(1981).
7. H.Kambara, S.Hishida, H.Naganawa: J.Antibiotics <u>35</u>, 67(1982).
8. K.Morimoto, N.shimada, T.Takita, H.Umezawa, H.Kambara: J.Antibiotics <u>35</u>, 378 (1982).
9. C.J.McNeal: Anal. Chem. 54, 43A(1982).
10. M.Barber, R.S.Bordoli, G.J.Elliott, R.D.Sedgwick, A.N.Tyler: Anal. Chem. <u>54</u>, 645A(1982).

106

3.4 Time-of-Flight Measurements of Metastable Decay

K.G. Standing, W. Ens, and R. Beavis

Physics Department, University of Manitoba, Winnipeg, Canada R3T 2N2

Abstract

Secondary $[(CsI)_n Cs]^+$ cluster ions, produced by 8 keV Cs^+ primary ions, were observed to decay within ~70 μs after production for n>7. Parent ions with n=14 to 18 decayed preferentially to ions with n=13, corresponding to a 3X3X3 structure. Secondary ions from large organic molecules, bradykinin in particular, were observed to decay on a similar time scale.

1. Introduction

It has long been recognized that linear time-of-flight mass spectrometers have certain advantages for the observation of metastable ions [1]. When an ion decomposes within the flight tube of such an instrument, the centre-of-mass of the resulting fragments continues on to the detector with unchanged velocity. Although the energy release in the decomposition is effectively amplified by the motion of the centre-of-mass [2], it is still normally small compared to the total ion energy. Thus the fragments appear in the mass spectrum at the approximate position of the parent ions, but the peak is broadened by the energy release. Decompositions occurring all along the flight tube may be detected, so the efficiency is relatively high.

In the Manitoba time-of-flight mass spectrometer, an ion of mass 1200 u reaches its full energy in 0.1 μs and takes 40 μs to reach the detector [3]. Thus metastable decays with lifetimes in this range may be expected to produce noticeable effects. To examine the fragments, a set of three grids (each 90% transmission at 1.3 cm spacing) was inserted in front of the detector at the end of the 1.6 m flight tube [1,4-6]. The detector was maintained at ground potential as before, so the energy of the ions striking it was unchanged. The entrance grid was also kept at ground, but a retarding potential V_R was applied to the central grid. Under these conditions the time-of-flight of neutral fragments is unchanged, but charged particles are delayed. The delay is larger for a charged fragment (mass m_f) than for the parent ion (mass m_p), since the fragment has only a fraction ($\simeq m_f/m_p$) of the parent energy; if the energy of the fragment is too low, it will be reflected. This enables examination of the mass spectrum at the time the ions arrive at the detector, while the normal spectrum (without retarding potential) measures the spectrum immediately after acceleration.

2. Measurement of CsI Cluster Ions

Large secondary cluster ions ejected from CsI surfaces by 4.7 keV Xe^+ ion bombardment have recently been observed in a sector-field mass spectrometer [7-9]. Positive ions $[(CsI)_n Cs]^+$ were detected up to n=70, and negative

107

clusters $[(CsI)_nI]^-$ up to n=4. The ion intensity decreased with increasing n, but striking anomalies were observed in the neighbourhood of particularly symmetrical cluster geometries; e.g. an intensity increase by a factor ~2 at n=13 and a decrease by an order of magnitude for n=14 and 15, corresponding to a 3x3x3 structure at n=13. The anomalies were interpreted [7,10] by a bond-breaking or a cleavage model, assuming that they arose in the ion production process.

We have observed similar clusters in our time-of-flight spectrometer [3] when electrosprayed deposits of CsI were bombarded with 8 keV Cs^+ ions. Positive clusters with n up to ~40 and negative clusters with n up to ~20 were detected (the latter with 28 keV Cs^+ bombardment). Fig. 1 shows a negative time-of-flight spectrum; the positive spectrum is reported elsewhere [11]. In marked contrast to the results mentioned above, the ion yield varied smoothly with n for both positive and negative ions; no significant anomalies were observed.

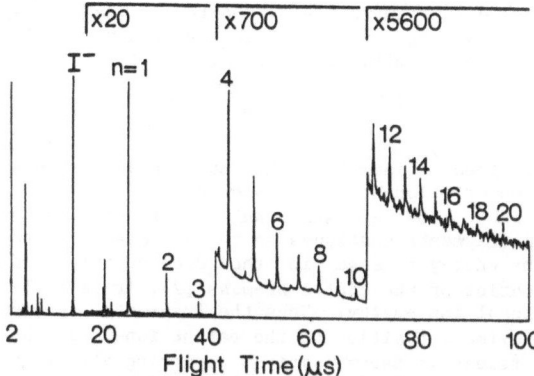

Fig. 1.
Time-of-flight spectrum
of $[(CsI)_n I]^-$ clusters

The time scales of the two measurements are significantly different, suggesting that metastable decay may be an important factor in explaining the apparent disagreement. In the sector-field spectrometer [12] a cluster ion with n=13 (mass ~3500 u) and energy 0.3 keV requires ~750 μs to traverse the instrument; if the ion decomposes within that time it is not observed, at least not at its original mass. By contrast, the fragments from an ion which decomposes in the flight tube of the time-of-flight spectrometer are still registered at the approximate position of the parent ion, as discussed above. Since the n=13 cluster is accelerated to its full energy of 10 keV in 0.17 μs, our measurement is made at an effective time ~0.2 μs after the ions are produced. Thus it appears that no anomalies in the intensity distribution are visible at that time.

In order to examine the mass spectrum at much later times, a retarding potential was applied to the grid in front of the detector, as described above. The flight time to the detector is ~70 μs, so this measures the intensity distribution at a time ~70 μs after the ions are produced. Time-of-flight spectra were measured for various retarding potentials between 7 and 10 keV applied to the central grid. Results for the positive clusters with n=4 to 7 are shown in Fig. 2. For clusters $[(CsI)_n Cs]^+$ up to n=4 the parent ions are predominant, but for n>4 fragmentation is seen to increase rapidly. Above n=7 no parent ions could be observed, only fragments. The measurements were taken at a pressure of ~10^{-7} Torr; an

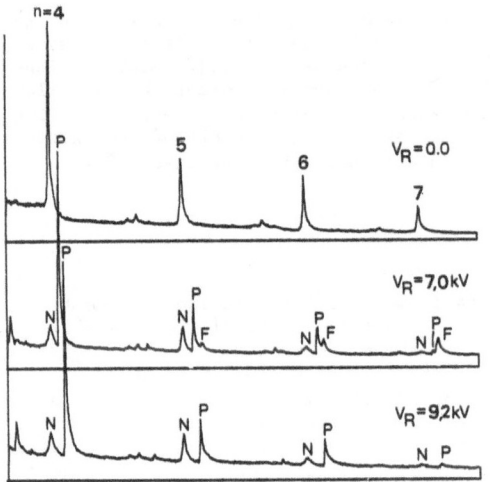

Fig.2. Spectra of 10 keV $[(CsI)_nCs]^+$ cluster ions with zero retarding voltage V_R and with V_R=7.0 and 9.2 kV. P labels the parent ion, N a neutral fragment and F a charged fragment

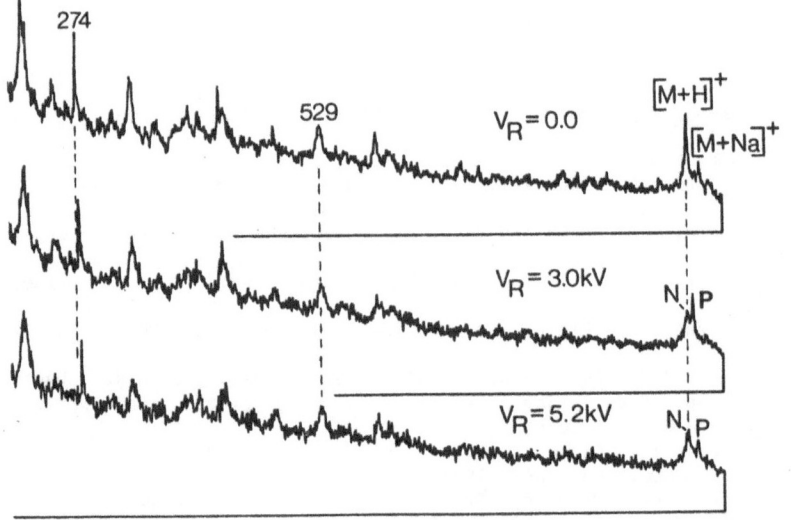

Fig.3. Positive spectra of bradykinin with 5.9 keV secondary ions and various retarding potentials (V_R). Most of the sample ions (e.g. 529) are accounted for by neutral fragments, N, from metastable decay (not delayed) while impurity peaks like 274 are more stable (delayed). The delayed parent ion (labelled P) in the bottom spectrum accounts for ~10% of the observed M+H ions

increase in pressure by a factor of 3 gave no observable change in fragmentation. Collisions in the flight tube therefore play no important role in the process. On the other hand, the results are consistent with metastable decay; evidently the larger clusters acquire enough internal energy during emission to make them unstable on a time scale ≪100 μs.

Most of the charged fragments observed from these decays correspond to ejection of one or more CsI molecules from the parent clusters. Decay of the clusters above n=13 was investigated in some detail [11]. Daughter ions with n=13 were found to predominate; n=14 and 15 clusters had very low yields. Thus our cluster distribution also shows an anomaly near n=13, but only at times >>1 μs after emission; the anomaly is a result of the pattern of metastable decay, not a consequence of the production process [13].

3. Decomposition of Organic Ions

Many of the ions produced from organic molecules by fission fragment or keV ion bombardment are also found to be unstable [3,6,14], the instability increasing with molecular size. Fig. 3 shows the spectra of the peptide bradykinin with various retarding potentials. Here ≲10% of the (M+H)$^+$ ions survived the ~40 μs flight time, and the fragment ions from the target also undergo further decomposition. Such spectra observed in a time-of-flight spectrometer at effective times <<1 μs after emission may therefore be quite different from those observed in a sector-field or quadrupole instrument several tens of microseconds later.

References

1. See for example W.F. Haddon and F.W. McLafferty, Anal. Chem 41, 31 (1969).
2. J.H. Beynon, Proc. R. Soc. Lond. A 378, 1 (1981).
3. B.T. Chait and K.G. Standing, Int. J. Mass Spectrom. Ion Phys. 40, 185 (1981).
4. R.E. Ferguson, K.E. McCulloh and H.M. Rosenstock, J. Chem. Phys. 42, 100 (1965).
5. W.W. Hunt, Jr., R.E. Huffman and K.E. McGee, Rev. Sci. Instrum. 35, 82 (1964).
6. B.T. Chait and F.H. Field, Int. J. Mass Spectrom. Ion Phys. 41, 17 (1981).
7. J.E. Campana, T.M. Barlak, R.J. Colton, J.J. DeCorpo, J.R. Wyatt and B.I. Dunlap, Phys. Rev. Lett. 47, 1046 (1981), and private communication from R.J. Colton.
8. T.M. Barlak, J.E. Campana, R.J. Colton, J.J. DeCorpo and J.R. Wyatt, J. Phys. Chem. 85, 3840 (1981).
9. T.M. Barlak, J.E. Campana, R.J. Colton, J.J. DeCorpo and J.E. Campana, J. Am. Chem. Soc. 104, 1212 (1982).
10. B.I. Dunlap, Bull. Am. Phys. Soc. 27.225 (1982).
11. W.Ens, R.Beavis and K.G. Standing, submitted to Phys. Rev. Lett.
12. R.J. Colton, J.E. Campana, T.M. Barlak, J.M. DeCorpo and J.R. Wyatt, Rev. Sci. Instrum. 51, 1685 (1980).
13. In the sector-field measurements [7–10], an ion with n=13 spends ~50 μs within the extraction lens before entering the spectrometer, so it appears likely that many of the large clusters observed were disintegration products originating from regions close to the target.
14. J.B. Westmore, W.Ens and K.G. Standing, Biomed. Mass Spectrom. 9, 119 (1982).

3.5 Design and Performance of a New Time-of-Flight Instrument for SIMS

P. Steffens, E. Niehuis, T. Friese, and A. Benninghoven

Physikalisches Institut der Universität Münster, Domagkstraße 75, D-4400 Münster, Fed. Rep. of Germany

1. Introduction

Sputtering of organic molecules results in the emission of parent-like secondary ions as $(M\pm H)^{\pm}$e.g., provided the molecules are in an appropriate chemical environment [1,2]. Yields in the 10% range can be obtained from amino acids, e.g. These yields and typical damage cross sections of some 10^{-14} cm^2 result in a probability of some percent for a surface molecule to be transformed into a parent-like secondary ion during sputtering. A considerable decrease of this probability is expected for larger molecules.

A complete utilization and efficient analysis of the generated secondary ions is crucial in organic and inorganic trace analysis, for the undisturbed analysis of surfaces that are very sensitive to ion bombardment, and especially for the secondary ion mass spectrometry very large organic molecules. These requirements demand a high transmission mass spectrometer detecting all secondary ions simultaneously after their e/m analysis, at least for one charge sign. Additional properties of such an instrument should be a high mass range, a high resolution and energy and angle focussing.

The main requirements, high transmission and simultaneous detection of the whole mass range, can not be accomplished either by quadrupole instruments, which have been mostly applied in SIMS investigations over the last decade, nor by scanning magnetic instruments, as they are now widely applied in SIMS combined with a regenerating liquid matrix ("FAB").

All the requirements mentioned above other than the high resolution, however, meet a time-of-flight instrument (TOF), provided some crucial aspects are considered. A time-of-flight mass spectrometer for SIMS can be of a straightforward design [3]. For an optimum utilization of a given number of sample molecules or the corresponding secondary ions with their characteristic energy and angular distribution, a special design of such an instrument is desired.

We have designed and run a time-of-flight mass spectrometer with a pulsed primary ion source. The instrument has a large energy and angular acceptance. It follows a geometry described by POSCHENRIEDER [4]. A high purity mass separated primary beam is required for secondary ion studies because any contamination in the primary beam will distort the secondary ion spectrum. The primary beam system actually is a time-of-flight spectrometer in itself because different species are separated in the drift path between the deflection system and the target. So the secondary ion mass spectrum is convolved with the primary ion spectrum.

The ion detection device must have a very high dynamic range to facilitate trace analysis and because ion yields may differ by several orders of
magnitude for different species. Single ion counting therefore is the
adequate mode of detection. In order to achieve compatibility with other
methods of surface analysis and to make operation convenient the target
should be at ground potential.

2. Design of the Instrument

The main components of the instrument, which was designed to fulfill all of
the above requirements [5], are the primary ion source and the deflection
condenser, a magnet for primary ion separation and bunching, a multiple
focussing time-of-flight analyser for the secondary ion separation and an
ion-electron-photon converter, which is used as a detector. The schematic
of the apparatus is shown in Fig.1.

2.1 Primary Beam System

An electron impact ion source with magnetic focussing of the radial electron beam is used to produce a continuous beam of argon ions. Acceleration to 6 keV and focussing is obtained by slit electrodes. A short bunch
of ions is cut out of the continous beam by applying a high voltage pulse

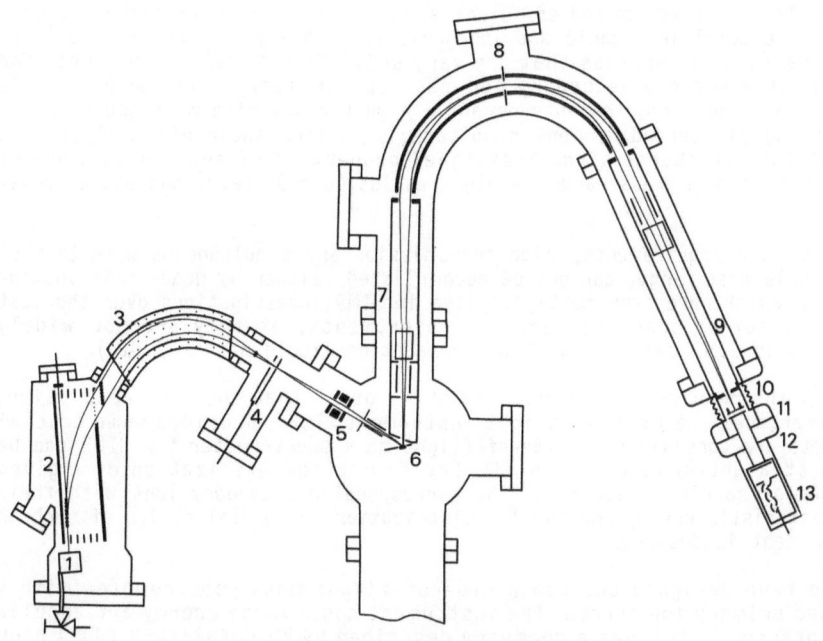

Fig.1 Schematic of the apparatus. (1) Primary ion source, (2) Deflector,
(3) Bunching magnet, (4) Start pulse multiplier, (5) Lens, (6) Target,
(7)+(9) Linear drift paths, (8) 163° toroidal condenser, (10) Postaccelera-
tion gap, (11) Channelplate, (12) Scintillator and light guide, (13) Photomultiplier

to the parallel plate deflection system. While the deflection voltage is switched on the ions are accelerated in the direction perpendicular to the undeflected beam. Thus the energy is increased to about 9 keV. Before the ions leave the deflection system through a hole in the grounded plate the HV pulse is switched off again, in order to compensate for the effects of the beam width and the fringe fields at the exit aperture. The rise and fall times of the 4 kV pulse are 0.03 μsec while the duration is approximately 1 μsec. The ion packet is tilted towards its direction of movement when it leaves the deflector. This gives an equal path length to the exit slit of the magnetic separator for all ions in the packet. Due to this bunching effect the instantaneous current in the pulse can be much larger than the ion current from the source.

A part of the ion packet hits the edges of the exit slit and generates secondary electrons. These are detected by a magnetic multiplier. In this way the pulse shape can be monitored and a start pulse for the registration of the secondary ion spectrum is obtained. The pulsed ion beam is focussed on a spot 1 mm by 10 mm in the target plane. In this plane a pulse width of 7 nsec was measured with a magnetic multiplier mounted perpendicular to the beam direction. The pulse length increases for non-normal incidence because of the rather large width of the ion beam. The actual pulse width of 10 nsec limits the resolution only in the lower part of the secondary ion mass spectrum.

The pulse repetition rate can be adjusted up to 40 kc and the average current can be as high as 1 nA. This is equivalent to 150 000 ions per pulse. For normal operation however the beam current must be reduced to less than 1 pA because the detector is used in the single ion counting mode.

2.2   Time-of-Flight Analyser

The mass of the secondary ions is determined by measurement of the time of flight through a combination of an electrostatic sector field with two linear drift spaces. This geometry, which had been first described by POSCHENRIEDER [4], combines stigmatic imaging with vanishing energy dispersion in time of flight. The sector field has a mean radius of 20 cm and a total deflection angle of 164 degrees. The total drift length is 157 cm. The toroidal deflector had originally been built by OETJEN and POSCHENRIEDER [6] , who also investigated its second-order focussing errors [7].

For an acceleration voltage of 5 kV the relative energy spread of the secondary ions is usually less than 1%. In this range the second-order energy focussing error does not yet limit the resolution of the instrument. The target width is the most critical parameter because it has a first-order effect on the time of flight. Lateral time focussing can be obtained by tilting the entrance or exit plane to 50 degrees versus the direction of the incident beam. If both planes are tilted by 25 degrees lateral directional and energy focussing can also be obtained. It was, however, found that the steep slope of the image plane increases the influence of small geometrical misalignments and instabilities of the deflection voltage on the transit time. Therefore the detector is mounted perpendicular to the beam direction and the tilt of the entrance plane is achieved by an inhomogeneous acceleration field.

2.3   Detection System

The detector is operated in the single-ion counting mode. This gives a much higher dynamic range than analog to digital conversion and there is nearly no mass discrimination in the lower mass range. For large molecules however the yield of the ion electron conversion drops below one. In order to increase the useable mass range a post-acceleration-voltage of up to 20 kV can be applied to the conversion electrode. With this voltage a detection efficiency of 50 % can be expected in the mass range up to 5000 amu [8].

A single channel plate is used as the first stage of the detector, which is followed by a scintillator. This combination is very useful, because it can also be used to monitor the beam profile. For normal operation however a photomultiplier is coupled to the scintillator via a light guide. The pulse amplifier and discriminator at the photomultiplier output are always at ground potential. Therefore they can be DC-coupled. The normalized pulses from the discriminator are accumulated in a multichannel scaler. This instrument has 16 k channels of 24 bits capacity. The time resolution is 10 nsec.

2.4   Vacuum System and Target Mounting

The target is mounted in the center of a spherical vacuum chamber, which allows the attachment of additional equipment for other surface analytical techniques. The chamber is evacuated by a 500 l/sec turbomolecular pump. Differential pumping of the ion source and the detection system is provided by two 100 l/sec turbomolecular pumps. Commercially available sample load locks and manipulators can be used. The whole system has UHV- compatible metal sealed flanges and is bakable up to 250° C.

The target, which is at ground potential, and in an horizontal position, is mounted approximately 1 cm below the extraction electrode. The extraction voltage can be increased up to 5 kV. Deflection plates are provided for the adjustment of the primary and the secondary ion beams.

3.   Performance of the Instrument

The performance of the instrument may be demonstrated by some preliminary results. Fig.2 presents low mass and high mass sections of the positive secondary ion spectrum in the mass range between 1 and 3000 amu emitted from a silver target covered with vitamin B12. Fig.3 presents in a logarithmic scale the low mass range up to 20 amu.

The highest detected masses are $(2M+H)^+$ and $(2M+Ag)^+$ at 2711 and 2818 amu respectively. The mass resolution is about $M/\Delta M = 500$. From both spectra it can be recognized that the background of metastable ions, limiting the dynamic range in a linear time-of-flight instrument, is drastically reduced by the geometrical arrangement of this instrument.

The qualities of the mass-selected pulsed primary ion source may be demonstrated by the logarithmic spectrum of Fig.3. The H^+ peak in the linear spectrum covers only one channel, i.e. the primary ion pulse is well below 10 ns. The mass-separated argon primary beam avoids all effects of faster or slower primary ions. This is demonstrated by the clean background.

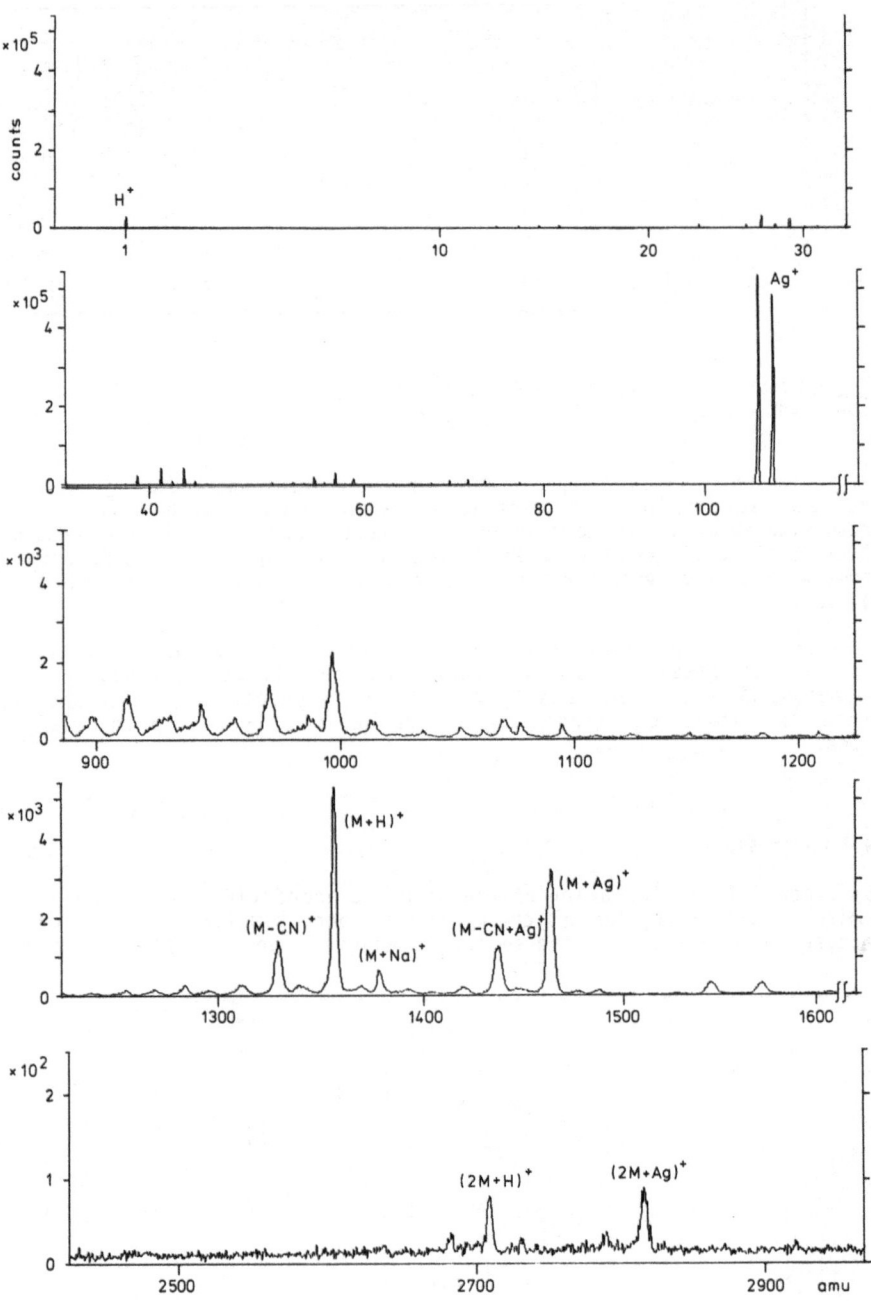

Fig.2 SI spectrum of vitamin B12. Total number of primary ions: $3 \cdot 10^8$

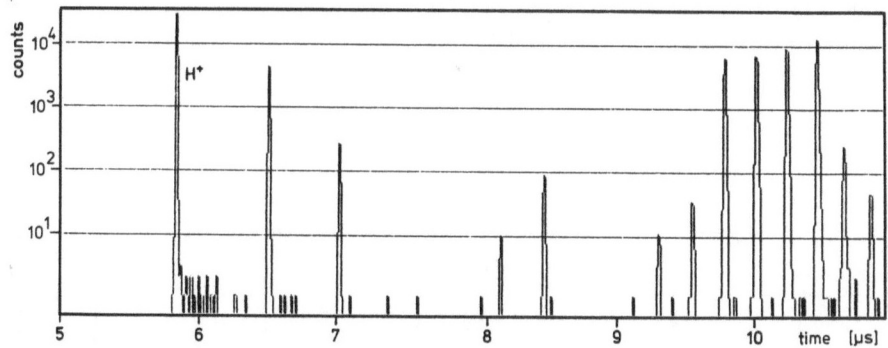

<u>Fig.3</u> Spectrum in the low mass range. Total number of primary ions: $8 \cdot 10^7$.
Logarithmic scale

The energy focussing capability of the instrument was tested by adding a voltage sweep either to the acceleration electrode or to the target. The resulting spectra are shown in Fig.4. Without energy focussing the resolution is seriously deteriorated. By the use of energy focussing secondary ions with a broad energy distribution can be mass analysed with good resolution.

From the experimental results available at present, no final information on the transmission of the instrument can be obtained. The high transmission, however, can be recognized from the spectrum in Fig.2 and the fact that it was obtained by a total number of only $3 \cdot 10^8$ primary argon ions.

4. Conclusion

By a careful consideration of the specific properties of secondary ion emission such as angular and energy distribution, simultaneous emission of metastable ions, etc., and the special strength and application of secondary

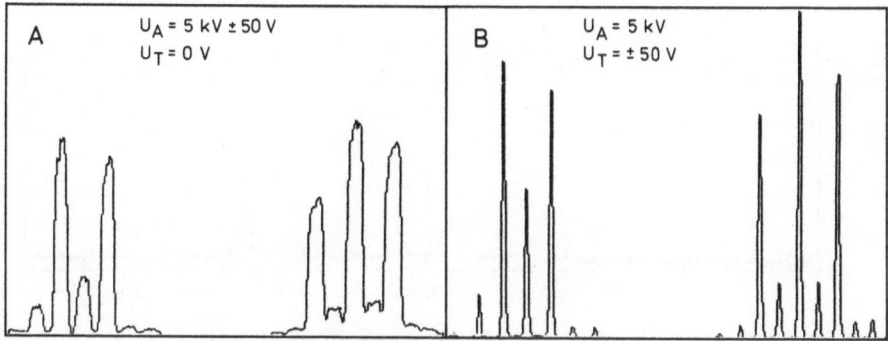

<u>Fig.4</u> Test of energy focussing properties. Relative energy modulation:±1%,
mass range 25-45. (A) without, (B) with energy focussing

ion mass spectroscopy, a high performance time-of-flight instrument can be constructed. The instrument is useful for trace analysis of organic as well as inorganic material, for the detection, identification and structural analysis of large organic molecules, especially if only small amounts of sample material are available, and for the undisturbed SIMS investigation of surface structures, which are very sensitive to ion bombardment.

References

1 A. Benninghoven, W. Sichtermann: Anal. Chem., 50, 1180 (1980)
2 A. Benninghoven: these proceedings
3 B.T. Chait , K.G. Standing: Int.J.Mass Spec.Ion Phys., 40, 185 (1981)
4 W.P. Poschenrieder: Int. J.Mass Spec. Ion Phys., 9, 357 (1972)
5 P.Steffens: Thesis Muenster 1983
6 G.H. Oetjen, W.P. Poschenrieder: IPP Report 9/15, January 1974
7 G.H. Oetjen, W.P. Poschenrieder: Int.J.Mass Spec.Ion Phys.,16,353(1975)
8 E. Niehuis: Diploma work, Muenster 1981

3.6 Secondary Ion Emission from Adsorption Layers on Nickel

D. Greifendorf, P. Beckmann, M. Schemmer, and A. Benninghoven

Physikalisches Institut der Universität Münster, Domagkstraße 75,
D-4400 Münster, Fed. Rep. of Germany

1. Introduction

The interaction of small molecules like C_2H_2 or C_2H_4 with a metal
surface results in the formation of various surface complexes.
The investigation of their secondary ion emission supplies information on the chemical composition of the surface and details
of the corresponding surface reactions, as well as information
on the sputtering and secondary ion formation process, as for
example on the energy transfer to adsorbed surface species by
primary ion induced impact cascades.

2. The Interaction of C_2H_4 and C_2H_2 with an Ni Surface

We investigated the interaction of H_2, C_2H_2, and C_2H_4 with a clean
nickel surface by combined thermal desorption mass spectrometry
(TDMS) and static secondary ion mass spectrometry (SIMS) [1-4] .
We found the secondary ion species MeA^+ or Me_2A^+ to be most specific for a metal (Me) bonded adsorbate A. In order to establish
this relationship for the most important surface complexes formed
during the interaction of ethylene or acetylene with a clean
nickel surface on one hand, and the corresponding secondary ion
species on the other hand, we investigated surface layers predominantly consisting of only one surface species (Table 1).
By combined SIMS/TDMS measurements we found a proportionality
between the ratios $NiC_2H_2^+/Ni^+$, $NiC_2H_4^+/Ni^+$, and Ni_2H^+/Ni_2^+, and
the relative surface coverage by the corresponding surface complex [5] . Since no thermal carbon desorption from nickel occurs,
the corresponding calibration for the ratio NiC^+/Ni^+ was not
possible.

Table 1 Nickel surface complexes, their preparation and the
corresponding "characteristic" secondary ion

Surface complex	Preparation	Characteristic secondary ion
Ni-H	H_2 exposure at 150 K	Ni_2H^+
Ni-C	C_2H_2 or C_2H_4 exposure at 300 K, followed by heating up to 450 K	NiC^+
Ni-C_2H_2	C_2H_2 exposure at 80 K	$NiC_2H_2^+$
Ni-C_2H_4	C_2H_4 exposure at 80 K	$NiC_2H_4^+$

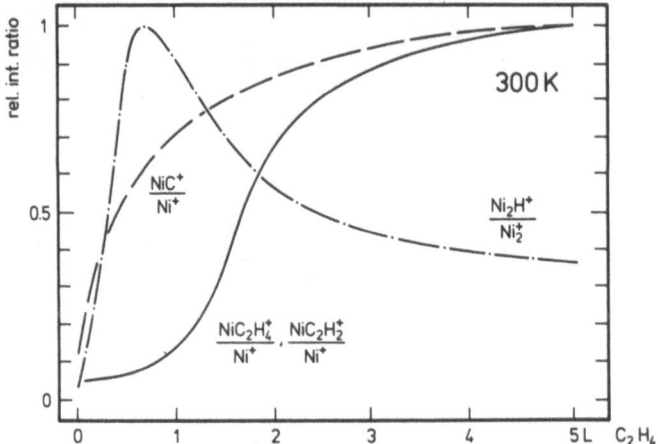

Fig.1 Changes of those secondary ion intensity ratios that are proportional to the relative coverage of a nickel surface with chemisorbed hydrogen, nickel carbide, and adsorbed acetylene or ethylene respectively, during C_2H_4 exposure of a clean nickel surface at 300 K. 1 L = 10^{-6} Torr·S.
Primary ions: Ar^+, 3 keV, $3 \cdot 10^{-9}$ A on 0.1 cm^2

Fig.1 presents the C_2H_4 exposure dependence of the secondary ion intensity ratios that are characteristic for the surface concentration of Ni-H, Ni-C, and Ni-C_2H_4 complexes. The corresponding experiment with acetylene gives quite similar results. The behaviour indicates a complete dissociation of ethylene on the clean nickel surface. Only after a high precoverage by hydrogen and carbon, the increase of the $NiC_2H_4^+/Ni^+$ intensity indicates the beginning of molecular ethylene adsorption.

In order to demonstrate that only the carbon precoverage is responsible for the adsorption of undissociated C_2H_4 or C_2H_2, the target was heated up to 500 K after ethylene exposure. This results in a complete desorption of hydrogen and hydrocarbons, so that the surface is only covered by Ni-C (nickel-carbide) complexes [6] . Fig.2 presents the dose dependence of the $NiC_2H_4^+/Ni^+$ ratio for a low (a), medium (b), and high (c) carbon precoverage of the nickel surface. The results show clearly that the deactivation of the nickel surface by the carbon precoverage determines the adsorption of undissociated C_2H_4.

For the interpretation of the results presented in Figs. 1 and 2, we made the assumption that the linear relationship between relative coverage by a surface complex on one hand, and the corresponding secondary ion intensity ratio on the other hand, is valid even if different surface complexes are present on the target at the same time. This assumption is strongly supported by the observation that during enhanced ion bombardment an exponential decrease of all relevant secondary ion ratios

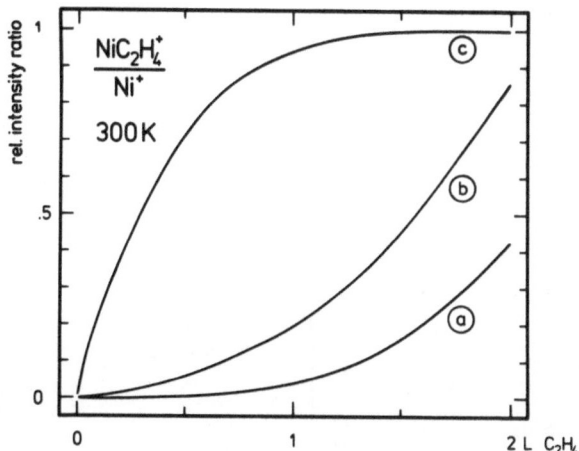

<u>Fig.2</u> Increase of the intensity ratio $NiC_2H_4^+/Ni^+$ during 2 L C_2H_4 exposure at 300 K. Surface preparation: cleaned Ni (a); 0,5 L C_2H_4, heated to 500 K (b); and 2 L C_2H_4, heated to 500 K (c)

is found, in complete agreement with this assumption and a simple desorption model.

2. Information on the Energy Transfer to Surface Species by the Impact Cascade

Surface species can be removed (sputtered) from the surface if a sufficient energy transfer from the impact cascade to surface atoms or molecules occurs. We expect a high ion bombardment in-duced desorption or damage cross section σ for weakly bonded sur-face complexes, and a low cross section for strongly bonded spe-cies [7] .

Determining the energy transfer E to a particle at the point (x_1, x_2) on the target surface relative to the point, where a primary ion hits the surface, and repeating this a large number N_0 of times, one gets a distribution $\frac{\partial}{\partial E} N(E, x_1, x_2)$ with

$$\int_0^\infty \frac{\partial}{\partial E} N(E, x_1, x_2) \, dE = N_0 \; . \tag{1}$$

Division by N_0 provides the density of probability with re-spect to the energy scale. It depends on (x_1, x_2).

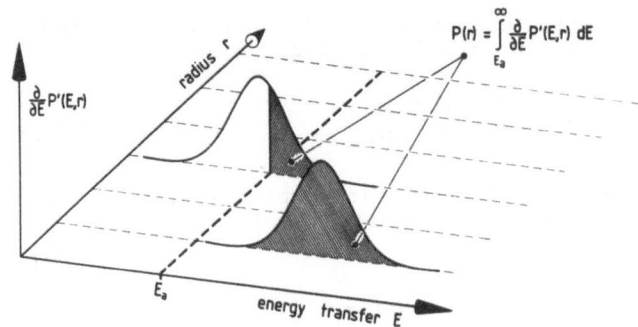

Fig.3 Probability density $\frac{\partial}{\partial E}P'(E,r)$ for the energy transfer E to
a surface species at a distance r from the point where the
primary ion hits the surface, as a function of r and the energy
E transferred to the particle. The dashed line corresponds to
the thermal activation energy of the surface species. The hatched
area corresponds to the probability P(r) for desorption (disappearance) of the particle, and is still a function of the
radius r

$$\frac{\partial}{\partial E}P''(E,x_1,x_2) := \frac{1}{N_0}\left[\frac{\partial}{\partial E}N(E,x_1,x_2)\right]$$

(2)

with
$$\int_0^\infty \frac{\partial}{\partial E}P''(E,x_1,x_2)\,dE = 1 \ .$$

If rotational symmetry is assumed, only a dependence on the
radius $r = \sqrt{x_1^2 + x_2^2}$ remains. This can be illustrated by adding
a third axis to the plot of $\partial/\partial E\,P'(E,r)$ (Fig.3). The dashed line
corresponds to the activation energy E_a for thermal (!) desorption
of a surface adsorbate. This line cuts the graph of the probability density for a given r into two parts. The area at the right
of this line gives the probability for desorption and is of course
still a function of the radius r:

$$P(r) = \int_{E_a}^\infty \frac{\partial}{\partial E}P'(E,r)\,dE \ .$$

(3)

P(r) is related to the disappearance cross section σ by the
following expression:

$$\sigma = \iint_{-\infty}^{\infty} P(\sqrt{x_1^2+x_2^2})\,dx_1\,dx_2 = \int_0^\infty\int_0^{2\pi} P(r)r\,d\varphi\,dr = 2\pi\int_0^\infty P(r)r\,dr \ .$$

(4)

In a first approximation we set $P(r) = 1$ for r smaller than a critical radius R_0 and $P(r) = 0$ for greater values:

$$P(r) := \begin{cases} 1 & \text{for } r \leq R_0 \\ 0 & \text{for } r > R_0 \end{cases} \quad . \tag{5}$$

The adsorbate is assumed to disappear completely from an area with the critical radius R_0 around the point of impact. Using (5) one obtains

$$\sigma = 2\pi \int_0^\infty P(r)\, r\, dr = 2\pi \int_0^{R_0} r\, dr = \pi R_0^2 \quad . \tag{6}$$

Table 2   Cross sections σ , obtained for 3 keV Ar^+ bombardment and thermal activation energies E_a, obtained by TDMS, for various adsorbed species on nickel

Adsorption-system	"characteristic" SI-signal	cross section cm^2	E_a eV/atom
O_2, 300 K	O^-	$1.8 \cdot 10^{-15}$	5.2
H_2, 150 K	$\dfrac{Ni_2H^+}{Ni_2^+}$	$6.0 \cdot 10^{-15}$	2.7
C_2H_4, 300 K	$\dfrac{NiC_2H_4^+}{Ni^+}$	$1.0 \cdot 10^{-14}$	1.6
C_2H_4, 80 K	"	$2.4 \cdot 10^{-14}$	0.4
H_2, 80 K	$\dfrac{NiH_2^+}{Ni^+}$	$4.5 \cdot 10^{-14}$	0.2

The combinations of these considerations with experimental values for σ and E_a supplies some insight into the ion bombardment induced impact cascades that arrive at the surface. Relation (6) is based on the simplifying assumption that the average energy transferred to a surface species A, located at a distance r from the point of primary ion impact, is $> E_a$ for $r \leq R_0$ and $< E_a$ for $r > R_0$. In this picture, the impact of a primary ion on a surface completely covered by the adsorbate A results in the complete depletion of a surface area $\sigma = \pi R_0^2$, whereas no adsorbed particles are desorbed outside this area. For a given target and constant bombardment conditions, R_0 depends on the binding energy of the adsorbate, i.e. for a given substrate on the nature of the adsorbed species. In reality, this surface depletion has a more statistical character.

In Fig.4 experimental values of σ obtained for 3 keV argon ion bombardment are plotted versus the corresponding activation energy E_a for thermal desorption, as they were determined by flash desorption experiments (TDMS), for 5 different adsorbates: chemisorbed O, chemisorbed H, physisorbed H_2, and physisorbed C_2H_4. The resulting curve may be interpreted as the average energy transferred by the ion-induced impact cascade to a

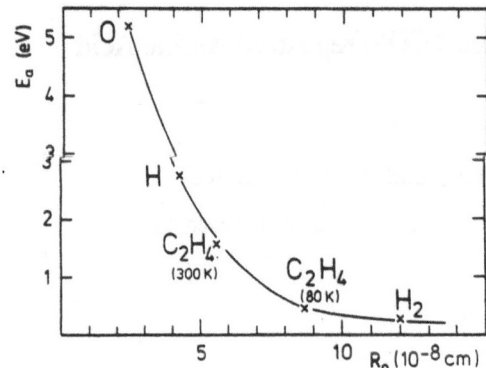

Fig.4 Activation energy E_a for thermal desorption plotted versus the "critical" radius R_O for several adsorbates on polycristalline nickel. The R_O values correspond to 3 keV Ar^+ ion bombardment

surface species A as a function of its distance from the point of primary ion impact on the surface. This interpretation, of course, is based on all the assumptions and approximations made above. On the other hand, the curve is based on direct experimental results.

References

1   A.Benninghoven, P.Beckmann, D.Greifendorf, K.H.Müller, M.Schemmer: Surface Sci. <u>107</u>, 148 (1981)

2   A.Benninghoven, O.Ganschow, L.Wiedmann: Nederl.Tijdschr. Vacuumtechn. 16 No. 2-4, 22 (1978)

3   A.Benninghoven, P.Beckmann, D.Greifendorf, M.Schemmer: Appl. of Surface Sci. <u>6</u>, 288 (1980)

4   A.Benninghoven, P.Beckmann, D.Greifendorf, M.Schemmer: Surface Sci. <u>114</u>, L 62 (1982)

5   M.Schemmer, PhD Thesis, Münster (1981)

6   M.Barber, J.C.Vickerman, J.Wolstenholme: J. of Catalysis <u>42</u>, 48 (1976)

7   A.Benninghoven, these proceedings

3.7 Secondary Ion Emission from UHV-Deposited Amino Acid Overlayers on Metals

W. Lange, D. Holtkamp, M. Jirikowsky, and A. Benninghoven

Physikalisches Institut der Universität Münster, Domagkstraße 75, D-4400 Münster, Fed. Rep. of Germany

1. Introduction

Normally samples for the investigation of the secondary ion emission from organic molecules are prepared outside the vacuum chamber of the mass spectrometer, by deposition of an appropriate solution on a substrate surface. Because of unavoidable contaminations originating from the ambient atmosphere and the solvent, and possible reactions between the solvent and the substrate material, this deposition technique does not result in a well-defined and reproducible chemical environment of the sample molecules on the target.

In order to investigate the influence of the chemical environment on the secondary ion formation from adsorbed amino acids under more defined surface conditions, a molecular beam source was used to deposit controlled amounts of material on atomically clean polycrystalline metal surfaces under UHV conditions. The results indicate strong matrix effects in the submonolayer and monolayer range of coverage.

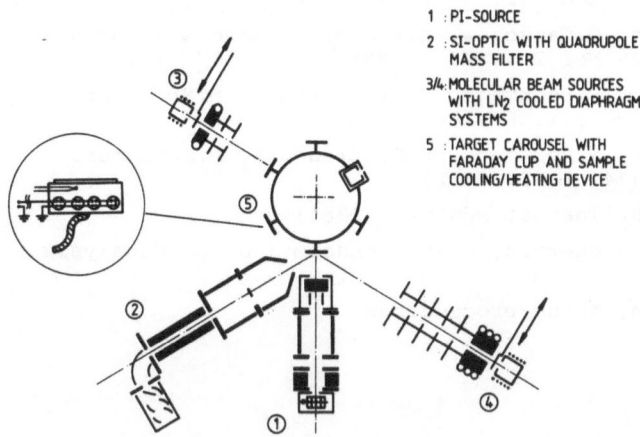

1 : PI-SOURCE
2 : SI-OPTIC WITH QUADRUPOLE MASS FILTER
3/4: MOLECULAR BEAM SOURCES WITH LN2 COOLED DIAPHRAGMA SYSTEMS
5 : TARGET CAROUSEL WITH FARADAY CUP AND SAMPLE COOLING/HEATING DEVICE

Fig.1 Secondary ion mass spectrometer with molecular beam devices

2. Instrumental

The investigations were carried out in two separate UHV systems for the study of layer growth and evaporation of amino acids and for SIMS experiments (Fig.1).

The basic components of the SIMS system were a quadrupole mass analyzer, an ion gun with a scanning/gating capability providing focussed ($\emptyset$ = 0.5 mm) ion beams (10^{-11} - 10^{-5} A, 0.5 - 5 keV, scanned area: 1 cm^2), and a target carousel carrying a Faraday cup for beam control and a heatable and coolable target holder. The base pressure in the instrument was < $5 \cdot 10^{-10}$mbar.

Atomically clean metal surfaces were prepared by heating (600 C, 1 h) and simultaneous ion bombardment ($5 \cdot 10^{-6}$ A cm^2, 3.5 keV Ar^+, 1 h). The cleaning was checked by SIMS and AES spectra. In all cases the remaining total surface contamination was found to be well below 1 %.

Amino acid molecules were evaporated from a molecular beam source consisting of a Knudsen-type evaporation cell containing the organic material, a nitrogen-cooled system of diaphragms, and a shutter allowing control of the evaporation time. The purpose of the cooled diaphragms was to maintain UHV conditions ($p < 2 \cdot 10^{-9}$ mbar) during degassing or evaporation of the amino acids. The temperature of the organic material inside the evaporation cell was controlled by Ni-NiCr thermocouples.

The majority of peaks in the SIMS spectra of UHV-deposited samples originate directly from the amino acid molecules and can be identified unambiguously. On the other hand, the conventional preparation from a solution generally results in additional peaks due to inorganic (salts) and organic (hydrocarbons) contaminations that complicate the interpretation, especially for low-intensity peaks.

A second UHV system was equipped with an Auger spectrometer, a quartz-crystal microbalance and a similar evaporation device as described in Fig.1. The Auger spectrometer had been modified for "static" mode operation, i.e. with the electron beam current reduced to about 10^{-9} A, so that radiation damages produced by the electron bombardment became negligible. This was accomplished by application of a pulse-counting technique combined with a digital simulation of the lock-in technique [1] . The AES system was used to monitor layer growth in the submonolayer range of coverage. In the multilayer range layer growth was directly controlled by the quartz microbalance with gold- or silver-coated crystals ($\emptyset$ = 0,8 cm, sensitivity: 10^{-8} g). This second UHV system was used for the calibration of the evaporation source and for the determination of the vapour pressures of the amino acids.

3. Layer Formation and Control

The rate of layer growth during evaporation of amino acids on the (cooled) microbalance could be recorded versus the temperature of the evaporation cell. Taking into account the geometry of the experimental arrangement, the results could be transformed into vapour-pressure curves (Fig.2), which in turn allowed a calibration of the SIMS experiments. (A sticking coefficient near 1 resulted from complementary measurements of reevaporation rates from heated quartz crystals [2].) By extrapolation of the vapour-pressure data it can be concluded that in terms of monolayers an appreciable reevaporation occurs already at room temperature. This has to be considered in SIMS experiments, especially if the coverage exceeds a monolayer. Due to this re-evaporation thick layers are stable only on cooled target surfaces.

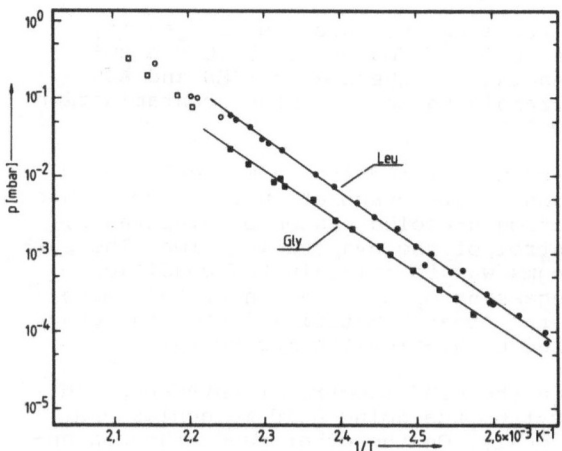

Fig.2 Vapour pressure of leucine and glycine. The open symbols present data from [3]

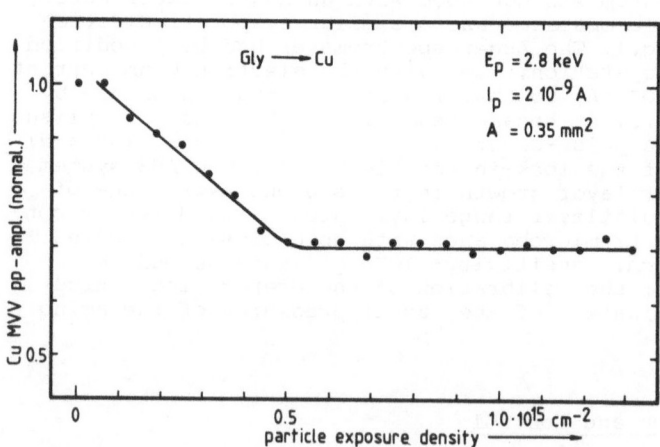

Fig.3 Change of the Cu MVV (60eV) AES signal during the deposition of glycine on a copper substrate at 310 K

The effect of reevaporation from uncooled targets could be observed by monitoring the layer formation with AES (Fig.3). For reasons of sensitivity the substrate signals (Cu,e.g.) were observed instead of sample specific lines (C,O,N) during the amino acid deposition. Obviously layer growth stops after an exposition time that corresponds to a vapour dose of approximately one monolayer. This finding suggests that monolayers of amino acids adsorbed on metal surfaces, Cu e.g., do not reevaporate at room temperature, due to the formation of strong bonds between these molecules and the corresponding metal atoms.

4. SIMS Results

SIMS control of layer growth on targets at room temperature confirm the Auger results. Characteristic SIMS signals reach a steady state when the surface has been exposed to approximately one monolayer equivalent of amino acid molecules.

4.1 Intensity Changes During Heating

More details concerning the bonding of the amino acids on the metal surface can be concluded from SIMS intensity changes during target heating. Figure 4 depicts the changes of ion emission during heating of targets covered by several monolayers of an amino acid. Before starting the temperature increase, the targets were cooled (200 K) in order to avoid a rapid evaporation of the multilayers. Whereas all fragment-ion yields of amino acid de-

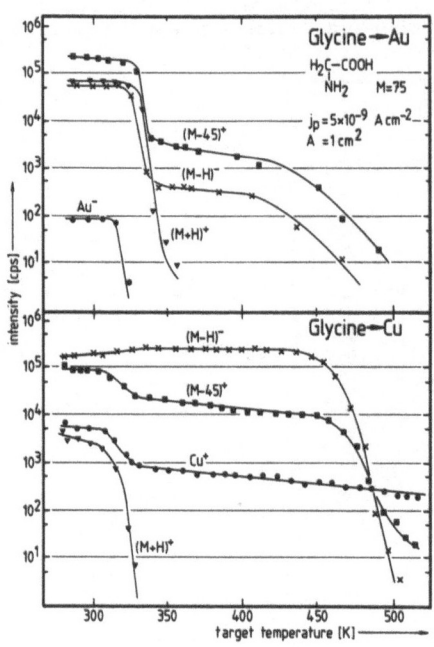

Fig.4 Development of some secondary ion intensities during heating of an Au and a Cu substrate precovered by some monolayer equivalents of glycine

posited on Au decrease by orders of magnitude in a low-temperature range (T < 350 K), only the positive ion intensities disappear in this temperature range for a Cu substrate. The negative yields from a Cu target decrease only in a temperature range (T > 450 K) where thermal decomposition of the amino acid molecules occurs [4] . The results of similar experiments with a variety of amino acids and substrate metals indicate that the results for the Cu substrate are typical for reactive metals (including Ag). On the other hand, a similar behaviour as for Au was found for Pt surfaces.

These experimental data indicate that the noble metals Au and Pt form weak bonds with amino acid molecules, even in the first monolayer, allowing the thermal desorption of this monolayer at a low temperature. For reactive metals, however, the fact that the (M-H)$^-$ yield remains almost constant in the low-temperature range supports the assumption of strong bond formation with amino acid molecules. Only weakly bonded molecules from amino acid clusters or thick layers are thermally desorbed in this case, leaving behind a strongly bonded monolayer coverage of amino acid on the metal surface. This explains the complete disappearance of the (M+H)$^+$ emission and the virtually constant emission of (M-H)$^-$ in the low-temperature range.

4.2   Intensity Variation with the Substrate Material

The different behaviour of positive and negative secondary ions during target heating indicates that the secondary ion emission is sensitive to the chemical environment of the adsorbed molecules. Therefore the influence of substrate material on the secondary ion emission was studied for a variety of metals. It was found that in the monolayer range the secondary ion yields differ by orders of magnitude for different substrates (Fig.5).

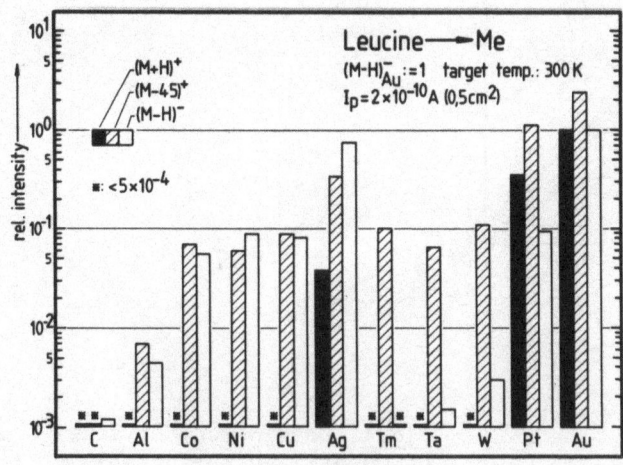

Fig.5   Relative intensities of the molecular ions (M±H)$^±$ and the typical fragment ion (M-45)$^+$ emitted from different metals covered by about one monolayer of leucine

Positive spectra are far more affected than negative spectra. Most of the positive mass lines are strongly reduced for reactive substrate metals.

In accordance with these results, the secondary ion intensity changes during layer growth are different for positive and negative ions and depend on the substrate material (Fig.7 in [5]). Just as during target heating both $(M+H)^+$ and $(M-H)^-$ behave quite similar on gold substrates. They reach a maximum at a dose corresponding to slightly more than one monolayer and then start to decrease, finally attaining values that are no longer typical for adsorbed molecules but for compact amino acid multilayers. On reactive metals initially there is observed an increase of the $(M-H)^-$ emission only. After its saturation the $(M+H)^+$ emission increases as well, if reevaporation is avoided. Thus the direct contact of an amino acid molecule with a reactive metal atom seems to prohibit the formation of the $(M+H)^+$ ion (and most of the other positive fragment ions).

In a qualitative manner the different behaviour of positive and negative molecular secondary ion emission can be explained if the bonding situation of amino acid molecules on metal surfaces is regarded. As amino acids are known to form stable complex compounds with transition metals [6] , strong bonds of a similar type are likely to be formed on metal surfaces as well (chemisorption).

The positive charge in these metal-amino acid complexes is concentrated near the metal atoms which have replaced active protons (cationization). As a consequence, large fragments of the amino acid molecules (not containing a metal atom) are most probably emitted as negative ions. Especially the formation of $(M+H)^+$ ions would need the attachment of two protons which is very unlikely to occur.

On the other hand it can be assumed that on the noble metals Au and Pt physisorption of intact molecules on the substrate surface is dominant. In this situation proton exchange between adjacent molecules (dimers, clusters) favour both the emission of positive and negative molecular ions. This explains the similar behaviour of both ion species during layer formation and heating. The real existence of proton exchange between adsorbed molecules has been proved by the investigation of coadsorbed deuterated glycine and not deuterated methionine on Ag. The SIMS spectra of this mixture definitely contained deuterated molecular ions and fragments of methionine [7] .

4.3 Ion Bombardment

Ion bombardment effects were investigated for several amino acids. Damage cross sections for $(M+H)^+$ and $(M-H)^-$ do not significantly differ for amino acid monolayers on noble metal substrates as Au e.g. (Fig.6). For a reactive metal as Ag, however, the damage cross section for $(M+H)^+$ exceeds that for $(M-H)^-$ by about a factor 3 (Fig.6).This may be explained within the frame of the qualitative model developed above, assuming weakly bonded amino acid mole-

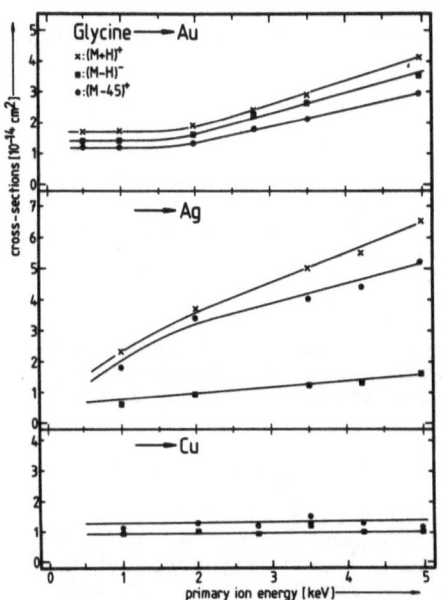

Fig.6 Ion bombardment in-
duced damage cross sections
for $(M\pm H)^{\pm}$ and $(M-45)^{+}$ for
different substrates covered
by about one monolayer of
leucine

cules on the Ag surface which are responsible for the $(M+H)^{+}$
formation, in addition to strongly bonded molecules, responsible
for the formation of $(M-H)^{-}$.

5. Conclusion

Well-defined, clean amino acid overlayers on metal substrates
in the mono- and multilayer range can be produced under UHV
conditions by a molecular beam technique. For the production of
stable multilayers, target cooling is required in order to pre-
vent reevaporation. The first monolayer of an amino acid is
strongly bonded to the metal substrate, avoiding a reevaporation
of the directly metal-bonded molecules at room temperature.

In the submonolayer and monolayer range the secondary ion
emission from amino acids depends strongly on the chemical nature
of the matrix metal. Noble metal substrates yield the most intense parent-like secondary ions. Positive ion yields are far
more sensitive to the chemical environment than negative secondary ion yields. This different behaviour indicates the formation
of polarized bond between reactive metal atoms and amino acids,
which favours the emission of negative $(M-H)^{-}$ ions.

For the formation of positive parent ions $(M+H)^{+}$ proton exchange between adjacent molecules is an important mechanism. It
occurs in adsorbed layers on the noble metals Au and Pt as well
as in multilayers, and gives rise to the emission of both $(M+H)^{+}$
and $(M-H)^{-}$ secondary ions.

References

1 A.Benninghoven, O.Ganschow, P.Steffens, L.Wiedmann: J.Electron
   Spectros.Rel.Phen. 44, 19 (1978)

2 D.Holtkamp, W.Lange, M.Jirikowsky, A.Benninghoven: submitted
   to Appl.Surf.Sci.

3 H.J.Svec, D.D.Clyde: J.Chem.Eng.Data 10, 151 (1965)

4 Handbook of Chemistry and Physics: R.C.Weast,Cleveland,
   Ohio, 58. Ed. (1977)

5 A.Benninghoven, these proceedings

6 R.J.Day, S.E.Unger, R.G.Cooks: Anal.Chem. 52, 253 (1980)

7 M.Jirikowsky, D.Holtkamp, P.Klüsener, W.Lange, A.Benning-
   hoven: to be published

3.8 Temperature Dependence of Secondary Ion Emission from Phenylalanine

Willy Sichtermann

Physikalisches Institut der Universität Münster, Domagkstraße 75,
D-4400 Münster, Fed. Rep. of Germany

1. Introduction

The investigation of numerous organic compounds [1] has shown
that the secondary ion spectra of differently prepared samples
in general do not coincide in their intensity pattern of
molecular and fragmentary ions. Obviously the chemical environment plays an important role and consequently contaminations as
well as admixed substances, pH value of the solution and the
substrate material may give rise to hazardous modifications of
the spectra. Therefore, the recognition of substance and the
determination of quantities may be seriously aggravated even in
case of low order amounts of accompanying materials.

Efforts have been made to standardize the sample preparation
e.g. to evaporate sample material under UHV conditions onto a
substrate cleaned by ion etching [2]. However, successful
applications of this method are restricted to sufficiently
available pure substances which also possess a certain vapor
pressure below their decomposition temperature. Analytical
investigations involving secondary ion mass spectrometry (SIMS)
for determination of certain components within e.g. urine and
other body fluids require wet preparation techniques preferably
after chromatographic preseparation. The aim of this work should
be regarded as a first step towards the quantitative investigation of arbitrary samples involving thermal desorption. Computer
aided evaluation of the quick motion picture of relevant intensities (e.g. Fig. 1) should allow an unambiguous determination
of sample composition.

2. Experimental

The construction of the utilized secondary ion mass spectrometer
which has been described elsewhere [3] has been extended by
introducing the facility of resistive sample heating. Then
investigations have been carried out by thermally removing layers
of the amino acid phenylalanine from an etched silver foil while
SIMS with low current density Ar^+ ions ($5 \cdot 10^{-9}$ A/cm^2, 3 keV) has
been used for probing the remaining part of the substance. Data
have been accumulated by cyclically sampling the intensity of the
required secondary ions during the temperature sweep with an
average rate of 0.5 K/s.

For sample preparation as usual 1 µl of the 10^{-3} mol/l solution of phenylalanine in 10^{-3} normal hydrochloric acid or in water has been deposited on the silver foil; thus on an average $6 \cdot 10^{14}$ molecules of the substance itself cover 1 cm² of the substrate. In case of a homogeneous distribution this results in a coverage of approx. one monolayer or - which is assumed - islands of multilayer structures caused by crystallization during the drying process.

The thermal desorption of phenylalanine prepared as mentioned above is completed at ca. 530 K with a sweep rate of 0.5 K/s. It should be noted that the substrate after an 1 h recrystallization treatment at 400 K in Ar atmosphere to avoid deformations during the heating has been used several times unless a severe contamination e.g. by Li gave rise to an exchange.

3. Results

On the basis of spectra of some preliminary samples characteristic secondary ions have been chosen for the thermal desorption experiment. The presentation of spectra has been omitted because any cross section of the thermal desorption curves represents the essential parts of the spectra.

The intensity development of the four prominent peaks $[M+H]^+$, $[M+Ag]^+$, $[M-45]^+$ and $[M-H]^-$ in case of the acidic preparation is shown in Fig. 1 in dependent on time and temperature. Obviously three phases of desorption are to be distinguished:
- phase 1 (310K<T<340K):decrease of $[M+M]^+$ and $[M-45]^+$, $[M+Ag]^+$ and $[M-H]^-$ constant,
- phase 2 (340K<T<380K):decrease of $[M+Ag]^+$ and increase of $[M-H]^-$,
- phase 3 (410K<T<490K): simultaneous decrease of $[M-45]^+$ and $[M-H]^-$ (ratio 0.75).

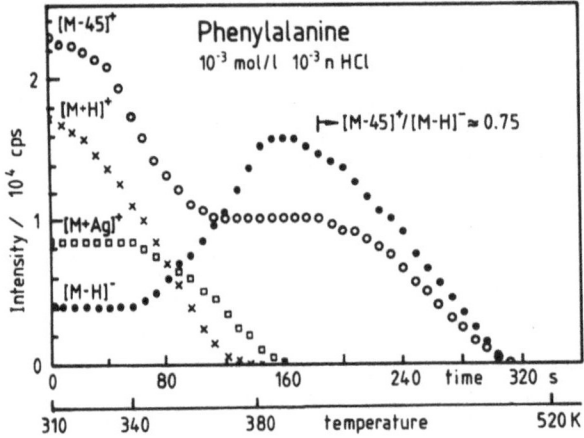

Fig. 1: Thermal desorption of acidic phenylalanine from silver. Intensities of representative secondary ions versus time and temperature. Primaries: Ar^+, 3keV, current density $5 \cdot 10^{-9}$ A/cm²

Presumably the emission of the four shown types of secondary ions originates from at least two different adsorption states, which desorb independently. Thus the intensities must be regarded as a combination of several parts, however, a separation needs more detailed information on the system e.g. whether the decrease of [M+Ag]$^+$ corresponds either to a desorption or conversion according to binding forces. Remarkable is that the ratio of [M-45]$^+$/[M-H]$^-$ which may be interpreted as fragmentation factor is constantly 0.75 over a temperature range of about 100 K.

The neutral preparation reveals only two kinds of secondary ions i.e. [M-H]$^-$ and [M-45]$^+$. The development of intensities during the application of thermal energy is shown in Fig. 2. It seems possible also to distinguish three phases of desorption:
- phase 1 (310K<T<390K): [M-H]$^-$ and [M-45]$^+$ constant, ratio = 1,
- phase 2 (390K<T<440K): increase of [M-H]$^-$, decrease of [M-45]$^-$,
- phase 3 (440K<T<530K): decrease of both, for T>480K ratio of
      [M-45]$^+$/[M-H]$^-$ = 0.25.

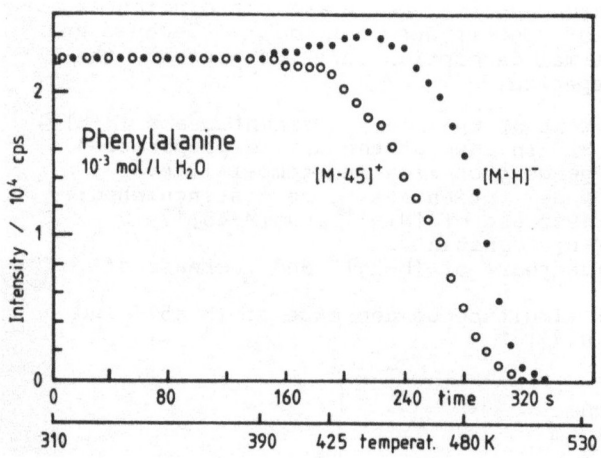

Fig. 2: Thermal desorption of neutral phenylalanine from silver. Intensities of [M-H]$^-$ and [M-45]$^+$ in dependent on time and temperature. Other conditions same as for Fig. 1

Because of the characteristic behavior i.e. constant ratio of [M-45]$^+$/[M-H]$^-$ for certain temperature ranges, the existence of two different states S_1 and S_2 for the phenylalanine is assumed. The value of the ratio itself may reflect the stability of the molecules as it is well known in case of cationization. This consideration is adapted to establish a mathematical formulation in order to explain the intensities of the two involved secondary ions dependent on the coverages θ_1 and θ_2 of S_1 and S_2:

$$[M-H]^- = \alpha_1 \cdot \theta_1 + \beta_1 \cdot \theta_2 \qquad (1)$$
$$[M-45]^+ = \alpha_1 \cdot \theta_2 + \beta_2 \cdot \theta_2 \qquad (1a)$$

The coefficients α_i, $\beta_i (i = 1,2)$ have been determined from the original experimental data under the following secondary conditions from the view of the primary ion beam:

$\Theta_1 = 1$, $\Theta_2 = 0$ for temperature range 310 K to 390 K, $\Theta_1 + \Theta_2 = 1$ for 390K to 425K and $\Theta_1 = 0$, $\Theta_2 < 1$ for 483 K to 523 K; the coefficients values are for the given experiment $\alpha_1 = \alpha_2 = 22380$ cps, $\beta_1 = 36358$ cps, $\beta_2 = 0.25 \cdot \beta_1$ with standard deviation of less than 1 %.

The coverages Θ_1 and Θ_2 now have been calculated as a function of the intensities $[M-H]^-$ and $[M-45]^+$ by means of (1) and (1a). The result of the calculation is shown in Fig. 3.

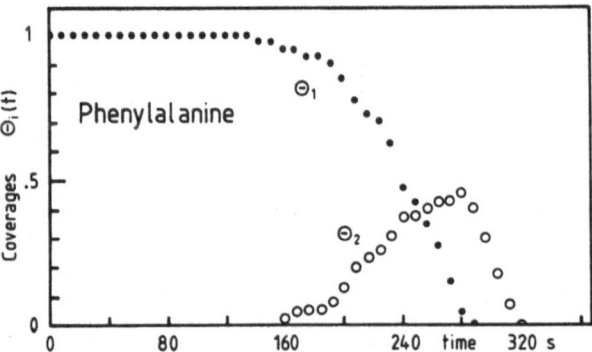

Fig. 3: Thermal desorption of neutral phenylalanine on silver. Coverages Θ_1 and Θ_2 of the two assumed adsorption states S_1, S_2.

After the numerical separation of the two states it is possible to apply the Polanyi-Wigner rate equation:

$$d\Theta_i/dt = -\Theta_i^{n_i} \cdot r_{ni} \cdot \exp{(-E_{ai}/kT)} \qquad (2)$$

with i = number of state, Θ_1 coverage by state S_i, n_1 = reaction order, r_{ni} = frequency factor, E_{ai} = activation energy, k = Boltzmann constant, T = thermodynamical temperature. This rate equation is applicable under the assumption that all parameters other than coverage are independent of temperature. However, a thermal desorption at constant temparature which would by-pass this hurdle may not easily be realized. Nevertheless, the relatively small temperature range of decreasing coverages (Fig. 3) and the constant intensity ratio for this range seems to guarantee the applicability of the above rate equation. At this point it should be noted that other limitations as wall effects, generation of the observed compounds after desorption etc. as in case of pressure measurements [4] are not involved when SIMS is used as a probe.

For the determination of reaction order n_i and activation energy E_{ai} the plot of

$$\ln \left(- \Theta_i^{-n_i} \cdot d\Theta_i/dt \right) \qquad (3)$$

against the reciprocal temperature is useful, because in case of correct choice of n_i the plot is a straight line and the activation energy can be determined as the slope of it. The values of Θ_i and $d\Theta_i/dt$ have been determined by a fourth order Chebychev-approximation according to Fig. 3.

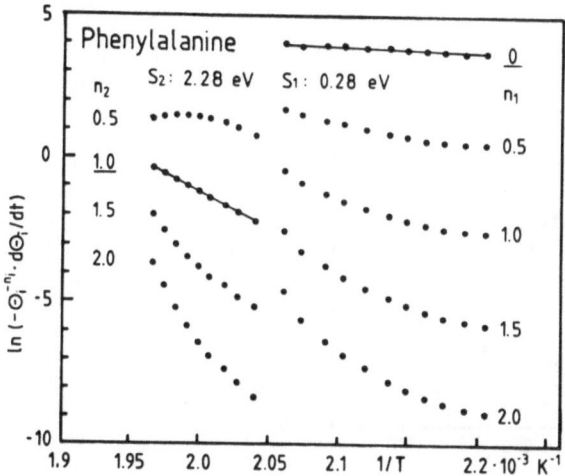

Fig. 4: Polanyi-Wigner Plot according to (2) and (3) for determination of reaction order and activation energy. Results for state S_1: $n_1 = 0$; $E_{a1} = 0.28$ eV; for S_2: $n_2 = 1$; $E_{a2} = 2.28$ eV.

As expected by means of Fig. 4 two states can be identified on the surface: S_1 which is only in contact with molecules of the substance, the other S_2 in contact with the silver. In accordance state S_1 is characterized by zero reaction order and low activation energy which is of the same order of magnitude to that reported for a vapor pressure experiment [2]. The state S_2 is of first order with an activation energy appropriate to binding forces expected for a metal-molecule compound.

4. Conclusions

First investigations of the amino acid phenylalanine with a combination of thermal desorption and secondary ion mass spectrometry show that also a wet prepared sample provides considerable information. Contrary to ion bombardment induced disintegration, different adsorption states are distinguished so that, up to a certain degree, the effects of sample preparation can be understood as a consequence of chemical environment.

Beside the evaluation of kinetic data, the results obtained
for both the acidic and the neutral preparation allow certain
reflections on the behavior of phenylalanine molecules depending
on their environment. For example the ratio $r = [M-45]^{+}/[M-H]^{-}$
varies between 0.25 and 1 and obviously represents the stability
of the molecules before emission. Considering the supposed
adsorption sites this corresponds to the experience in FD (field
desorption) and SIMS that weak molecules e.g. sugar are stabilized by cationization. Probably the general influence of matrix
embedding on the emission properties of the organic substance can
be elucidated in this combination of methods.

In consequence it turns out to be valuable to support quantitative measurements by application of thermal desorption. Finally,
the expected sequence of spectra will ease the direct evaluation
of quantities for each component of a mixture without the need of
preseparation.

Acknowledgement

The author wishes to thank V. Anders for making accessible the
original data for the calculations presented here.

References

1. A. Benninghoven and W. Sichtermann, Anal. Chem. 50 (1978) 1180
2. W. Lange, M. Jirikowsky, D. Holtkamp and A. Benninghoven in
   Springer Series of Chemical Physics 19 (1981) 416
3. A. Eicke, W. Sichtermann and A. Benninghoven, Org. Mass
   Spectrom. 15 (1980) 289
4. L.A. Petermann in Progress in Surface Science 3 (1973) 1
5. W.D. Lehmann and H.R. Schulten, Anal. Chem. 42 (1977) 1744

3.9 Matrix Effects on Internal Energy in Desorption Ionization

K.L. Busch, B.H. Hsu, Y.-X. Xie, and R.G. Cooks

Department of Chemistry, Purdue University, West Lafayette, IN 47907, USA

Desorption ionization (DI) mass spectrometry encompasses a family of techniques in which energization of a condensed phase leads to ejection of ions into the vacuum with subsequent mass analysis and detection [1]. In contrast to the gas–phase ionization methods of electron, chemical, and photo- ionization, DI techniques are by their nature sensitive to the physical and chemical nature of the matrix from which the ions are ejected. Success in the analysis of nonvolatile and thermally fragile molecules has been enough to thrust these techniques into routine use in just a few years. DI spectra of pure compounds can often be interpreted in terms of known gas–phase ion chemistry. This provides evidence for the desorption, from the surface, of intact ions with some degree of internal excitation; the fragmentation processes undergone are then defined by the nature and the amount of the internal energy. Since the ion is in an isolated state, fragmentation processes should be the same as those undergone by ions of the same structure formed directly in the gas phase by chemical ionization [2]. MS/MS experiments have confirmed this premise for particular DI conditions. Ions are isolated by a first stage of mass analysis, and then activated by collision. The masses and relative abundances of the resultant fragment ions (determined by the second stage of mass analysis) match those of the DI spectra. Metastable ion transitions have also been directly observed in DI spectra.

There remain samples for which DI analysis is not successful, and the problem can often be traced to the matrix in which the sample is found. We report results of a series of experiments in which the matrix is varied in order to delineate some of its effects, and to determine how they can be turned to advantage in an analytical experiment. In direct mixture analysis, for instance, the sample to be analyzed is but one component in an undefined mixture. In other experiments, the sample may be purposefully mixed into a matrix of known composition for analysis. We have previously used ammonium chloride matrices in the SIMS analyses of biological and organometallic samples [3,4]. Desorption of structurally informative ions from amino acids, nucleosides, and vitamins is facilitated when the sample is admixed with ammonium chloride. It has also been shown that for compounds in which intermolecular reactions take place, products of such reactions can be reduced in relative abundance when the sample is dispersed in an ammonium chloride matrix. More significantly, the signal-to-noise ratios of the organic ions of interest did not decrease [5] on dilution in the matrix. Another example of matrix design is the dissolution of samples in glycerol, a technique used in electrohydrodynamic ionization [6] and fast atom bombardment mass spectrometry [7]. Derivatization reagents (such as acids or bases) are often mixed with samples in order to create ionic species primed for desorption into the vacuum [8].

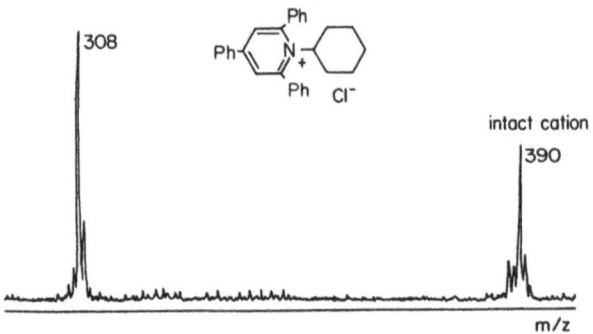

DILUTION 1:50,000 NH$_4$Cl
from solution onto silver foil
4.5 keV Ar$^+$ 2 x 10^{-9} A

Figure 1

308

intact cation
390

m/z

Several aspects of the matrix effect in SIMS will be presented here. First, experiments with ammonium chloride matrices have demonstrated marked effect on the fragmentation patterns of the spectra, with decreased fragmentation occurring in the presence of the salt matrix. Second, a series of experiments has confirmed the surprising persistence of organic ions in the SIMS spectrum of a predominately inorganic matrix. Studies of a quaternary ammonium salt in an ammonium chloride matrix have shown that even at 1:50,000 dilution, a clear and interpretable spectrum is obtained (Fig. 1). In this example, the component of interest is present at the 20 part-per-million level, yet the S/N for the organic ions is excellent, and the fragmentation pattern has not changed greatly from that seen in the spectrum at a 1:20 dilution. A decrease in the absolute number of ions desorbed is balanced by an increase in the primary ion current. This spectrum could be observed for several minutes before refocussing of the primary ion beam onto a new portion of the surface was necessary. Detection of preformed ions of carbonium and phosphonium salts has also been achieved despite high dilution of the organic in an ammonium chloride matrix.

The effect of a solid matrix on the fragmentation of a desorbed ion is illustrated in Fig. 2. The upper spectrum is that of neat candicine chloride, a quaternary ammonium salt. Upon 20 fold dilution in ammonium chloride, the center spectrum is obtained, in which the ratio between the intact cation at m/z 180 and the fragment ion (loss of trimethylamine) at m/z 121 reflects a decrease in the extent of fragmentation. It can be seen that the matrix-assisted spectrum begins to resemble that obtained (lower figure) in an MS/MS experiment when the intact cation is desorbed from a surface by photon impact and selectively excited.

Desorption of organic ions from liquid matrices very often produces effects similar to those of solid matrices. An example is given in Fig. 3, which shows the positive ion SIMS spectrum of a pyridinium salt. The upper spectrum is that obtained from the neat compound, and the lower that of the same compound dispersed in polyethylene glycol. The ratio between the intact cation at m/z 137 and the fragment ion at m/z 80 changes on dilution to favor the intact cation. The general trend is that of decreased fragmentation when desorption occurs from the matrix. A

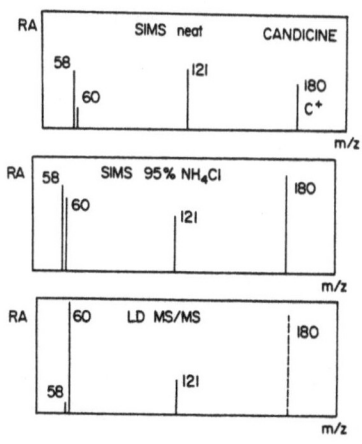

Figure 2

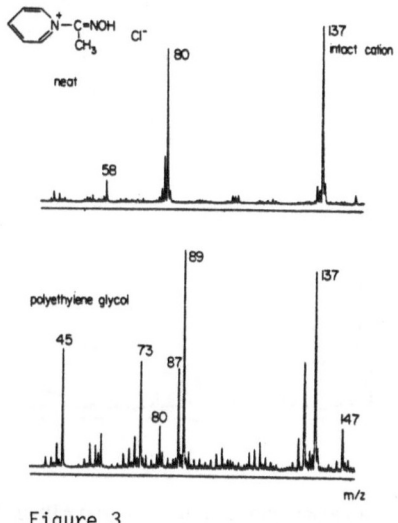

Figure 3

number of such observations in DI can be drawn together under the aegis of a desorption/desolvation mechanism as shown in Fig. 4. Central to this proposal is the idea that polyatomic clusters are desorbed from energized surfaces [9]. With the appropriate matrix, these clusters consist of ions solvated by molecules of the matrix. The cluster carries some amount of internal energy depending upon the ease with which it was desorbed. Excess internal energy is used to jettison solvent molecules. As these evaporate into the vacuum, they can carry away some portion of the total energy of the complex. This process of desorptive relaxation can yield cations with a reduced internal energy, and thus the degree of fragmentation is decreased.(Fig. 5)

Much recent work in DI has been concerned with the desorption of ions from liquid matrices. The earliest examples of this phenomenon were reported in SIMS and FAB work. At the 1980 Münster Conference, COOKS reported the desorption of organic compounds from polyphenyl ether [10]. In addition to observation of the usual protonated and cationized species, it was noted that the formation of cluster ions from organic salts was diminished when the sample was dispersed within the matrix. In 1981, SURMAN and VICKERMAN used polyphenyl ether as the solvent in FAB experiments [11]. At high fluxes of bombarding atoms, they reported a rapid decrease in secondary ion yields of (M+1)+ ions from the amino acids valine and phenylalanine. This decrease could be avoided by suspension of the organic sample in polyphenyl ether. SURMAN and VICKERMAN postulate a steady evaporation of the liquid solvent with a surface constantly replenished with sample molecules diffusing from the bulk of the solution. BARBER has investigated the use of glycerol as a liquid matrix in FAB [7]. Although early reports omit mention of any solvent use [12], in later reviews, BARBER describes in some detail a model of sample/solvent interactions in which the main advantage accrued is an extended sample lifetime due to diffusion and enhanced sensitivity of detection [7]. The desorption of clusters leading to internally relaxed ions may help explain the latter observation.

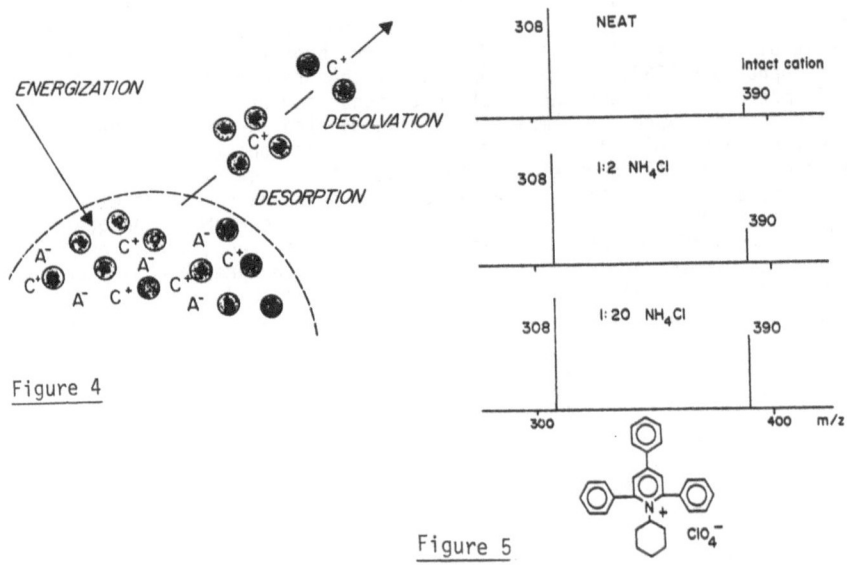

Figure 4

Figure 5

Two additional examples from other ionization methods serve to buttress this concept. In EHMS, samples in a glycerol matrix are pumped directly into the high vacuum source of a mass spectrometer. Cluster ions of many glycerol molecules can be seen in the spectra thus obtained; organic ions are most often seen either unsolvated, or complexed with one or two molecules of the glycerol, indicating a large degree of solvent loss by evaporation. Evidence that relaxed organic ions are formed (presumably accompanying desolvation) is evident in the fact that doubly-charged cations are observed as base peaks for the diquaternary ammonium salts in EHMS [13], whereas in SIMS and other techniques, the same ions when they occur at all, are present in much reduced abundance [14]. The Coulombic repulsive energy inherent in these dications is added to any energy deposited as the result of desorption, and these compounds can thus be used as probes sensitive to the energetics of the desorption process. GIESMANN and BAROFSKY have used doubly charged compounds to investigate the effects of liquid matrices in field desorption mass spectrometry and have drawn similar conclusions [15].

Similarities to the proposed desolvation processes can also be found in the thermospray method investigated by VESTAL as an interface between a liquid chromatograph and a mass spectrometer [9]. Effluent from the chromatograph is nebulized and then passed through a heated constriction into the high vacuum of the mass spectrometer source. Droplets of solvent containing the organic ions of interest are sequentially desolvated until only an organic ion remains. Often, the presence of alkali ions leads to the formation of cationized species such as $(M+Na)^+$ as the residual ion. The fragmentation patterns of molecules investigated with this thermospray interface are similar to those observed in DI spectra. Experiments with the doubly-charged salts might reveal relaxation processes such as those described above. Other work with direct LC/MS has also been interpreted in similar terms [16].

In the direct analysis of chromatograms by particle and photon impact, it is desirable that the matrix have a minimum effect on the spectrum obtained. In work with paper and thin—layer chromatography [17], and electrophoresis [18], use has been made of the preferential desorption of preformed ions to insure the relative transparency of the chromatographic matrix itself. When the sample to be investigated is itself uncharged, then the analyst will turn to derivatization reactions to create the desired charge. Although each reagent added to the matrix is a potential interferent, the analytical advantages often outweigh such problems, especially when the DI method is used along with the MS/MS technique for the reduction of the chemical noise introduced into the system. As general acid/base derivatizations are replaced with more selective redox coupled reagents, the problems of matrix effects should diminish still further and the selectivity of DI be further increased [19].

One further experimental variable in DI is time resolution. The analytical potential of time resolved desorption methods will depend strongly on the matrix. Experiments have shown that there are long delays between energization by nanosecond laser pulses and ejection of ions from the condensed phase into the vacuum [20]. This delay is thought to correspond to the time for the isomerization and transfer of energy through the condensed phase to the actual site of desorption of the organic molecule. The matrix can be expected to influence this rate and to suggest experiments which allow more detailed characterization of the DI mechanism.

In conclusion, control of the matrix is one key to successful DI analysis. Innovations are likely to continue at the rapid pace of the past few years; understanding may develop more slowly. The debate about the nature of the secondary ion yield enhancement of metals treated with oxygen has continued for many years now yet this matrix effect is simple when compared with the various recipes for matrices used in DI analysis of organics.

ACKNOWLEDGEMENT
    This work was supported by the National Science Foundation. We thank Prof. A. R. Katritsky for samples.

REFERENCES

1. K.L. Busch and R.G. Cooks: Science, in press.
2. H. Grade and R.G. Cooks: J. Am. Chem. Soc. 100, 5615 (1978).
3. L.K. Liu, K.L. Busch, and R.G. Cooks: Anal. Chem. 53, 109 (1981).
4. J.L. Pierce, K.L. Busch, R.G. Cooks, and R.A. Walton: Inorg. Chem. 21, 2597 (1982).
5. S.E. Unger, R.J. Day, and R.G. Cooks: Int. J. Mass Spectrom. Ion Phys. 39, 231 (1981).
6. B.P. Stimpson, D.S. Simons, and C.A. Evans, Jr.: J. Phys. Chem. 82, 660 (1978).
7. M. Barber, R.S. Bordoli, G.J. Elliott, R.D. Sedgwick, and A.N. Tyler: Anal. Chem. 54, 645A (1982).
8. K.L. Busch, S.E. Unger, A. Vincze, R.G. Cooks, and T. Keough: J. Am. Chem. Soc. 104, 1507 (1982).

9. C.R. Blakely, J.J. Carmody, and M.C. Vestal, J. Am. Chem. Soc. 102, 5931 (1980).
10. K.L. Busch, S.E. Unger, and R.G. Cooks, Proceedings of the Conference on Ion Formation from Organic Solids, University of Münster, October 1980, Springer-Verlag, New York, in press.
11. D.J. Surman and J.C. Vickerman: J. Chem. Res. (S), 170 (1981).
12. M. Barber, R.S. Bordoli, R.D. Sedgwick, and A.N. Tyler: J. Chem. Soc. Chem. Comm., 325 (1981).
13. K.D. Cook, Presented at the Thirtieth Annual Conference on Mass Spectrometry and Allied Topics, Honolulu, HI, 1982, paper MOD7.
14. T.M. Ryan, R.J. Day, and R.G. Cooks: Anal. Chem. 52, 2054 (1980).
15. U. Giessmann and D.F. Barofsky, Presented at the Thirtieth Annual Conference on Mass Spectrometry and Allied Topics, Honolulu, HI, 1982, paper WPB12.
16. P.J. Arpino and G. Guichon, J. Chromatog. 251, 153 (1982).
17. S.E. Unger, A. Vincze, R.G. Cooks, R. Chrisman, and L.D. Rothman: Anal. Chem. 53, 976 (1981).
18. A. Ba-isa, K.L. Busch, R.G. Cooks, A. Vincze, and I. Granoth: Tetrahedron, in press.
19. K.L. Busch and R.G. Cooks, in preparation.
20. D. Zakett, A.E. Schoen, R.G. Cooks, and P.H. Hemberger: J. Am. Chem. Soc. 103, 1295 (1981).

3.10 Mass Spectrometry of Secondary Ions: Polymers, Plasticizers and Polycyclic Aromatic Hydrocarbons

J.E. Campana, M.M. Ross, S.L. Rose, J.R. Wyatt, and R.J. Colton

Chemistry Division, Naval Research Laboratory, Washington, DC 20375, USA

1. Introduction

The mass spectrometry of secondary ions encompasses a number of techniques depending on the identity of the primary particle. Table 1 presents the various primary particles used to produce secondary ions and the name of the associated mass spectrometric technique. These new techniques were developed primarily for the study of nonvolatile and thermally labile compounds. Two of these techniques, SIMS and FAB, and their application to the analysis of polymers, plasticizers and polycyclic aromatic hydrocarbons are the subject of this contribution.

Table 1. Primary particles and techniques associated with the mass spectrometry of secondary ions

Primary Particles	Instrumental Technique
ions	secondary ion mass spectrometry (SIMS)
neutral atoms	fast atom bombardment (FAB)
MeV particles	plasma desorption mass spectrometry (PDMS) or fission fragment mass spectrometry (FFMS)
photons	laser desorption mass spectrometry (LDMS)

Molecular SIMS utilizes an energetic ion beam to desorb molecules directly from surfaces [1,2]. In earlier molecular SIMS studies the samples were prepared by placing a dilute solution of the sample onto an acid-etched Ag foil [3,4]. In fact, a thin layer of the sample compound would deposit on the uppermost surfaces of the acid-etched foil as the solution evaporated and receded into the etched pores of the substrate [4]. If, on the other hand, the substrate was not etched and the concentration of the solution was high, the adsorbed organic film would grow too thick and the resultant insulating film would accumulate charge under ion bombardment and consequently quench the secondary ion emission. Therefore, the acid-etched surface provided a substrate onto which an extended range of solution concentrations could be deposited without degrading the secondary ion signal [4].

In the meantime, several other sample preparation methods were developed to simplify the solution-deposition procedure. For example, COOKS and co-workers studied adduct ion formation (cationization) of several organic compounds with various alkali, transition and noble metals by several different sample preparation methods. In one case the organic sample was burnished (rubbed) onto a metal foil; in another case the organic was mixed with a metal salt and then burnished onto a metal foil and finally some samples were mixed with metal powders and pressed into pellets. Not only did the SIMS spectra show dramatic differences in the adduct ion formation efficiencies for various metals, but the sample preparation methods had an equally dramatic effect on the SIMS spectra, i.e., the metal salt/organic mixture yielded higher organic molecule-metal adduct ion intensities than did the organic films on the metal foil or the pressed pellets [2,5].

Enhanced adduct ions were also observed when samples were sputtered from a metal-supported ammonium chloride matrix [6]. Although the reasons for the enhancement are not completely understood, the NH_4Cl matrix may reduce molecule-molecule interactions. In addition, since the NH_4Cl matrix is not inert but chemically active, it can protonate molecules more basic than ammonia [6].

Other workers have studied molecular ion emission from a variety of different matrices including the rare-gas solids [7,8] and frozen molecular solids [9-12]. These matrices yield cluster ions particularly intense if the sample molecules are not well diluted.

Finally, the recent use of a liquid matrix in FAB has lead to several significant accomplishments, particularly in the analysis of biomolecules [13]. In the FAB technique, the sample to be analyzed is dissolved in a liquid matrix, such as glycerol, at some optimum concentration in order to provide a reservoir of sample molecules in the liquid matrix. During neutral bombardment or FAB, the sputtered surface is constantly replenished with intact sample molecules that diffuse from the bulk liquid to the surface. Therefore, a dynamic primary atom beam, with densities of 10^{10} to 10^{11} atoms/cm², can be used to produce intense and long-lived molecular ion emission [13] without detrimental bombardment-induced surface damage effects.

The methods of sample preparation affect the chemical and physical properties of the sample molecules, which can have a profound influence on the ion formation/emission process. For example, the efficiency of the various ionization processes in SIMS falls off in the order of direct emission of preformed ions > adduct ion formation > electron transfer. The secondary ion intensity of molecular ions from organic salts is generally about two orders of magnitude higher than organic molecules detected as adduct ions [2]. The detection limit of the organic salts is in the picogram range [14]. Therefore, if organic molecules can be derivatized to form salts during sample preparation, detection limits could be greatly improved. For example, depositing an amino acid from an acidic solution enhances the emission of [M+H]$^+$, while depositing the amino acid from a basic solution enhances the emission of [M-H]$^-$ ions [15].

The mass spectrometry of secondary ions offers the mass spectrometrist several alternate sample preparation methods. However, we have experienced several severe problems in the analysis of certain compounds such as polymers and the nonpolar polycyclic aromatic compounds (PAC) for which few of the above sample preparation methods were particularly effective. We have investigated several novel approaches to the analysis of polymers, plasticizers, and PAC which will be reported here.

2. Experimental

A high performance SIMS instrument, developed in our laboratory [16], was operated in the static mode to reduce bombardment-induced surface damage. This static beam condition was accomplished by reducing the primary beam flux and rastering it over the sample surface to further decrease the effective beam density. Dynamic emittance matching was used to enhance the transmission of the secondary ion beam [17]. A beam density of approximately 10 nA/cm^2 was used to analyze the samples cast as films on acid-etched silver foil.

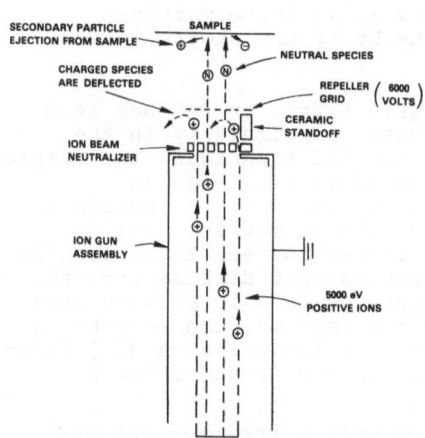

Fig. 1. Ion beam neutralizer

A simple modification of the commercial ion gun provided ion-to-neutral conversion and is shown schematically in Fig.1. The device can operate by three processes: (a) resonance neutralization

$$A^+ + n(e^-) \rightarrow A* + (n-1)e^- \tag{1}$$

followed by Auger de-excitation
$$A* + n(e^-) \rightarrow A + e + (n-1)e, \tag{2}$$

(b) Auger neutralization
$$A^+ + n(e^-) \rightarrow A + (n-2)e^- + e^- \tag{3}$$

(a) and (b) occuring when ions approach the metal surface[18] or
(c) capture of low-energy secondary electrons generated in
processes (a) and (b), where A is the species to be neutralized
and n is the total electrons in the metal. A metal with a low
work function, such as Ta or Mo is used as the neutralizer.
The neutalization efficiency of the device increases with the
magnitude of the difference between the work function of the
metal (neutralizer) and the first ionization potential of the
ion to be neutralized. This technique is similar to one
described previously for an ion microprobe[18]. Our work used
Ta or Mo as the neutralizer with many holes drilled in it to
maximize the conversion surface area and beam transmission
through the device, further allowing use of the beam raster
system. Finally, because the ion-to-neutral conversion process
is not 100%, it is desirable to remove the ions from the
neutral beam by electrostatic deflection as shown in Fig.1.

3. Polymers

The mass spectrometry of secondary ions is an especially
attractive surface analytical technique for the characterization of polymer surfaces. The potential of the technique is
exemplified by the recent detection of ultra-high mass inorganic secondary ions on conventional (m/z > 18,000) [20-23] and
commercial (m/z > 25,000) [24] mass spectrometers. In addition,
SIMS recently has been shown useful for the nondestructive
evaluation and characterization of polymeric thin films [25].

3.1 Polystyrene

Spin-cast samples of polystyrene (mw=500) were prepared by
dropping a dilute solution of the polymer on a Ag substrate re-
volving at 2000 rpm, resulting in a thin uniform film. Figure
2 shows the SIMS spectrum obtained by Ar^+ bombardment and two
different solvent systems.

The spectra are very similar for toluene (Fig.2a) and for cylo-
hexane, a theta solvent for polystyrene (Fig.2b). Major ion
intensities recorded correspond to the phenylium ion (m/z 77),
the benzyl or tropylium ion (m/z 91), the protonated styrene
monomer (m/z 105) and the Ag substrate (m/z 107 and 109). Other
major ion signals at higher m/z values correspond to structures
previously postulated in the GC/MS of polystyrene pyrolosates
[26]. Thick films of polystyrene were also prepared by the

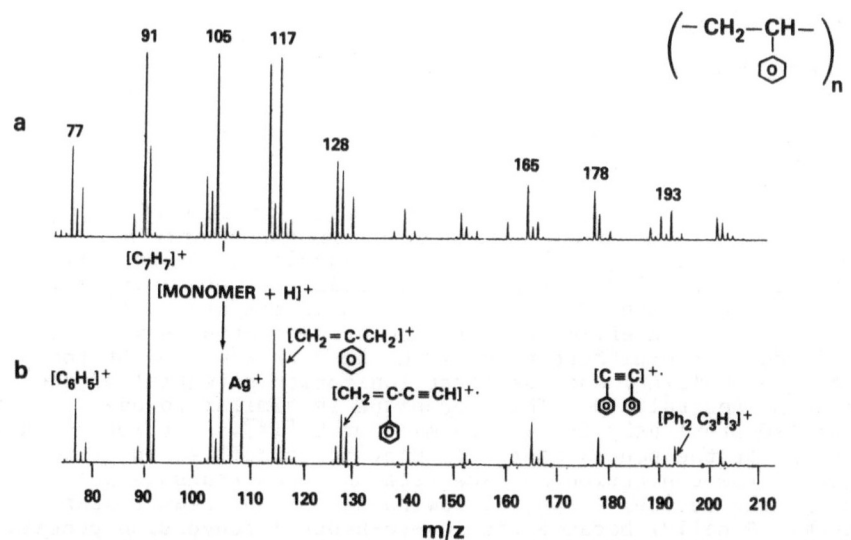

Fig. 2. SIMS spectrum of polystyrene under Ar⁺ bombardment,
spin-cast from (a) toluene and (b) cyclohexane

spin-cast method. These thick, insulating films could not be
mass-analyzed during ion bombardment due to sample charge
buildup resulting in secondary ion energy broadening and more
typically the complete suppression of the secondary ion current
[27].

Neutral atom bombardment using the device described in Section 2 was used to investigate the thick samples. The results
of Ar and Xe bombardment on the mass spectra are shown in Fig. 3.

The mass spectra are very similar to the mass spectra from thin
films by ion bombardment. Xe bombardment yields a much more
intense spectrum due to the increased momentum transfer over Ar

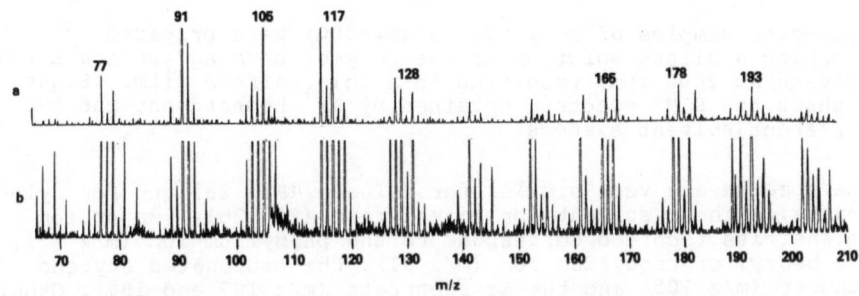

Fig. 3. FAB spectrum of polystyrene spin-cast from cyclohexane
solution under (a) argon and (b) xenon bombardment

bombardment (Fig. 3b). The absence of Ag^+ ion signals indicates a thick film coverage. Sample charge buildup has been reported to be an problem in the analysis of polystyrene films (cut from $100\mu m$ thick sheets) by quadrupole-based SIMS [28]. Neutral atom bombardment appears to reduce charging effects (especially in sector instruments); therefore, FAB is useful for bulk polymers (insulating samples).

This has been demonstrated in our laboratory by the characterization of polytetrafluoroethylene sheet stock and a pressed pellet of poly(n-butyl)-methacrylate under neutral bombardment, the spectra of which could not be obtained by ion bombardment.

3.2 Poly(alkyl)-methacrylates

A series of poly(alkyl)-methacrylates consisting of the methyl, ethyl, isobutyl n-butyl and lauryl substituents (Table 2) were characterized as thin-films solvent cast on silver by Ar ion bombardment [25]. The relative ion intensities at four m/z values were sufficient to distinguish between the various homologs (Table 3).

For example, in the methyl and ethyl methacrylate, the methyl ion and the ethyl ion were the most intense ion species, respectively. In the case of the butyl isomers it is interesting

Table 2. Polymer Characterization: Polyalkylmethacrylates

NAME	ACRONYM	R	
POLYMETHYLMETHACRYLATE	PMM	$-CH_3$	
POLYETHYLMETHACRYLATE	PEM	$-CH_2CH_3$	
POLYISOBUTYLMETHACRYLATE	PIB	$-CH_2CH(CH_3)_2$	
POLY(n-BUTYL)METHACRYLATE	PNB	$-(CH_2)_3CH_3$	
POLYLAURYLMETHACRYLATE	PLM	$-(CH_2)_{11}CH_3$	

Table 3. Poly(alkyl)methacrylate characterizations

m/z	SPECIES	
15	$[CH_3]^+$	PMM m/z $15 > 55 > 29 > 43$
29	$[C_2H_5]^+$	PEM m/z $29 > 43 > 15 \approx 55$
43	$[C_3H_7]^+$	PIB m/z $29 > 43 \approx 55 > 15$
55	$[C_4H_7]^+$	PNB m/z $43 > 29 \approx 55 > 15$
		PLM m/z $55 > 43 > 29 > 15$

to note that for the polyisobutylmethacrylate the ethyl ion is predominent over the propyl ion; however for the poly(n-butyl)-methacrylate the propyl ion is more intense than the ethyl ion. The $[C_4H_7]^+$ ion is most intense in the spectrum of polylauryl-methacrylate.

The SIMS spectra of the same series of polymers, analyzed in the bulk form and under more destructive beam conditions (0.1 to 1 μA/cm^2), have been reported [29] . Although similar ion species are observed in that work, there are notable ion intensity differences at certain m/z values due to the differences in the experimental conditions. Our study also demonstrated the potential of static SIMS as an "apparent" nondestructive method of analysis [25] in that very little surface damage occured during the first several minutes of analysis as monitored by mass spectral ion intensity changes as a function of time.

4. Plasticizers

Plasticizers are common components and contaminants of polymer systems. Therefore, plasticizers compatible to the polymer systems studied were investigated by SIMS. Recognition of the presence of plasticizers is essential to the interpretation of the complex secondary ion mass spectra of polymer systems.

4.1 Phthalic Acid Based Plasticizers

A series of dialkyl phthalates and tere-phthalic acid were investigated by molecular SIMS using a 4 keV Ar$^+$ beam. Tere-phthalic acid was burnished into an acid-etched Ag substrate. The dialkyl phthalates were applied to acid-etched Ag substrates as dilute solutions (0.5-5% by weight in spectro-

Table 4. Relative ion abundance of major ions in the SIMS spectra of phthalates

STRUCTURE	149$^+$	[148 + Ag]$^+$	[M + H]$^+$	[M + Ag]$^+$	[2M + Ag]$^+$
COOH — COOH TERE-PHTHALIC ACID	100	10	30	10	—
DIALKYL PHTHALATES					
METHYL	40	70	<1	<1	100
ETHYL	100	57	<1	<1	57
ISODECYL	100	6	32	18	<1
2-ETHOXYETHYL	100	<1	8	2	<1

metric grade acetone) and allowed to air dry. Table 4 lists
some of the phthalates analyzed and summarizes the relative intensities of major ions observed in the mass spectra. The
carboxonium ion of phthalic acid (m/z 149) was the major ion in
the spectra of all compounds except dimethyl phthalate.
The dialkyl phthalates with short-chained R groups gave low-
intensity protonated molecules. Ions corresponding to the
molecule–metal adducts were also observed where the $[M + Ag]^+$ was
more intense for the dialkyl phthalates with bulky R groups. A
second and novel adduct ion, corresponding to $[2M + Ag]^+$, was
observed for the dimethyl and diethyl phthalates while it was not
significant for the others. This suggests that steric hindrance
by the alkyl group is a factor in the metal complex formation of
$[2M + Ag]^+$. Metal chelate formation has been previously reported
in SIMS studies [30]. The $[2M + Ag]^+$ adduct ion also shows a time
dependent formation which is demonstrated for dimethyl phthalates
in Fig. 6. This phenomenon is currently being studied in
detail.

TIME DEPENDENT METAL COMPLEX FORMATION OF DIMETHYL PHTHALATE

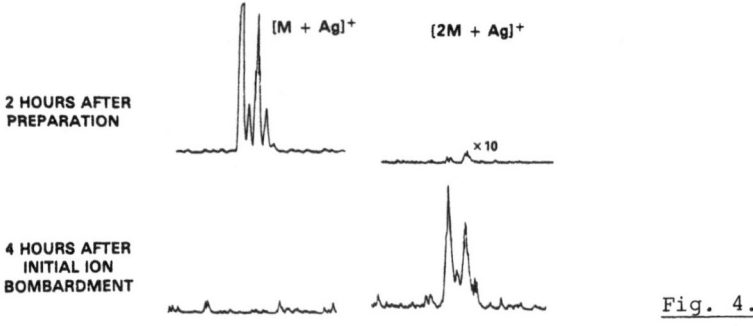

Fig. 4.

5.0 Polycyclic Aromatic Compounds

Activated carbons or charcoals have long been used as adsor-
bents for airborne or aqueous pollutants. In addition, carbonaceous particulate matter, or soot, produced by combustion of
hydrocarbon fuels is known to contain a multitude of adsorbed
compounds such as the PACs, many of which are toxic, carcinogenic
or mutagenic. Analysis of such adsorbed species on carbons
usually involves the extraction of the species by an appropriate
solvent followed by chromatographic fractionation, and finally
the separation and identification of the components of the
extract with a combined chromatographic-mass spectrometric
technique. Although this analytical scheme often yields a large
amount of information about the adsorbed compounds, the
experimental procedure is quite tedious and time-consuming.

One objective of this research is to develop new analytical
techniques for the direct and rapid identification of organic
compounds adsorbed on carbon. Specifically, we plan to use

SIMS methods to analyze the organic adsorbates [31]. Therefore, the first part of this section on the PACs is devoted to
the investigation of the secondary ion formation and emission
mechanisms of organic molecules on carbon substrates as
compared with organic SIMS results from metal surfaces [12].
The second part deals with the use of liquid metal surfaces,
such as gallium, as a sample substrate for dynamic SIMS analyses.

5.1   Carbon Substrates

The organic compounds studied include several mono- and polycyclic aromatic compounds such as benzene, methyl-substituted
benzenes, naphthalene, anthracene, phenanthrene, pyrene, acen-
aphthylene, etc., which are commonly found as combustion by-
products on airborne carbon particulate matter. The organic
compounds were burnished on Ag or adsorbed from chloroform
solutions on carbon particles that were burnished on Ag. The
results for the substituted benzenes are presented in Table 5.

Table 5.   SIMS Results for the Substituted Benzenes

NAME/STRUCTURE	RELEVANT PHYSICAL CONSTANTS[1]	CONCENTRATION RANGE FOR DETECTION	SIMS RESULTS	
			SOLUTION/SILVER	ADSORBED ON CARBON/SILVER
BENZENE	M.W. = 78.1 B.P. = 80.1 ΔH$_a$ = 9.8 V.P. = 104.2	–	–	81, M^{++}?, [Ag+M]$^+$?
TOLUENE (CH$_3$)	M.W. = 92.1 B.P. = 110.8 ΔH$_a$ = 11.6 V.P. = 31.5	NEAT LIQUID ONLY (SATURATED)	–	107, M^{++}?, 199, 201, [Ag+M]$^+$
m-XYLENE (CH$_3$, CH$_3$)	M.W. = 106.2 B.P. = 139 ΔH$_a$ = 16.2 V.P. = 9.3	NEAT LIQUID- 0.1 ml/g C	–	107, M^{++}?, 213, 215, [Ag+M]$^+$
MESITYLENE (CH$_3$, H$_3$C, CH$_3$)	M.W. = 120.2 B.P. = 164.7 ΔH$_a$ = 15 V.P. = 2.8	NEAT LIQUID- 0.1 ml/g C	–	109, M^{++}?, 227, 229, [Ag+M]$^+$

[1] M.W. = MOLECULAR WEIGHT IN GRAMS/MOLE
B.P. = BOILING POINT IN °C AT 760 Torr
ΔH$_a$ = HEAT OF ADSORPTION, IN kcal/mole, ON GRAPHITIZED CARBON BLACK
V.P. = VAPOR PRESSURE, IN Torr, AT 300°K AND 760 Torr

These compounds are normally difficult to analyze without
cryogenically cooling the sample, due to their high volatility.
Consequently, secondary ions characteristic of the organics
were not detected when Ag alone served as the sample sub-

strate. Adsorption of toluene, xylenes, and mesitylene on the carbon burnished on Ag permitted detection of the $[M + Ag]^+$ adduct ion from each of these compounds. Since toluene, but not benzene, is detected, the minimum adsorption energy necessary to permit detection is taken to be ~10 kcal/mole. Since the heats of adsorption of organic compounds on carbon are relatively high, the adsorbate is held in place long enough to be detected in a SIMS experiment. When the carbon matrix is saturated with toluene or mesitylene, the ion emission lasts for ca. 0.5-1.0 hour.

Table 6. SIMS Results for the PAC's: Phenanthrene, Anthracene, Fluoranthene and Pyrene

NAME/STRUCTURE	RELEVANT PHYSICAL CONSTANTS	CONCENTRATION RANGE FOR DETECTION	SIMS RESULTS	
			SOLUTION/SILVER	ADSORBED ON CARBON/SILVER
PHENANTHRENE	M.W. = 178 B.P. = 338 V.P. = 1.4 × 10⁻³	SOLUTION/SILVER: 0.2µg < [M] <2µg ON CARBON/SILVER: M⁺[Ag +M]⁺:>10µg M⁺ ONLY: <1µg	M⁺· 178 [Ag + M]⁺ 285 287	M⁺· 178 285 287 [Ag + M]⁺
ANTHRACENE	M.W. = 178 B.P. = 340 V.P. = 1.3 × 10⁻³	SOLUTION/SILVER ~0.2-2µg ON CARBON/SILVER 1 mg/50 mg C	M⁺· [Ag +M]⁺ 178 285 287	M⁺· 178
FLUORANTHENE	M.W. = 202 B.P. = 384 V.P. = 2 × 10⁻⁶	1 mg/50 mg C	–	M⁺· [Ag + M]⁺ 202 300 311
PYRENE	M.W. = 202 B.P. = 395	1 mg/50 mg C	–	M⁺· [Ag + M]⁺ 202 300 311

The SIMS results obtained with some of the PACS are presented in Table 6. These compounds are usually difficult to analyze with SIMS due to their nonpolar nature and low adsorption energies on metal surfaces. As expected, ion emission from these compounds on silver proved to be unpredictable and irreproducible. However, when these organics are adsorbed on carbon, $M^{+\cdot}$ molecular ions and $[M + Ag]^+$ adduct ions are readily detected. We believe that the high surface area and porosity of the carbon causes the samples to be distributed over a three-dimensional surface which can emit molecular and adduct ions over a long time period during sputtering. In fact, $[M + Ag]^+$ ions are observed for over one hour from a carbon charged with ~2µg of a PAC. In addition, under dynamic sputtering conditions, (10-100nA/cm²) we have measured a detection limit of ~1 ng for phenanthrene on carbon.

5.2 Liquid Metal Substrates

FINE and co-workers [32] previously observed that bombarding the surface of gallium with an electron beam caused the mass transport of impurities to the beam where they were desorbed. Using this idea, we purposefully dispersed samples of the PACs on the gallium surface and bombarded the surface with a dynamic ion beam (5×10^{-5} A/cm^2 of O_2^+) from a CAMECA IMS 300 ion microscope. As expected, the sample particles could be seen with the ion microscope moving toward the primary ion beam where they are desorbed and finally detected by the mass analyzer. Sputtering of microgram quantites of the PACs on gallium produced a long-lived molecular ion $M^{+\cdot}$ signal, which lasted for more than 30 minutes. All attempts to detect $M^{+\cdot}$ or $[M + Ag]^+$ ions from the organic deposited on Ag were unsuccessful due to the surface damage caused by the dynamic ion beam. No molecular-like ions from a glycerol matrix were observed since the PACs are not soluble in glycerol.

Several properties of liquid metal and glycerol substrates are summarized below. Basically, the two substrates are quite different in the way the sample molecules are introduced and maintained during analysis. The gallium acts as a mobile support over which any compound can (theoretically) be dispersed. The number and types of compounds which can be analyzed using a solvent substrate will be limited by the chemical properties of that solvent. When glycerol or other fluids are used, the sample to be analyzed must be readily dissolved. The surface of the gallium is relatively inert and very smooth, and its low vapor pressure allows a low background pressure consistent with UHV operation. The vapor pressure of glycerol is much higher than that of the liquid metal; therefore, the type of vacuum equipment used in the SIMS or FAB experiment must be considered. Pertaining to SIMS, the high conductivity of the gallium substrate is advantageous to avoid sample charge buildup. The lower conductivity of glycerol is not a serious problem because salts can be added, if necessary, to increase its conductivity, but these additives can contribute interference peaks in the mass spectrum.

In conclusion, gallium has been shown to be an effective sample substrate for the nonpolar PACs analyzed with dynamic SIMS. The use of the ion microscope in these studies is particularly relevant since we are striving to detect low concentrations of adsorbed compounds on individual carbon particles. In addition, the use of a mobile gallium substrate should allow a wide variety of nonvolatile and thermally labile compounds to be analyzed by dynamic SIMS.

References

1. R. J. Colton, J. Vac. Sci. Technol., 18, 737 (1981).
2. R. J. Day, S. E. Unger, and R. G. Cooks, Anal. Chem., 52, 557A (1980).
3. A. Benninghoven, D. Jaspers and W. Sichtermann, Appl. Phys., 11, 35 (1976).

4. R. J. Colton, J. S. Murday, J. R. Wyatt, and J. J. DeCorpo, Surf. Sci., 84, 235 (1979).
5. R. J. Day, S. E. Unger, and R. G. Cooks, Anal. Chem., 52, 354 (1980).
6. L. K. Lin, K. L. Busch, and R. G. Cooks, Anal. Chem., 53, 109 (1981).
7. R. G. Orth, H. T. Jonkman, D. H. Powell, and J. Michl, J. Am. Chem. Soc., 103, 6026 (1981).
8. R. G. Orth, H. T. Jonkman, and J. Michl, J. Am. Chem. Soc., 104, 1834 (1982).
9. M. Barber, J. C. Vickerman, and J. Wolstenholme, J. Chem. Soc., Faraday Trans., 1, 76, 549 (1980).
10. G. M. Lancaster, F. Honda, Y. Fukuda, and J. W. Rabalais, J. Am. Chem. Soc., 101, 1951 (1979).
11. H. T. Jonkman and J. Michl, J. Am. Chem. Soc. 103, 1007 (1981).
12. R. G. Orth, H. T. Jonkman, and J. Michl, J. Am. Chem. Soc., 103, 1564 (1981).
13. M. Barber, R. S. Bordoli, G. J. Elliott, R. D. Sedgwick, and A. N. Tyler, Anal. Chem., 54, 645A (1982).
14. S. E. Unger, T. M. Ryan, and R. G. Cooks, Anal. Chem. Acta, 118, 169 (1980).
15. A. Benninghoven and W. K. Sichtermann, Anal. Chem., 50, 1180 (1978).
16. R. J. Colton, J. E. Campana, T. M. Barlak, J. J. DeCorpo, J. R. Wyatt, Rev. Sci. Instrum., 51, 1688-89 (1980).
17. J. E. Campana, J. J. DeCorpo, J. R. Wyatt, Rev. Sci. Instrum., 52, 1517-20 (1981).
18. G. Borchardt, H. Scherrer, S. Weber, S. Scherrer, Int. J. Mass. Spectrom. Ion Phys., 34, 361-73 (1980).
19. M. Kaminsky, Atomic and Ionic Impact Phenomena on Metal Surfaces, Academic Press, N.Y., 236-300 (1965).
20. J. E. Campana, T. M. Barlak, R. J. Colton, J. J. DeCorpo, 1046-49 (1981).
21. J. E. Campana, T. M. Barlak, J. R. Wyatt, J. J. DeCorpo, R. J. Colton, J. Vac. Sci. Technol., 20, 1068-69 (1982).
22. T. M. Barlak, J. R. Wyatt, R. J. Colton, J. J. DeCorpo, J. E. Campana, J. Amer. Chem. Soc., 104, 1212-15 (1982).
23. R. J. Colton, T. M. Barlak, J. R. Wyatt, J. J. DeCorpo, J. E. Campana, J. Vac. Sci. Technol., 20, 421-22 (1982).
24. J. E. Campana, J. R. Wyatt, R. J. Colton, R. H. Bateman, B. N. Green, to be published.
25. J. E. Campana, J. J. DeCorpo, R. J. Colton, Applications Surface Sci., 8, 337-342 (1981).
26. A. Alajbeg, P. Arpino, D. Deur-Jiftar, G. Guiochon. J. Anal. Appl. Pyrolysis, (1), 203-212 (1980).
27. W. Reuter, M. L. Yu, M. A. Frisch, M. B. Small, J. Appl. Phys., 51, 850-55 (1980),
28. D. Briggs, A. B. Wootton, Surface Interface Anal. 4, 109 (1982).
29. J. A. Gardella, D. M. Hercules, Anal. Chem. 52, 226 (1980).
30. R. J. Day, S. E. Unger, R. G. Cooks, J. Am. Chem. Soc., 101, 499, (1979).
31. M. M. Ross and R. J. Colton, submitted for publication.
32. J. Fine, S. C. Hardy, and T. D. Andreadis, J. Vac. Sci. Technol. 18, 1310 (1981).

3.11 SIMS Studies of Polymer Surfaces

D. Briggs

ICI Petrochemicals and Plastics Division, Welwyn Garden City, Herts, UK

1. Introduction

In principal SIMS has several advantages over other surface analytical techniques currently used for studying polymer surfaces. These include greater molecular specificity, greater surface sensitivity and much greater spatial resolution. Realising these advantages is the object of our current research [1,2]. In the present context, however, we restrict discussion to the molecular specificity aspects of SIMS. It should be noted that FABMS probably has advantages over SIMS for general studies of this type since the problem of charge neutralisation is avoided [3], but that the goal of molecular specificity at high spatial resolution requires SIMS with electron neutralisation.

2. Experimental

The VG Scientific system, incorporating the MM12-12 quadrupole mass spectrometer, used for these studies has been previously described [1]. Unless otherwise specified all the spectra described herein were obtained using 4keV Ar^+ ions with a current density of 1-2nA cm^{-2} (area irradiated ca. 0.3 cm^2) and a neutralising beam of 700eV electrons with a current density of <5nA cm^{-2} (area irradiated ca.1 cm^2). Spectra were obtained in 300s or less with a maximum total exposure time of ≈500s (corresponding to a maximum ion dose of ≈6 x 10^{11} ions cm^{-2} under the highest current conditions used).

3. Materials

Poly(ethyleneterephthalate)(PET) and cellulose triacetate(CTA) were studied as free standing films cut from reels of the pure material immediately prior to examination. Polyethylmethacrylate(PEMA), polyhydroxyethyl-methacrylate(PHEMA) and a 1:1 copolymer of EMA and HEMA were studied as thin films spin-cast from chlorobenzene and dimethylformanide respectively onto specially cleaned glass coverslips. Nylon-6 (poly(ε-caprolactam)) was in the form of a thin film prepared by multiple evaporation of a solution of the purified polymer in formic acid onto a stainless steel mount. Conducting (ultrathin) films of paraffin wax (a short chain saturated hydrocarbon) were similarly cast from dilute hexane solutions. Identically prepared films were examined by XPS (using MgKα radiation) in the same instrument and found to be contaminant free.

4. Results and Discussion

Three factors which complicate the detailed analysis of SIMS data from polymer surfaces need to be mentioned, having been discussed in detail elsewhere [1]. The first is secondary ion emission stimulated by the neutralising electron beam. Under the conditions used this is not a problem for the polymers discussed here: only low mass clusters are observed for electron excitation alone and these are 1-2 orders of magnitude less intense than in SIMS. However, this is not true for all polymers and the situation changes markedly as the electron beam density increases. The second complicating factor is the sample surface potential. This is obviously affected by the relative magnitude of ion and electron current densities. It can also be altered by application of a bias voltage to the sample holder and this is used routinely to maximise secondary ion peak intensity by shifting the secondary ion energy distribution to coincide with the 'window' of the energy filter on the entrance to the quadrupole mass spectrometer. Since the secondary ions do not have identical energy distributions relative intensities are affected by change in surface potential. This is illustrated in Fig.1 for the conducting hydrocarbon film - in this case the surface potential is exactly the applied voltage [1]. The third factor is the rate of surface damage which also affects relative peak intensities. This is not considered to be important in this study because of the very low fluences used.

The hydrocarbon spectrum in Fig.1 is expected to be similar to that from polyethylene (essentially $-(CH_2-CH_2)_n-$) and this is observed [2]. The spectrum is also similar to conventional electron compact (EI) spectra of volatile linear hydrocarbons with a maximum at C_3 and a smooth decrease with high C_n fragments.

In nylon-6 the hydrocarbon chain is punctuated by the peptide linkage
$-\overset{\text{O}}{\underset{\text{H}}{\overset{\parallel}{C}} - \overset{\mid}{N}} -$ (Fig.2). If the monomer unit is denoted M then a prominent
peak occurs at M+H (114 amu). Characteristic peaks (i.e. not frequently observed) are also seen at 55amu ($C_2H_3CO^+$) and 30 amu ($CH_2NH_2^+$)

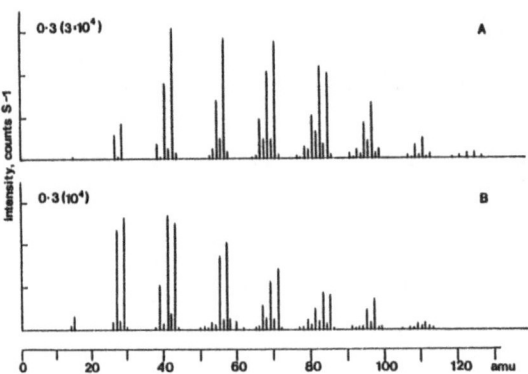

Fig. 1. Positive ion mass spectra from conducting paraffin wax film. Target bias of 12.5V(A) and 7.5V(B)

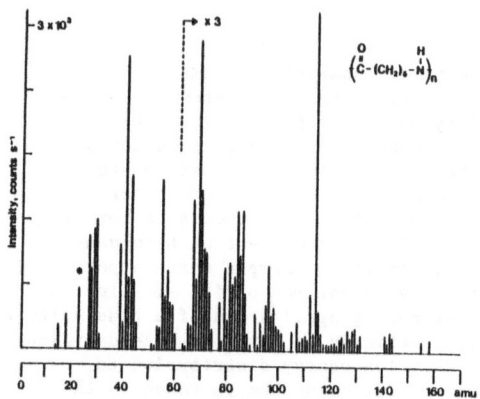

Fig. 2. Positive ion mass spectrum from nylon 6.

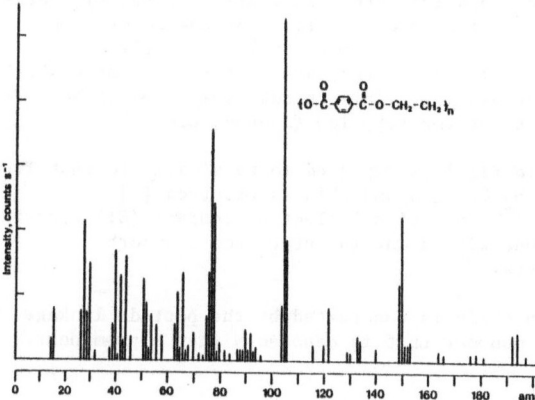

Fig. 3. Positive ion mass spectrum from PET.

and these are prominent in the EI mass spectrum of ε-caprolactom. A peak at 69 amu ($C_5H_9^+$) is common to polymers with β-methyl substituents along the carbon chain (e.g. polypropylene, polymethacrylates) and is thought to be due to the dimethylcyclopropyl ion [2]. However, in this case the linear $C_5H_9^+$ ion is an obvious candidate. Although negative ion spectra are usually uninformative, nylon-6 has a prominent peak at 26 amu due to CN^-.

Fig.3 shows the spectrum from PET. Peaks at 193 and 191 amu correspond to $(M+H)^+$ AND $(M-H)^+$. The fragmentation sequence can be envisaged as:

which accounts for the high mass signals.

158

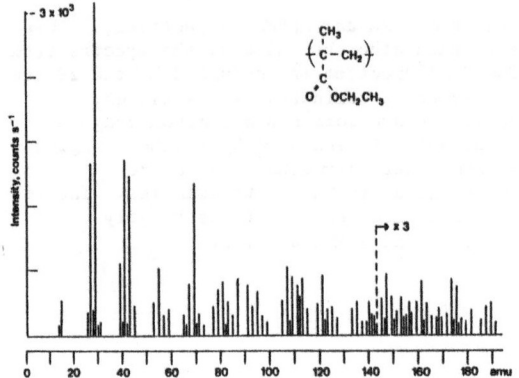

Fig. 4. Positive ion mass spectrum from PEMA

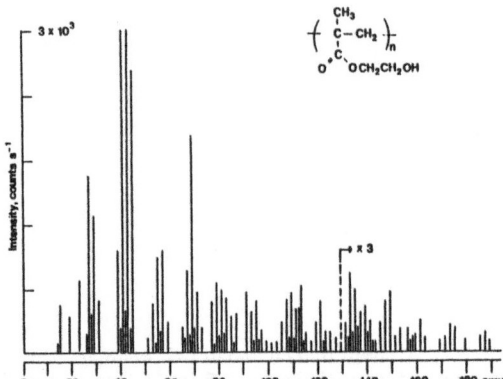

Fig. 5. Positive ion mass spectrum from PHEMA

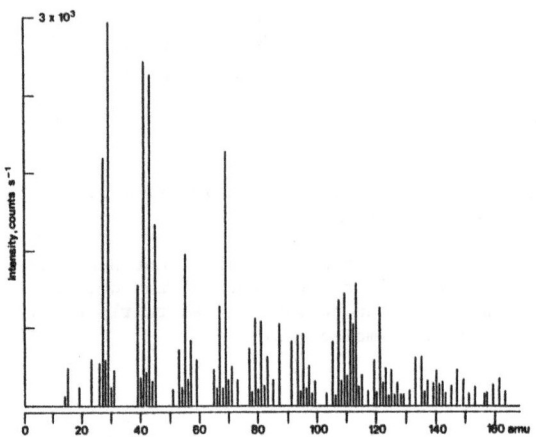

Fig. 6. Positive ion mass spectrum from 1:1 EMA/HEMA copolymer

Figs. 4 and 5 show the spectra from PEMA and PHEMA respectively. These are strikingly similar not only to each other but also to the spectra from other polymethacrylates [4]. The PEMA spectrum is dominated by the 29 amu signal ($C_2H_5^+$) whilst the PHEMA spectum is dominated by the 41, 43, 45 amu triplet. Peaks at 41 and 43 amu are common hydrocarbon fragments but in this case their relative intensity is enhanced by further fragment-ation of the ion (C_2H_4OH)$^+$. As with other polymethacrylates the prominant peaks are due to the ester group in these two examples. The 1:1 copolymer incorporating EMA and HEMA is seen (Fig.6) to be roughly equivalent to the addition of the PEMA and PHEMA spectra.

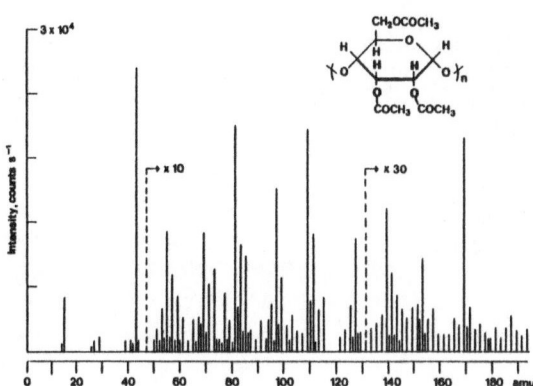

Fig. 7 shows the spectrum of CTA, a complicated material made by the acetylation of the natural polysaccharide cellulose. This is dominated by the 43 amu peak (CH_3CO^+) and the 15 amu peak (CH_3^+) is much more intense than usual, indicating preferential fragmentation of the side chain. A series of other prominant peaks can be assigned as follows:

which points to the stabilising effect of the aromatic heterocyclic ion which can be derived from the saccharide ring.

This data can be summarised by observing that most prominent secondary ions from polymers are closely related to the structure of the original monomer and that fragmentation patterns can be anticipated fairly easily by resource to EI mass spectra from related molecules.

Acknowledgement

We thank Dr B D Ratner (University of Washington, Seattle) for preparing the thin film samples of PEMA, PHEMA and the EMA/HEMA copolymer.

References

1   D. Briggs and A. B. Wootton, Surf. Interface, Anal 4, 109(1982).

2   D. Briggs, ibid 4, 151(1982).

3   D. Briggs, A. Brown, J. A. van den Berg and J. C. Vickerman, in this
    volume.

4   D. Briggs, to be published.

3.12 A Comparative Study of Organic Polymers by SIMS and FABMS

D. Briggs
ICI Petrochemicals and Plastics Division, Welwyn Garden City, Herts, UK

A. Brown, J.A. Van Den Berg, and J.C. Vickerman
Chemistry Department, UMIST, Manchester, UK

1. Introduction

Recent SIMS studies of polymer film surfaces [1,2] have identified and investigated several practical problems which may limit the applicability of the SIMS technique to these materials. These problems include (a) the high rate of ion beam damage, (b) the need for charge neutralisation leading to (c) the uncertainty of surface potential and (d) the possibility of electron stimulated desorption (ESD) of secondary ions.

These studies were able to demonstrate the usefulness of SIMS in providing stable "fingerprint" spectra of several polymer surfaces. The spectra were readily interpreted using conventional EI mass spectrometry fragmentation mechanisms. However, spectral decay was observed and it was found to be difficult to assess the relative importance of the two effects of surface potential drift and ion beam damage. Furthermore, the physical form of the material may also be an important parameter in SIMS experiments through its effect on the stability and uniformity of surface charge. For example, the secondary ion spectrum of poly(ethylene oxide)(PEO) powder has been obtained but that of polyacrylonitrile(PAN) powder could not be obtained [3] using the techniques previously described [1,2].

Fast atom bombardment mass spectrometry (FABMS) offers the possibility of distinguishing between these effects. It appears to suffer little from surface charging problems and has been applied equally successfully to organic compounds [4,5] and non-conducting inorganic materials [6].

In this paper we compare the spectra obtained by SIMS (with electron beam neutralisation, EN) and FABMS on PEO, PAN, poly(ethyleneterephthalate)(PET), and polystyrene(PS). A further investigation is made into the problems of beam damage and surface charging by time dependent studies on PS using both ion and atom bombardment.

2. Experimental

The experiments were carried out in a UHV SIMS system fitted with a Vacuum Generators MM 12-12 quadrupole mass spectrometer with a mass range of 1200 daltons. The novel primary source provides a mass-filtered beam of 2keV argon ions or atoms with emission controlled beams fluxes in the range $10^{-10} - 10^{-8}A$ (Ar^+) or $\approx10^9 - 10^{11}$ particles s^{-1} (Ar). This is delivered into a spot size of area $0.4cm^2$ at the sample. Rapid switching is possible between ion and atom bombardment. Secondary ion collection and energy selection is performed by a modification of the optics described by Wittmaack[7]. A small sample bias of <5v was generally used to optimise signal intensity. No bias was applied to the quadrupole axis during these experiments.

One of the problems in using FABMS is to accurately measure the particle flux at the target. In this work a simple calibration procedure was devised using a Cu target as a standard. Assuming the sputter yields from ion and atom bombardment are the same, a calibration of current equivalent in atom bombardment was obtained by comparing Cu^+ yield with that obtained under ion bombardment of known flux. For the initial spectra and for the time dependency studies an atom flux of $\approx 5 \times 10^{10}$ particles cm^{-2} s^{-1} was used. A primary current of 1×10^{-8} A cm^{-2} was used for those experiments involving argon ions with a compensating electron beam of 10^{-9}–10^{-8} A cm^{-2} and $\approx 100eV$ energy.

PEO and PAN were in powder form whereas PET and PS were cut from reels of free standing film, 50 and 100 μm thick respectively. Powder samples were compressed into circular recesses (1cm diameter) in the Cu sample holder while the film samples (1cm x 1cm) were attached to the holder with double-sided tape.

3. Results and Discussion

3.1 Powders

PEO. In Figs. 1 and 2 the spectra produced by ion and atom bombardment of the PEO powder are compared.

Significantly they are virtually identical in terms of relative intensity of the major fragment ions. The base peak at 45 daltons is attributed to the protonated repeat unit $[-CH_2-CH_2-O-]H^+$. Other intense ions correspond to simple fragments of the polymer chain. Higher mass fragments of the type $[-CH_2-CH_2-O-]_nH^+$ could be detected from n=2 (89 daltons) to n=6 (265 daltons) in the FABMS spectrum.

PAN. One of the problems encountered in SIMS/EN analysis of materials has been the inability of certain non-conducting samples to yield a stable spectrum. A case in point is the powder form of PAN. As noted above this was probably due to surface charging effects since it is often difficult to obtain a stable and uniform surface potential on powdered materials by this method. However, the application of FABMS to this polymer readily yielded a spectrum as shown in Fig.3. It shows intense characteristic ions at 27 daltons ($C_2H_3^+$), 39 daltons ($C_3H_3^+$) and a group of ions in the region 51-55 daltons probably closely derived from the repeat unit $(-CH_2-CH(CN)-)$.

3.2 Films

PET. The SIMS/EN spectrum of PET has been reported previously by Briggs[2]. In this study a FABMS spectrum of the PET film revealed characteristic ions at 193, 149, 132, 121 and 76 daltons and an identical fragmentation sequence as that reported previously [2]. Although occurring with similar relative intensities the overall ion yields with FABMS were a factor of ≈ 5 higher than SIMS for an equivalent primary beam flux. This increase in signal intensity offered by FABMS suggests that such surface charging as may occur is small, rapidly attained, stable and uniform. Optimum ion extraction conditions would than be more easily attained giving rise to higher sensitivity in FABMS.

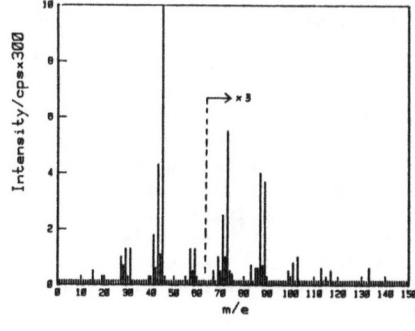

Fig. 1. SIMS/EN spectrum of poly(ethylene oxide) (PEO) using 4keV Ar$^+$ ($\approx$1 nAcm^{-2}) and 700eV electrons ($\approx$3 nAcm^{-2}). FSD = 3 x 10^3 cps

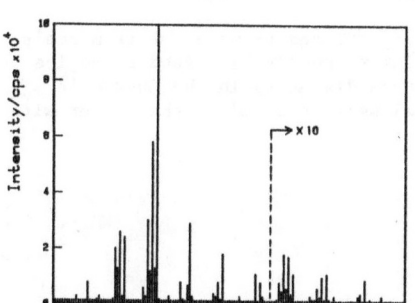

Fig. 2. FABMS spectrum of PEO using 2keV Ar atoms ($\approx$5 x 10^{10} particles cm^{-2}s^{-1}). FSD = 10^4 cps

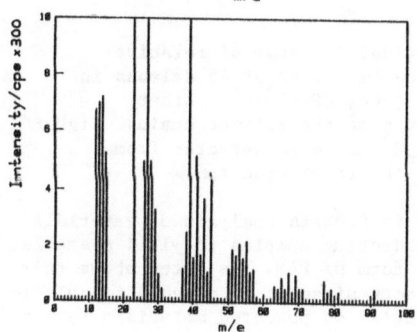

Fig. 3. FABMS spectrum of polyacrylonitrile (PAN) using 2keV Ar atoms ($\approx$5 x 10^{10} particles cm^{-2}s^{-1}) FSD = 3x10^3 cps

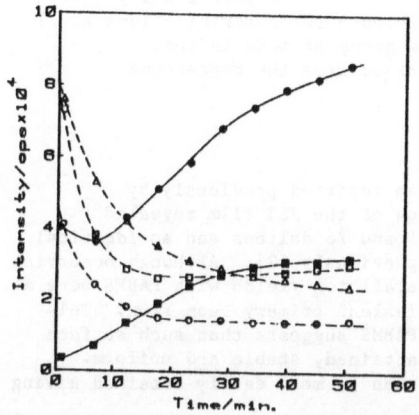

Fig. 4. Variation of the 27($\bullet$), 55($\blacksquare$), 51($\triangle$), 91($\square$) and 115 daltons ($\circ$) peaks from polystyrene(PS) with atom bombardment. (FABMS conditions $\approx$5 x 10^{10} particles cm^{-2}s^{-1} ,2keV Ar)

164

PS. In the case of PS film FABMS again yielded a similar spectrum to that reported previously using SIMS [1]. Characteristic fragment ions occur at 115 daltons ($C_8H_7^+$), 91 daltons (base peak, $C_7H_7^+$ the cyclic tropyllium cation) and 77 daltons ($C_6H_5^+$). A ca. 5 fold enhancement of signal intensity was again observed from FABMS as compared to SIMS.

3.3 Spectral Time Dependence

A serious problem in SIMS studies of polymers and other non-conducting organic materials has been the severe loss in spectral intensity and variation in relative signal intensities with primary beam fluence. This problem has been addressed previously [1] but the relative importance of the two effects of surface potential drift and ion beam damage remain unresolved.

In this study it has already been shown that the effects of surface charging are minimal for FABMS. With the instrumental capability of rapid switching between FABMS and SIMS/EN it is therefore possible to attempt to separate out these deleterious effects.

The time dependence of the PS secondary ion spectrum was studied under the conditions of (a) SIMS with $1 \times 10^{-8}A \ cm^{-2}$, 2keV Ar^+ and electron flood neutralisation for 1 hour and (b) FABMS with $\approx 5 \times 10^{10}$ particles $cm^{-2} \ s^{-1}$, 2keV Ar for 1 hour.

The results for ion bombardment are consistent with those of an earlier study [1] with the $C_7H_7^+$ ion (91 daltons) losing $\approx 15\%$ of its initial intensity in the first 100s of bombardment.

The results for atom bombardment are shown in Fig. 4. For the 115, 91 and 51 dalton ions the initial decay rates show very similar values to those obtained in the previous SIMS study [1] and above. For example, the $C_7H_7^+$ ion loses 13% of its intensity after 100s.

The combination of SIMS and FABMS results provides strong evidence for:-

(a)   secondary ion formation by identical mechanisms for both atom and ion bombardment and

(b)   the spectral time dependence being primarily due to <u>damage</u> processes since atoms and ions yield similar decay rates.

The latter conclusion is supported by the additional data for 27 and 55 daltons ions in Fig. 4. These ions were found to increase significantly as a function of atom dose. They are taken as direct evidence of surface chemical damage. Presumably these ions reflect permanent fragmentation of the PS aromatic structure. Ion bombardment data obtained in this work shows only a small increase in the 27 and 55 daltons ions after prolonged ion bombardment. This may be attributed to the difficulties associated with attaining optimum ion transmission conditions due to the lack of precision in the control of surface charging in SIMS/EN.

The spectral time dependence of PEO, PAN, PET and PS was further investigated using a <u>low</u> atom flux of $\approx 2 \times 10^9$ particles $cm^{-2} \ s^{-1}$. For PEO and PAN the spectral intensity remained constant over a 20 min. period but for PET and PS a significant loss of intensity (10-20%)

resulted. These results again demonstrate the importance of using low primary beam intensity to preserve spectral intensity and minimise beam damage [1]. For PET and PS a damage rate 100-1000 times higher than that estimated from simple sputtering calculations is obtained. This is in agreement with earlier work reported by Benninghoven et al. on sputtering of thin organic films from metal substrates [8]. It is significant to note that of the four polymers studied those containing aromatic units, PET and PS, incur damage most readily. This might be expected since their chemical structure offers more potential fragmentation routes than do the more simple PEO and PAN structures.

A further contributing factor to the stability of the PEO and PAN spectra may be an "annealing" effect whereby beam damaged surface units are replaced by labile sub-surface material. Consquently the stability of secondary ion spectral intensity is likely to depend on an equilibrium between surface damage and annealing processes and surface potential drift. In FABMS the latter process appears to be of much lower importance than with SIMS/EN.

References

1 D. Briggs and A. B. Wootton, Surf. Interface Anal, 4, 109(1982).

2 D. Briggs, ibid, 4 151(1982).

3 D. Briggs, unpublished data.

4 D. J. Surman and J. C. Vickerman, J. Chem. Soc., Chem. Comm. 1981, 324.

5 M. Barber et al. Nature, 293, 270(1981).

6 D. J. Surman and J. C. Vickerman, Appl. Surf. Sci., 9, 108(1981).

7 K. Wittmaack, Rev. Sci. Instr., 47, 157(1976).

8 W. Lange et al. in: 'Secondary Ion Mass Spectromety SIMS III' (Eds: A. Benninghoven et al.) Springer-Verlag, (1982) p.416.

3.13 Use of a Cesium Primary Beam for Liquid SIMS Analysis of Bio-Organic Compounds

W. Aberth and A.L. Burlingame

Department of Pharmaceutical Chemistry, University of California, San Francisco, CA 94143, USA

1. Introduction

In 1976 BENNINGHOVEN, JASPERS and SICHTERMAN first reported on obtaining secondary ion mass spectra from a number of amino acids using an argon primary beam to bombard a metal target covered with approximately a monolayer of sample [1]. In their pioneering work, they demonstrated that important molecular and structural information can be obtained from complex organic molecules which are nonvolatile and thermally labile, and the technique, labeled molecular SIMS, is in current and widespread use. Molecular SIMS, however, yields an inherently small secondary ion signal of short duration making the technique generally inaccessible to high resolution tandem type mass spectrometers of relatively low transmission efficiency.

In 1981 BARBER, SURMAN and co-workers reported on the development of "fast atom bombardment" (FAB) where the sample is supported in a glycerol or related viscous fluid matrix and bombarded with a fast (2-10 keV) neutral atom beam [2, 3]. The FAB technique is capable of producing a more intense secondary ion beam than molecular SIMS and thereby expanding the application of the technique to include the high resolution and high mass instruments.

In the initial publications describing FABMS, it was emphasized that a neutral primary beam was essential to the technique. If one could effectively replace the neutral primary beam with an alkali ion beam, however, several advantages would be achieved: 1) alkali ions can be produced by surface evaporation at temperatures between about 850 and 1200°C with insignificant amounts of outgassing, thus reducing pumping load requirements of the ion source chamber; 2) the use of ion optics and deflectors would provide better control over the beam size and position; 3) a gun could be designed sufficiently small to allow its being positioned entirely inside the standard source chamber of our Kratos MS-9 and MS-50S mass spectrometers; 4) the ion beam intensity can be controlled over a wide range by adjustment of the ion evaporator temperature. A cesium ion gun was therefore constructed and used to evaluate whether an ionic primary beam can be used effectively with a liquid matrix target (Liquid-SIMS). We chose cesium for the ion source for two reasons. Since its mass (133 d) is close to that of xenon we could make a more realistic performance comparison between the Cs^+ primary beam and our Kratos Xe FAB source. Also, it has been demonstrated by FAB that secondary ion efficiency increases with increasing mass of the primary beam and cesium is the heaviest practical alkali available.

Instrumental

A cross-sectional schematic of the Cs^+ gun, drawn to scale, is shown in Fig. 1. The design is based on that of CONFORTI et al. [4] with some modifications required to reduce its overall length to 2.7 cm. Electrode 1 is the Cs^+ evaporator source [5]. It consists of a porous tungsten plug in which the emitter material (a cesium alumina silicate) has been placed. The Cs^+ emitter is indirectly heated by an Al_2O_3 potted molybdenum filament. It is supported by electrode 2 which serves both as a heat shield for the emitter and to shape the electrostatic focusing field above the emitter surface. Both electrodes 1 and 2 are maintained at the same potential. The extractor electrode 3 is maintained at -300 V relative to electrodes 1 and 2. It serves to extract and focus the Cs^+ ions evaporated from the emitter. Both electrodes 2 and 3 are fabricated of molybdenum to reduce outgassing which can "poison" the emitter surface. Electrode 4 serves as an einzel lens to focus the Cs^+ ions on the probe target and is operated between 500 and 1500 volts below the emitter potential. Electrode 5 is maintained at the source block and sample target potential. It is typically 6000 V below the Cs^+ emitter potential, thus yielding a 6 keV Cs^+ primary beam. The Cs^+ beam is focused to a spot size between 1 and 2 mm in diameter on the sample target which is about 2 cm beyond electrode 5. Typical total Cs^+ beam currents used are between 0.1 and 1 µA obtained at emitter temperatures of between 920-1000°C. The total lifetime of the Cs^+ emitter has not yet been fully evaluated but is inversely a function of the Cs^+ emission. We have operated one emitter for approximately 300 hours at the 1 µA current level. In any case, the Cs^+ emitters are relatively inexpensive and can be readily replaced.

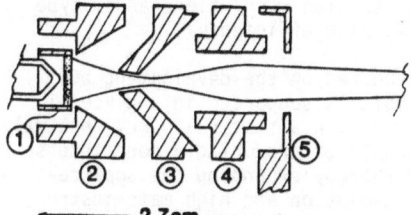

Fig.1 Schematic cross section of the cesium ion gun. (1) Cs^+ emitter and heater; (2) emitter support and field shaping electrode; (3) ion extractor electrode; (4) einzel lens; (5) final ion-accelerating electrode.

2.7cm

Mass analysis was performed on a Kratos MS-50S mass spectrometer equipped with a 23 kG magnet and negative ion switching. The instrument is capable of scanning up to m/z 3000 at 8 kV. Typical operating conditions were: scan rate, 30 s/decade; dynamic resolution, M/ΔM 3000. The spectrometer was interfaced to our laboratory's LOGOS-II/Xerox Sigma 7 data system, permitting real-time assignment of masses [6]. A standard Kratos FAB gun and ion source were used to obtain the neutral primary beam FAB spectra. The gun utilized xenon gas to produce an 8 keV beam of Xe atoms. A gun source current of between 40 and 50 µA was typically used.

Figure 2 is a schematic showing the mounting arrangement of the FAB and Cs^+ guns for producing comparative spectra. The sample matrix is placed on a copper probe tip with a surface plane at an angle of 40° with respect to a plane perpendicular to the MS beam center line. The Cs^+ primary beam is directed at an angle of 20° with respect to the probe tip surface and the Xe^0 beam strikes this surface at an angle of approximately 40°. Although the FAB primary beam is normally directed 20° to the target plane, we found that the larger striking angle had no significant effect on the efficiency

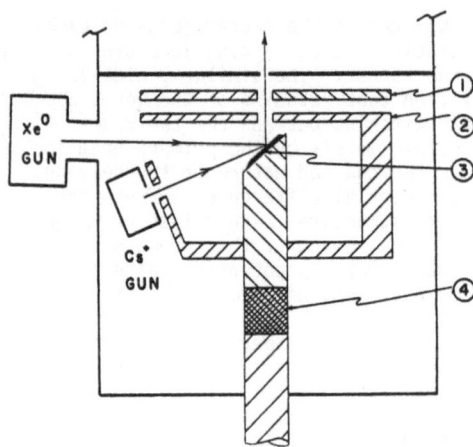

<u>Fig. 2</u>. Schematic of MS source showing mounting arrangement of Xe^0 and Cs$^+$ primary beam guns. (1) Beam-centering plates; (2) source block; (3) sample target; (4) insulator in direct insertion probe.

of secondary ion production. Both Xe^0 and Cs$^+$ primary beams could be directed at the sample target with no additional adjustments to the instrument except for a 15° rotation of the target about the MS axis.

Samples were prepared by either dissolving or making a slurry of the solid in glycerol. Alternatively, 1 µL of a solution of the sample (1-10 µg/µL) dissolved in water was added to about 3 µL of glycerol on the probe tip, and the excess water pumped away in the vacuum lock of the mass spectrometer. Calibration was achieved using either Ultramark 1621 (PCR, Gainesville, FL) or KI dissolved in glycerol. A wide variety of samples were comparatively examined by Xe FAB and Cs$^+$ liquid SIMS. These samples included peptides, lipids, nucleotides, steroid sulfates, carbohydrates and drug conjugates (sulfates, glucuronides) [7].

Results and Discussion

As described above, the size and placement of the Cs$^+$ ion gun within the ion source permits concurrent operation of the source with the commercial fast atom gun (mounted externally on the ion source housing). In switching between the FAB and SIMS modes, it was observed that the source tuning (Y and Z lens, ΔV, beam centering) was essentially unchanged. Since the quantity of sample and mode of preparation are not variables, a direct comparison of the nature of the spectra from the two modes of operation for a given sample can be made. Some retuning of the mass spectrometer is required when switching between positive and negative ion modes.

A comparison of the spectra, often both positive and negative, obtained using the Xe^0 and Cs$^+$ primary beams shows that the Cs$^+$ liquid SIMS generally yields a sensitivity improvement of the molecular ion species by a factor of about 3 as well as a relative reduction in fragmentation components by about a factor of 2 over that of Xe^0 liquid SIMS [7]. Part of the reason for the difference in results is probably due to the better focusing capability of the Cs^+ gun, thereby concentrating the bombarding ions over a smaller but more effective region of the target. The similar yields in secondary ion current, and the observation that retuning of the ion source focusing controls is not required when switching between the two modes of operation, show that charging of the sample matrix by the primary beam ion is not significant. This may be due to the charged nature of the selvedge region which would tend to equilibrate potential differences between the sample matrix and the surrounding support structure. Regardless of whether a neutral or ionic primary beam is used, the region of the target surface must have reasonable conductivity in order that its potential value be close to that of the probe tip for effective instrument performance.

We have observed with liquid targets that when either the Cs^+ or Xe^0 primary beam is turned on there is an initial intense secondary ion signal which diminishes in a few seconds to a stable signal of perhaps 1/5 the beginning amplitude. A probable explanation for these results is that the high sample liquid-vacuum interface concentration on the liquid surface initially produces an intense secondary ion signal. As sputtering progresses, the surface concentration decreases and the secondary signal then reflects the rate at which the sample solute diffuses to the liquid surface being sputtered. A further support of this explanation is the observation that the secondary sample ion signal tends to increase in proportion to primary Cs^+ beam current at low Cs^+ current levels. As the Cs^+ beam intensity increases, there is a leveling off of the secondary signal until no further increase is possible.

Conclusions

We have carried out a series of experiments which establish that the charge or neutrality of the incident energetic atomic beam effecting energy deposition has no significant effect on the secondary ion mass spectra produced from labile, polar organic substances dissolved in a viscous liquid matrix. With adequate attention to ion source design, the use of a primary ion beam with a liquid matrix target may result in superior performance for structural studies of biological compounds in magnetic sector instruments. Other researchers are demonstrating that ion beams can be effectively used with liquid matrix targets [8-10]. In light of these results, the use of the term FAB to describe the state of the sample and its matrix on the electrode surface (viscous liquid as opposed to solid) is misleading. SIMS is the established, preferred acronym for describing the phenomenological process of sputter ionization [11]. For the particular application to liquid matrix targets, it would be more appropriate to modify the term SIMS to specify this, i.e., Liquid SIMS. If necessary, this term could be further modified to indicate the type of primary beam used, i.e., Xe^0 Liquid SIMS or Cs^+ Liquid SIMS.

Acknowledgement

We wish to thank Fred C. Walls for his expert technical assistance and Kenneth M. Straub for his mass analysis work.

This work was supported by National Institutes of Health Division of Research Resources Grant RR00719/RR01614. Purchase of the negative ion unit for the Kratos MS-50S was made possible by a grant from the Academic Senate, University of California, San Francisco, 1981.

References

1. A. Benninghoven, D. Jaspers and W. Sichtermann, _Appl. Phys._ _11_, 35 (1976).
2. M. Barber, R.S. Bordoli, R.D. Sedgwick and A.N. Tyler, _J. Chem. Soc. Chem. Commun._ 325 (1981).
3. D.J. Surman and J.C. Vickerman, _J. Chem. Res. (S)_ 170 (1981); _ibid._, J. Chem. Soc. Chem. Commun. 324 (1981).
4. G. Conforti, F. Del Giallo, F. Pieralli and G. Ventura, _Int. J. Mass Spectrom. Ion Phys._ _36_, 343 (1980).
5. Antek, P.O.Box 51311, Palo Alto, CA 94303.
6. H. Kambara, F.C. Walls, R. McPherron, K.M. Straub and A.L. Burlingame, Presented at the 27th Annual Conference on Mass Spectrometry and Allied Topics, Seattle, WA, June 3-8, 1979.
7. W. Aberth, K.M. Straub and A.L. Burlingame, _Anal. Chem._, in press, Oct. 1982.
8. K. Haroda, M. Suzuki, N. Tokeda, A. Tatematsu and H. Kambara, _J. Antibiotics_ _35_, 102 (1982).
9. P. Williams and D.M. Adams, Presented at the 30th Annual Conference on Mass Spectrometry and Allied Topics, Honolulu, HI, June 6-11, 1982.
10. C.N. McEwen, Presented at the 30th Annual Conference on Mass Spectrometry and Allied Topics, Honolulu, HI, June 6-11, 1982.
11. A.R. Krauss and V.E. Krohn in "Mass Spectrometry", P.A. Johnstone, Sr. Reporter, Vol. 6, Specialist Periodical Reports, Burlington House, London, pp. 118-152.

3.14 Fast Atom Bombardment Study of Glycerol Mass Spectra and Radiation Chemistry

F.H. Field

The Rockefeller University, New York, NY 10021, USA

Neat glycerol was irradiated with beam of Ar atoms and ions produced by a Capillaritron gun operated at 5 KeV. The FAB spectrum produced changed markedly as the irradiation continued. Studies were made with irradiations lasting 12 minutes and 35 minutes. Both the absolute and relative intensities of the ions in the glycerol spectrum change, and new ions are formed. The glycerol sample after the irradiation contains clear crystals. The irradiated sample was subjected to negative chemical ionization analysis, which provided further evidence for the formation of radiation products from the glycerol and also evidence about the identities of the products. The major new products formed are

$$CH_2OH$$
$$HO-C-CH_2OH$$
$$CH_2OH$$

(MW=122)

$$CH_2OH$$
$$HO-C-CHOH$$
$$HO-CH_2-CH_2OH$$

(MW=152)

$$HO-CH_2 \quad CH_2OH$$
$$HO-C - C-OH$$
$$HO-CH_2 \quad CH_2OH$$

(MW=182)

$$CHOH$$
$$C-OH$$
$$CH_2OH$$

(MW=90) .

The radiation yield has been measured semiquantitively. Approximately 100 molecules of products are formed per incident Ar atom/ion. This comprises an interesting radiation chemistry, which presumably involves a non-Franck-Condon initial excitation.

3.15 The Use of FAB for the Solution of Difficult Mass Spectral Problems

C.E. Costello, A.M. Van Langenhove, S.A. Martin, and K. Biemann

Department of Chemistry, Massachusetts Institute of Technology,
Cambridge, MA 02139, USA

Introduction

FABMS promises to be a highly useful addition to the repertoire of mass
spectral ionization methods available for structural studies. We describe
here the experimental procedures being used in our studies and a few typical
examples of the application of FABMS to current research problems.

Experimental Procedures

At the beginning of our investigation of this new method, we carefully evaluated the instrumental and chemical protocols for the measurements. These
results have been reported in detail elsewhere [1], but are sumarized briefly
here. Among the parameters studied were the target´s shape and material, the
nature of the ionizing gas, sample preparation, concentration and matrix.
The instrument employed in our work is a Varian MAT 731 double-focusing mass
spectrometer of Mattauch–Herzog geometry, which has a mass range of 1–2000 u
at 8 kV accelerating potential. The EI/FI/FD combination ion source was fitted with an Ion Tech B12N 100 μA neutral source. The highest sensitivity was
achieved using xenon as the neutral beam and a 60° angle of incidence to the
stainless steel target which was mounted on the tip of the FD probe. Low –
resolution FAB spectra were recorded by oscillographic scans and high-resolution spectra were recorded on photographic plates.

For samples which have been separated from complex mixtures, the final
isolation steps have proven to be the most critical to the success of the
FABMS experiment, because the ionization process is matrix-dependent. Addition of acid [(+)FAB] or base [(−)FAB] may enhance signal intensity. After
the initial FABMS measurements, some simple derivatizations (e.g. acetylation, methylation) may be carried out in the glycerol matrix to provide information about reactive sites.

During the determination of structure of unknown compounds, it is particularly useful to obtain complete FAB mass spectra at high resolution, so
that elemental composition assignments may be made. The glycerol used as
the FAB matrix contributes to the spectrum observed, producing an abundant
series of protonated oligomer and related fragment ions, as well as lower
abundance peaks at every nominal mass. At high resolution, our results show
that these low abundance peaks are in fact clusters of ions having as many
as ten or more different compositions. Exact mass measurements, therefore,
must be conducted at a resolution sufficient to remove contributions of ions

arising from the matrix to the intensity of peaks whose exact mass is to be determined. The principal glycerol ions may, however, be used as reference peaks for the mass scale calibration [2]. Alternatively, FAB spectra of well—characterized compounds or EI spectra of reference materials such as Fomblin may be overlaid on the sample spectra or placed in adjacent exposures on the photoplate and used for the extrapolation [2].

FABMS Applications

Metal Complexes. We have used FAB (Fig. 1) to determine the number of ligands and the oxidation state of technetium in Tc-HIDA [3], an important radiopharmaceutical imaging agent. It was thought to have the structure: sodium bis-[N-(2,6-dimethyl-phenylcarbamoylmethyl)iminodiaceto]technetate(III).

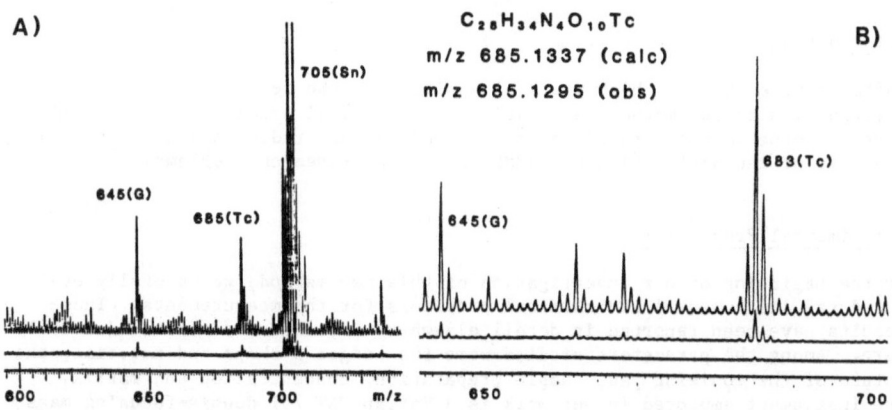

$C_{28}H_{34}N_4O_{10}Tc$

m/z 685.1337 (calc)

m/z 685.1295 (obs)

Fig. 1. (+)FABMS of the radiopharmaceutical Tc-HIDA: (A) after Sep-Pak® concentration of the eluate from the generator, (B) after HPLC of the Sep-Pak® concentrate and collection of the radioactive fraction

Exact mass measurement of the peak at m/z 685 by peak matching with the (glycerol$_7$ + H)$^+$ cluster at m/z 645 at a resolution of $M/\Delta M$=10,000 gave the value shown, corresponding to the elemental composition $C_{28}H_{34}N_4O_{10}Tc$, indicating the complex contained two ligands and had a net charge of -1, and thus required the addition of two protons to form the species observed.

Peptide Syntheses. During the synthesis of β-melanocyte-stimulating hormone (β-MSH), which was being prepared by D. Carr and M. Rosenblatt for use in the analysis of β-lipoprotein, some question arose as to whether the product was the desired peptide, or one which had a higher molecular weight due to incorporation of a blocking group such as trifluoroacetyl. The target peptide has a molecular weight of 2202 and the structure: H$_2$N-Asp-Glu-Gly-Pro-Tyr-Arg-Met-Glu-His-Phe-Arg-Trp-Gly-Ser-Pro-Pro-Lys-Asp-COOH. Fig. 2A shows the portion of the (+)FABMS in the molecular ion region. This result confirmed that a compound of the desired molecular weight had been prepared.

For a peptide of this sequence, the molecular weights and compositions of fragments predicted from a trypsin digest would be (A) 735, $C_{31}H_{45}N_9O_{12}$, (B) 718, $C_{31}H_{46}N_{10}O_8S$, (C) 670, $C_{32}H_{46}N_8O_8$, and (D) 133, $C_4H_7NO_4$. Fig. 2B shows

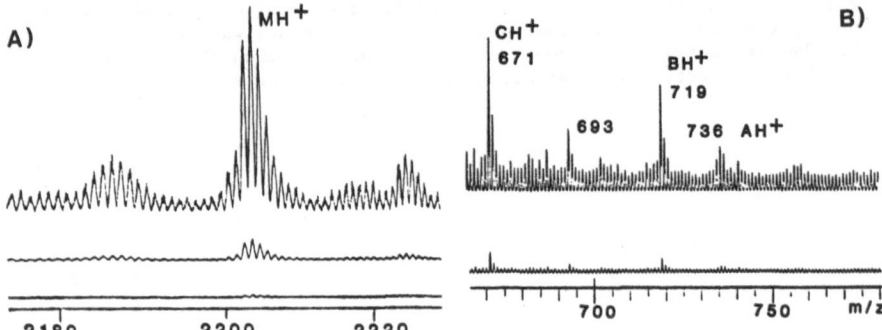

Fig. 2. (+)FABMS of β-MSH and its trypsin digest products: (A) intact peptide, molecular ion region, (B) peptide digest fragments, m/z 650-750 region

the m/z 650-750 region of the (+)FABMS, which includes the molecular protonated and sodium cationized ions of the expected trypsin fragments A, B, and C.

Muramyl peptides. For an immunostimulant prepared by C.M. Gupta, the (+)FAB
mass spectrum had a base peak corresponding to (MH)$^+$ and a peak of nearly
equal abundance corresponding to loss of water from the molecular protonated
ion. Ions arising via other fragmentation pathways were of only very low abundance. The (-)FAB spectrum, however, had a much higher overall signal intensity and many diagnostic fragments were observed, as shown in Fig. 3.

Thirty micrograms of another muramyl peptide, called Factor S because it
induces sleep in humans and other mammals, was isolated from 5000 liters of
human urine by J.R. Pappenheimer and M.L. Karnovsky. In the (+)FABMS of this
sample (Fig. 4A), ions at m/z 922 and 940 were observed and were postulated

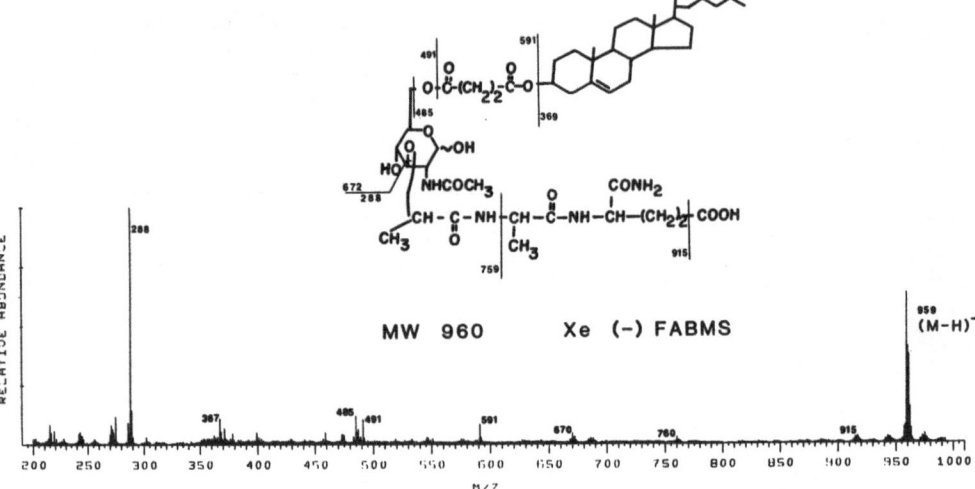

Fig. 3. (-)FABMS of a synthetic muramyl peptide, a potential immunostimulant

to be the MH^+ and $(MH - H_2\overline{O})^{\overline{+}}$ ions. In order to determine the number of free carboxyl groups in this compound, dilute HCl in methanol was added to sample in the glycerol matrix. The peaks at $\underline{m/z}$ 922 and 940 then shifted to $\underline{m/z}$ 950 and 968, respectively, indicating two free acid groups (Fig. 4B).

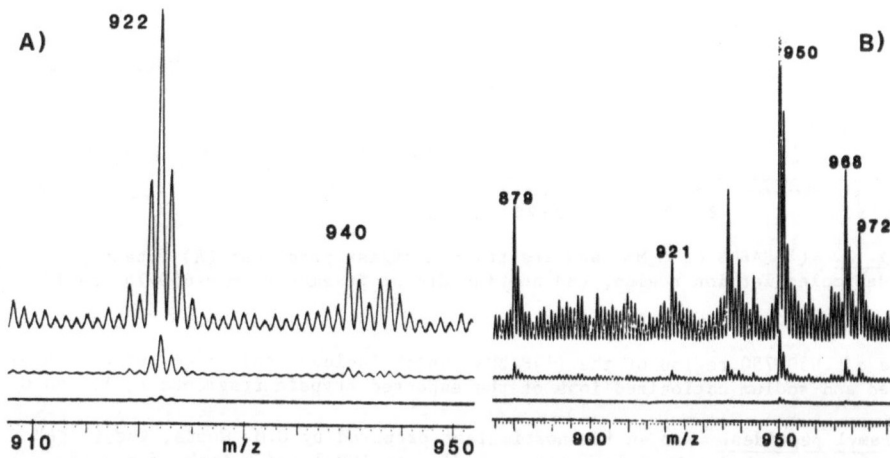

Fig. 4. Molecular ion region in the (+)FABMS of Factor S: (A) before esterification, (B) after esterification

Conclusion. As the examples provided illustrate, FABMS is well suited to the solution of problems which frequently arise in structural studies. Attention to experimental details can help to maximize the information obtained by this technique, since one may thus minimize interferences, selectively enhance the abundance of sample ions, carry out some functional group analyses, and observe both positive and negative ion spectra.

Acknowledgements. This work was supported by the National Institutes of Health (Grant No. RR00317 from the Division of Research Resources) and by the Office of Naval Research (Grant No. ND0014-78-C-0421).

References

[1] S.A. Martin, C.E. Costello and K. Biemann, Anal. Chem., in press.
[2] A. Van Langenhove, C.E. Costello, H.-F. Chen, J.E. Biller and K. Biemann, 1982 ASMS Meeting, Honolulu, HI, Abstracts, p 558.
[3] C.E. Costello, J.W. Brodak, A.G. Jones, A. Davison, D.L. Johnson, S. Kasina, and A.R. Fritzberg, J. Nucl. Med., in press.

3.16 Biological and Medical Applications of Organic SIMS

M. Junack, A. Eicke, W. Sichtermann, and A. Benninghoven

Physikalisches Institut der Universität Münster, Domagkstraße 75
D-4400 Münster, Fed. Rep. Of Germany

The detection and identification of involatile, thermally labile
organic compounds in complex solutions play an important role
in biology and medicine. Obviously SIMS is a very effective
technique for the analysis of these organic compounds. Parent
ions and characteristic fragment ions in the spectra enable
detection, identification and structural information for extremely small amounts of material down to the subnanogram range.
Possible interferences in the spectra, however, demand a preseparation of complex mixtures by liquid chromatography or thin-
layer chromatography in most cases. On-line or off-line combinations of SIMS with these separation techniques are possible.

For an assessment of the feasibility and power of SIMS in
this field we carried out a number of investigations that may
be divided into the following groups:

- Determination of the secondary ion spectra and yields for
  relevant groups of compounds.
- Detection of small amounts of these compounds after preseparation from a complex mixture.
- Quantitative determination of trace compounds in body fluids.

The first step of any SIMS application in biology and medicine
is the determination of the secondary ion spectra, in particular
the yields of parent-like and fragment ions. We carried out
this kind of investigation for many groups of organic compounds
as amino acids, peptides, barbiturates, sulphonic acids, nucleosides, nucleotides, folic acids, desoxynojirimycins, steroids,
benzodiazepines, etc. [1-4] . The standard procedure for sample
preparation is the deposition of 1 µl of an appropriate solution
on 1 cm^2 of an etched silver surface. As a typical example Fig.1
shows the positive and negative secondary ion spectrum of the
benzodiazepine nitrazepam. In the spectrum $(M+H)^+$ and $(M-H)^-$
peaks appear with high intensities, in addition to $(M+Ag)^+$
and $(M+Na)^+$ ions, which originate from cationization of the
sample molecules by the substrate atoms or Na impurities. For
some compounds, the presence of Cl ions results in the appearance
of $(M+Cl)^-$ and $(M-H+AgCl)^-$ ions.

The fragmentation of organic compounds can mostly be explained
by predictable bond cleavages, splitting off functional groups
and proton rearrangement, etc. A typical fragmentation of the
benzodiazepines is the elimination of an H and the C(2) atom
with its substitutes to form a pyrimidine ring. For nitrazepam

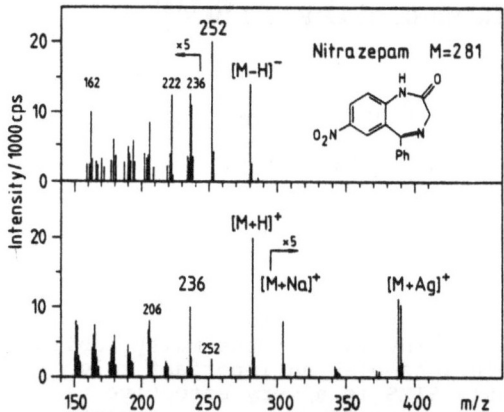

Fig.1 Negative and positive secondary ion spectrum of nitrazepam on Ag. Coverage: 3 nmol/cm^2. Primary ion bombardment:
3 keV Ar$^+$, $2.5 \cdot 10^{-10}$ A on 0.1 cm^2

this results in the elimination of CO and H and the appearance
of a strong peak at (M-29) = 252 amu.

An important application of SIMS in real sample analysis is
its use as LC detector. This has been demonstrated by the identification of products of the bacteria Streptomyces violaceoruber
which have been investigated with SIMS after separation and purification by thin-layer chromatography [5] . The bacteria produce the antibiotic granaticin A, but during incubation the
granaticin itself is metabolized to at least two new products.
The negative secondary ion spectrum of granaticin A shows molecular ions M$^-$ at 444 amu and characteristic fragments at 339,
356 and 399 amu (refer to Fig.2a). One metabolite is granati-
cinic acid which is identified by the M$^-$ peak at 462 (refer to
Fig.2b). The fragments are mainly the same as for granaticin,
but in addition a peak at 417 amu corresponding to the loss of
the carboxyl group is observed. Fig.2c presents the spectrum
of the second metabolite. Its structure is not completely known,
but the appearance of similar fragment ions as in the case of
granaticin indicates in addition to other results that this product is a derivate of granaticin, too. According to the possible
interpretation of the peak at 607 amu as M$^-$ or (M-H)$^-$, the molecular weight of this second metabolite must be 607 or 608
respectively.

As a further example for the analysis of a mixture and the
advantages of preseparation procedures, a urine control solution
with 5 opiates added in a concentration of 10 mg/l (= 3.10-5 mol/l)
has been investigated. Mass spectra were recorded from samples
prepared with the original solution and with different eluates
of an extraction of the drugs by an extrelut column. The spectrum
of the original solution is shown in the upper part of Fig.3.
Only low intensities of the molecular ions of the drugs (meperidine, morphine, codeine, methadone, ethylmorphine) are obtained,

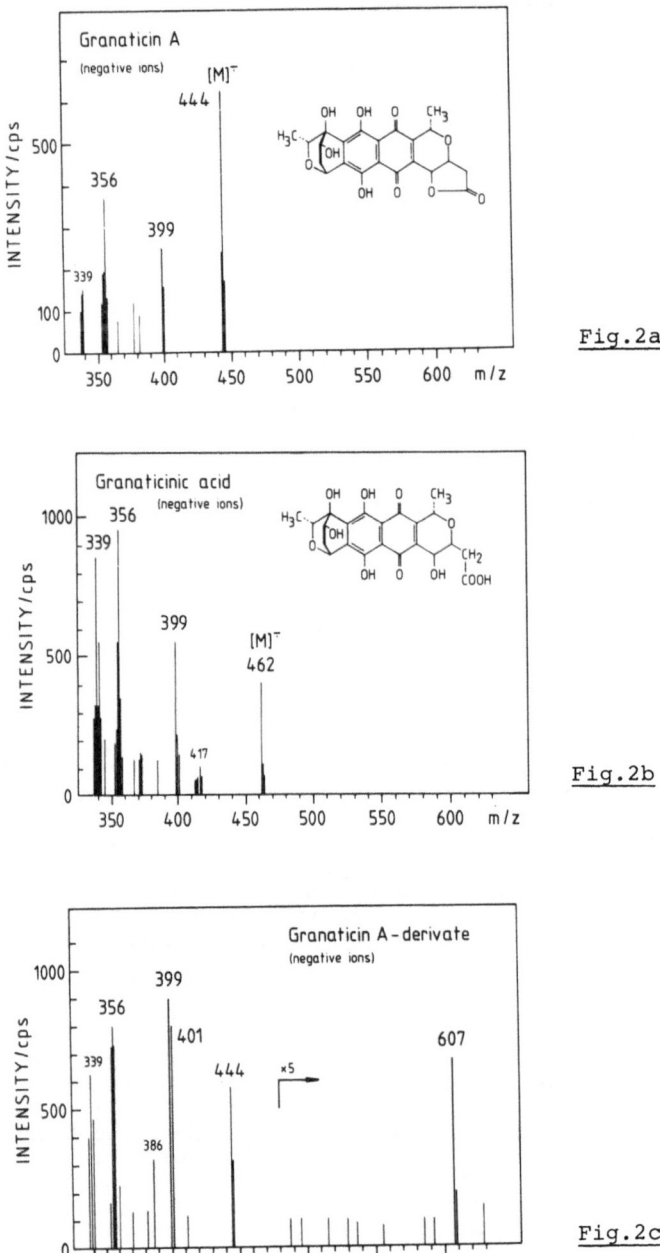

Fig.2a

Fig.2b

Fig.2c

<u>Fig.2</u> Negative secondary ion spectrum of granaticin A(a), grana-
ticinic acid(b), and an unknown granaticin A metabolite(c).
Primary ion bombardment: 3 keV Ar^+, $4 \cdot 10^{-10}$ A on 0.1 cm^2

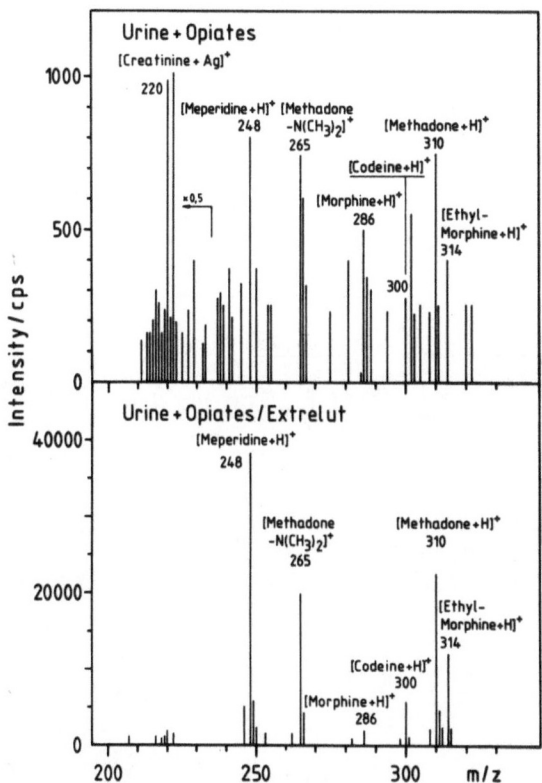

Fig.3 SIMS spectra of a urine control solution with added opiates (meperidine, morphine, codeine, methadone, ethylmorphine, each $3 \cdot 10^{-5}$ mol/l). Original solution, diluted 1:5 with water (upper spectrum), and after extraction with chloroform from an exrelut column (lower spectrum). Primary ion bombardment: 3 keV Ar^+, $5 \cdot 10^{-10}$ A on 0.1 cm^2

probably due to the relatively high concentration of other components in the urine sample. This is indicated by the strong appearance of the creatinine parent ions $(M+H)^+$ and $(M+Ag)^+$ (refer to Fig.3). After extraction the opiates are detected with considerably higher intensities, because of an enrichment in the eluate and the separation of disturbing components. The lower part of Fig.3 gives the spectrum of the eluate at pH 2 which contains predominantly meperidine, methadone and ethylmorphine.

The purification is demonstrated by the absence of the creatinine peaks. Morphine and codeine are eluated after buffering the column at pH 10. The results of this mixture analysis show in addition that components of fractions containing several compounds may be detected by SIMS directly, without further separation.

As a first application to quantitative analysis of body fluids we investigated the excretion of the steroid methandrostenolone in urine after oral intake [3] . The detection was possible after extraction with a reversed phase C-18 cartridge. The concentrations calculated by comparison with a sample containing an internal standard are in good agreement with values published in literature.

The examples reported here demonstrate the capacities of SIMS in the detection and identification of organic trace compounds in biological samples, the importance of preseparation and enrichment procedures for the analysis of complex mixtures and the feasibility of quantitative results through the application of internal standards.

References

1 A.Benninghoven, W.K.Sichtermann: Anal.Chem. 50, 1180 (1978)

2 A.Eicke, W.Sichtermann, A.Benninghoven: Org.Mass Spectrom. 15, 289 (1980)

3 W.Sichtermann, A.Eicke, M.Junack, A.Benninghoven: Fres.Z. Anal.Chem. 311, 410 (1982)

4 A.Eicke, V.Anders, M.Junack, W.Sichtermann, A.Benninghoven: Anal.Chem. 55 (1983), in press

5 M.Junack, E.Kormann, A.Eicke, W.Sichtermann, A.Benninghoven, H.Pape: Fres.Z.Anal.Chem. 311, 411 (1982)

3.17 Low and High Resolution FAB Applications in Positive and Negative Ionization Mode

Uwe Rapp and Monika Höhn

Finnigan MAT GmbH, Barkhausenstraße 2, D-2800 Bremen, Fed. Rep. of Germany

Introduction

In the field of alternative ionization techniques, FAB (or liquid SIMS) has attracted an extremely high degree of attention, because a variety of application problems could be solved which, up to then, were not attainable with techniques such as CI, DCI or even FD.

Besides the production of low—resolution spectral data, analysts are demanding all the other auxiliary techniques which make mass spectrometry such a valuable tool for structure elucidation. One of these features, available with double focussing sector field mass spectrometers, is exact mass determination and elucidation of elemental compositions.

Usually, medium to high resolution conditions are used together with suitable reference compounds such as perfluorokerosene (PFK) under electron ionization conditions. When alternative ionization techniques are used, the lack of universally employable reference compounds becomes a severe drawback.

Accurate mass determinations by magnetic scans are possible only if the mass range measured is adequately covered with reference peaks. Typically, with an equally spaced reference file, accuracies of about 1-3 millimass units are obtainable at medium (3000) to high (10,000) resolution. When the "density" of the references is diminished, the accuracy of exact mass determinations decreases because of the nonideal magnetic scan function.
There are only limited data available, which clearly indicate a not fully satisfactory accuracy when only 1-5 so-called "lock masses" are used for defining the magnetic scan function under medium/high resolution conditions.

In this paper a completely different approach will be discussed. Principally there are two ways to scan a sector field type mass spectrometer: either the magnet or the accelerating voltage may be scanned to obtain a mass spectrum. Theoretically, the acceleration voltage should be scanned from full voltage to zero. Practically this is not possible, since the ion source, under normal conditions, is defocused at low voltages, thus reducing, more or less drastically, sensitivity.

If the scan range is limited to certain values, the accelerating voltage scan becomes a very useful tool. As it deals with voltages and not magnetic fields, an extremely precise scan function is obtainable, thus making possible a scan requiring only a few reference peaks for exact mass determinations.

Results

In order to show the capabilities of FAB, the following few examples of normally difficult substances are presented, where FAB spectra are easily obtainable. The substances are: Nikomycin Z, a peptide-nucleoside antibiotic, Mezlocillin® a commercially available penicillin derivative and an aromatic sulfonic acid, a dyestuff synthesis product. All three substances were measured in positive and negative ionization modes to show the complementary information yielded.

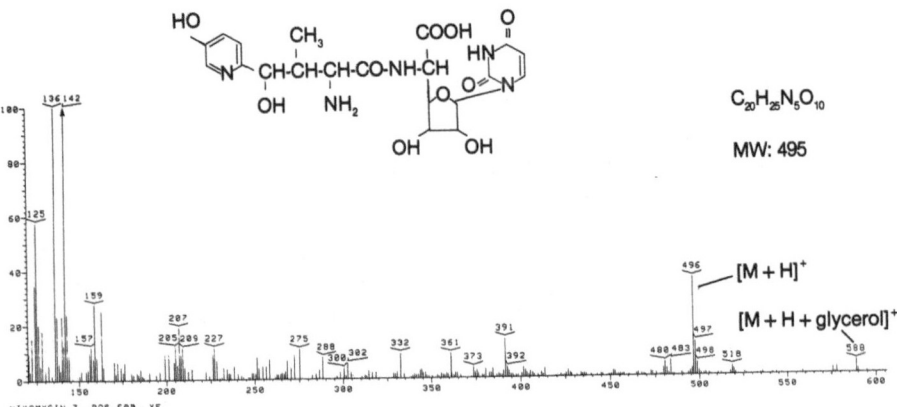

Fig. 1. Positive ionization FAB spectrum of the antibiotic Nikomycin, glycerol background subtracted

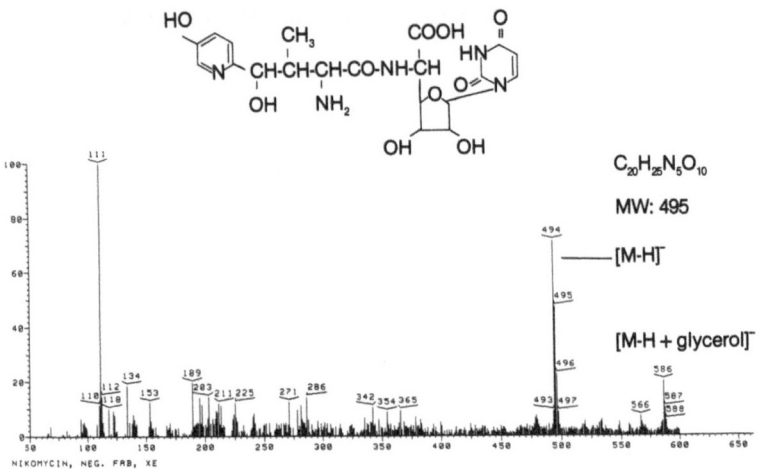

Fig. 2. Negative ionization FAB spectrum of the antibiotic Nikomycin, glycerol background subtracted

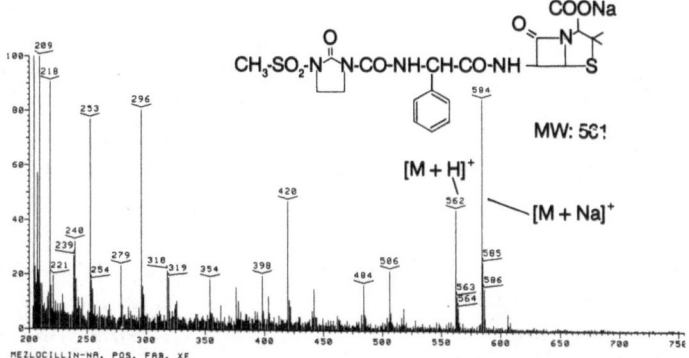

Fig. 3. Positive ionization FAB spectrum of the penicillin Mezlocillin®

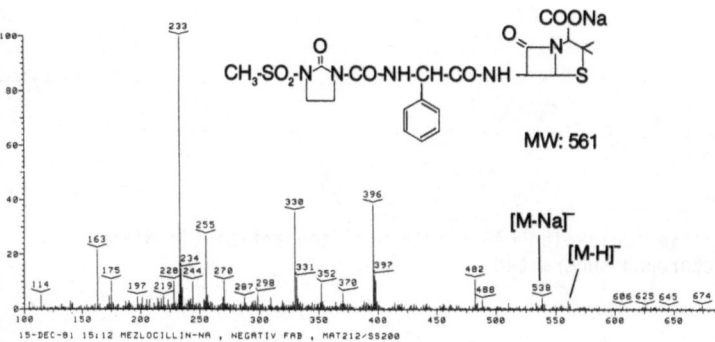

Fig. 4. Negative ionization FAB spectrum of the penicillin Mezlocillin®

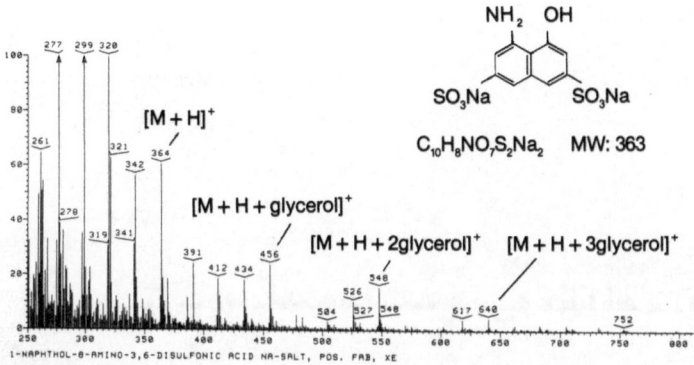

Fig. 5. Positive ionization FAB spectrum of 1-naphthol-8-amino-3,6-disulfonic acid Na salt

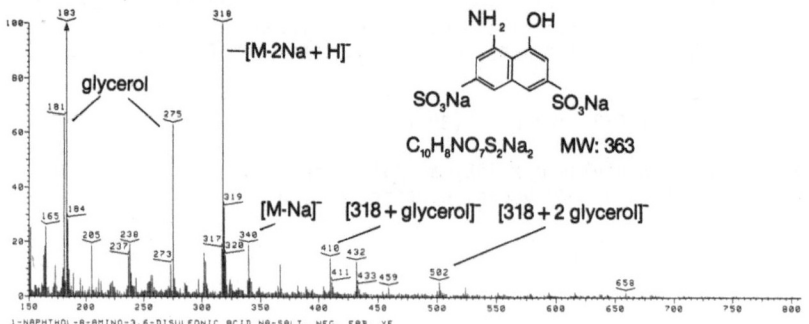

Fig. 6. Negative ionization FAB spectrum of 1-naphthol-8-amino-3,6-disulfonic acid Na salt

It is noteworthy that positive and negative mode information are indeed supplementary. Therefore, to take advantage of the most information for structural determination, both modes should be used.

In rating the value of positive and negative mode of operation it can be concluded that the wealth of information obtainable, strongly depends on the sample's structure. The sulfonic acid (Figures 5 and 6) yields a very complex positive mode spectrum by the formation of many adduct ions of the sample with the matrix, such as M+H+(glycerol)$_n$. In negative mode the spectrum looks clearer and may make interpretation easier. This behaviour is typical of sulfonic acids or salts under FAB conditions.

The Nikomycin spectra seem to be both equally valuable, which is also true for the Mezlocillin® data.

The second chapter now deals with exact mass determinations using the acceleration voltage scan (ESCAN).

Leucine enkephalin, a penta peptide with the sequence Tyr-Gly-Gly-Phe-Leu was used for evaluation. A resolution of 3000 (10% velley) was applied and the scan speed of the ESCAN was 4 s for a complete scan. As reference substance a KI/glycerol mixture was measured together with the peptide.

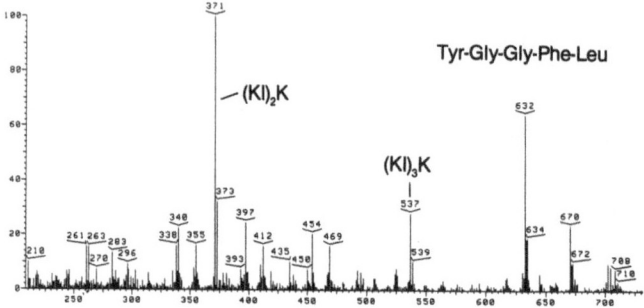

Fig. 7. Positive ionization FAB spectrum of leucine encephalin using KI/glycerol as matrix

Figure 7 shows the low resolution spectrum under these conditions. The molecular weight information (MW:555) is chemically shifted to m/z 632 and 670 because of the KI presence. An addition and substitution of hydrogen by potassium occurs:

555+K(39)+K(39)-H(1) = 632   and 555+K(39)+2K(78)-2H(2) = 670.

A further substitution yields m/z 708, a smaller peak in the spectrum. The ions corresponding to the KI/glycerol cluster are indicated with their formula.

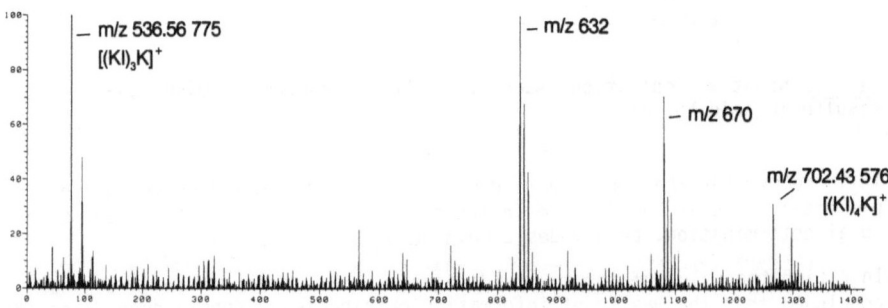

Fig. 8. Raw data acquisition spectrum of leucine encephalin at 3000 resolution, using KI/glycerol as matrix

Figure 8 gives the raw data acquisition spectrum at 3000 resolution where the magnet mass for starting the accelerating voltage scan was set to m/z 527 in order to establish m/z 536 (KI)$_3$K as the first reference mass. The mass range scanned was m/z 527 to m/z 840, although it is not completely displayed here.

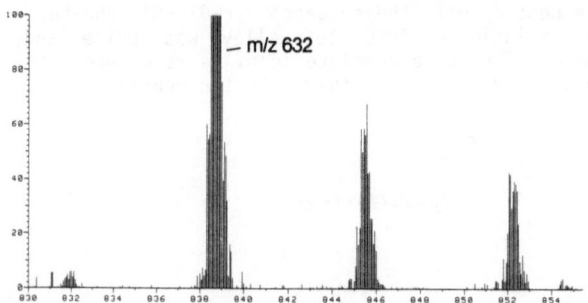

Fig. 9. Peak profiles of m/z 631 - 634 of leucine encephalin at 3000 resolution, using KI/glycerol as matrix [(M-H+2K) group]

Figure 9 makes a closer inspection of the m/z 632 peak profile possible. There are about 25 data points per peak available, guaranteeing an exact peak centroid calculation.

The elemental composition list, together with the exact mass values, are given in figure 10. Single spectrum results of consecutive spectra and averaged data. The latter show for m/z 632 0.2 millimass units and 0.0 millimass units deviation corresponding to values better than 0.5 ppm.

```
SSX:  ELIST FOR MHENESHR.CTD;1 USING ELEMENTS.ELM;1
AUG 27 82    12:20:06    VERSION V04.0         PAGE:    1

                    SPECTRUM NUMBER    2

        MASS INTENSITY     DIFF    C/C•    H    N    O    K
                 BASE              12/13    1   14   16   39

    632.1896    100.00     0.7     28/0    36   5    7    2

    633.1982     37.70     1.4     28/0    37   5    7    2



SSX:  ELIST FOR MHENESHR.CTD;1 USING ELEMENTS.ELM;1
AUG 27 82    12:20:06    VERSION V04.0         PAGE:    2

                    SPECTRUM NUMBER    3

        MASS INTENSITY     DIFF    C/C•    H    N    O    K
                 BASE              12/13    1   14   16   39

    632.1848    100.00    -4.1     28/0    36   5    7    2

    633.1921     48.50    -0.1     27/1    36   5    7    2
    633.1921     48.50    -4.6     28/0    37   5    7    2



SSX:  ELIST FOR MHENESHR.CTD;1 USING ELEMENTS.ELM;1
AUG 27 82    12:20:06    VERSION V04.0         PAGE:    3

                    SPECTRUM NUMBER    4

        MASS INTENSITY     DIFF    C/C•    H    N    O    K
                 BASE              12/13    1   14   16   39

    632.1923    100.00     3.4     28/0    36   5    7    2

    633.1894     82.25    -2.9     27/1    36   5    7    2
```

Fig. 10a. Exact mass values together with elemental composition proposals using KI/glycerol as reference

```
SSX:  ELIST FOR MHENESHR.AVE;2 USING ELEMENTS.ELM;1
AUG 27 82    12:23:23    VERSION V04.0         PAGE:    1

                    SPECTRUM NUMBER    1

        MASS INTENSITY     DIFF    C/C•    H    N    O    K
                 BASE              12/13    1   14   16   39

    632.1887    100.00    -0.2     28/0    36   5    7    2

    633.1922     58.01     0.0     27/1    36   5    7    2
    633.1922     58.01    -4.5     28/0    37   5    7    2

            ***  ELIST PROCESSING COMPLETE  ***
```

Fig. 10b. Averaged data of exact mass determinations with elemental composition proposals

Part 4

Laser Induced Ion Formation

4.1 Laser Induced Ion Formation from Organic Solids (Review)

Franz Hillenkamp

Institut für Biophysik, Universität Frankfurt,
D-6000 Frankfurt 70, Fed. Rep. of Germany

1. Introduction

The first reports on laser ion sources for mass spectrometric investigations of solid surfaces appeared in the literature in the early sixties. In these investigations lasers of various types had been used, either for the evaporation only, followed by an independent ionization,e.g. by classical electron impact, or for the simultaneous evaporation and ionization. A substantial number of publications came out in the late sixties and early seventies. VASTOLA, FENNER and DALY, HONIG and READY and coworkers were the most active groups in this field. Most of their work had however been concentrated on inorganic specimens, including bioorganic molecules and biological samples. First reports of such attempts began to appear in the literature from 1976 [1, 2, for a more general application. Such attempts were started in the early- to mid seventies by several different groups in various countries. An excellent review of all this work published until 1979 with an exhaustive list of references has been compiled by CONZEMIUS and CAPELLEN [1] .

This review concentrates on the Laser Desorption Masspectrometry (LDMS) of organic specimen, including bioorganic molecules and biological samples. First reports of such attempts began to appear in the literature since 1976 [1, 2, 3, 4, 5] . Since then an increasing number of groups have published results, the most pertinent of which will be discussed here. The greatly differing techniques used by the various groups has made - and to a certain extent still makes - intercomparisson of results difficult. Nevertheless the work of the last two years has brought about considerable progress in understanding the mechanisms of laser-induced ion formation, though much still needs to be done.

2. Materials and Methods

The results reported by the different groups have all been obtained with experimental arrangements that differ greatly in the types of lasers used, the wavelengths, time regime, and irradiances on the sample, in the sample geometry and sample preparation, the mass spectrometers and the detection systems. The most pertinent information on these different systems is compiled in Tab. 1.

All parameters involved scan a very large range of magnitude and it was certainly not expected that most spectra nevertheless show some striking similarities. Particular attention is drawn to the following facts:
- The covered wavelength range from 250 nm to 10.6 µm, though only a very small fraction of the electromagnetic spectrum, scans almost the total range of photon energies needed for very different primary interaction processes from ionization (at least in the liquid and solid phase), through electronic excitation to a mere population of vibrational and rotational states.

Table 1. Experimental parameters of the different groups

		University Bonn [5,12,13,15,16]	FOM Amsterdam [4,17,18,19,39]	
Laser / Frequency-multiplier	Type	CO_2	CO_2	TEA-CO_2
	Time regime	continuous	continuous	pulsed
	Laser energy or power	3 W	3 W	< 1 J
	Mode pattern	multimode	multimode	multimode
	Wavelengths used	10.6 µm multiline	10.6 µm multiline	10.6 µm multiline
Sample Stage	Sample geometry[1]	mostly "thick" layer	"thin" layer	"thin" layer
	Substrate	metal	quartz, metal	quartz, metal
	Irrad. geom.[2]	0^0	0^0	0^0
	Ion extraction geometry[2]	90^0	90^0 and 0^0	90^0
	Ion polarity	+, some -	+, some -	+, some -
	Spot diameter on sample	0.2 - 4 mm	0.2 - 0.5 mm	0.3 - 0.5 mm
	Sample Irrad. time	several seconds to minutes	usually 0.7 s	150 ns
	Irradiance on sample[3]	$20 - 7 \cdot 10^3$ Wcm^{-2}	$10^3 - 10^4$ Wcm^{-2}	$10^7 - 10^8$ Wcm^{-2}
Mass Spectrometer	Type	scanning quadrupole; 0.1-0.2 s scan time	magnetic sector	
	Detected mass range	$M_r/Z \leqslant 800$	1:1.2 - 1:1.6 for scan in range $M_r/Z \leqslant 1500$	
	Transmission[4]	10^{-5}	10^{-5}	
	Resolution	$M/\Delta M \sim 20$ at $M_r/Z = 365$	$M/\Delta M \sim 300$ at $M_r/Z = 600$	
Detector		SEM + multichannel analyzer	Channelplate - vidicon-optical multichannel analyzer	
	Major substances investig.	Oligosaccharides NAD Quart. ammonium salts	Oligosaccharides Aminoacids, small peptides Nucleotides Glucosides	

For footnotes see page 195

Table 1. Experimental parameters of the different groups (continued)

		Johns Hopkins Univ., Baltimore [14, 20, 21]	FOM - Amsterdam [4]	Univ. Vienna [24, 25]
Laser - Frequency multiplier	Type	TEA-CO_2	Nd-glass	Nd-YAG
	Time regime	pulsed	free run	free run 50 Hz prf
	Laser energy or power	0.1 - 0.7 J	1.0 J	0.1 J
	Mode pattern	multimode	multimode	multimode
	Wavelengths used	10.6 μm multiline	1060 nm	1060 nm
Sample Stage	Sample geometry[1]	"thick" layer	"thin" layer pellet, 2 mm diam.	"thin" layer
	Substrate	bulk org. polymer (Vespel)	metal	glass
	Irrad. geom.[2]	0°	0°	45°
	Ion extraction geometry[2]	90°	90°	90°
	Ion polarity	+	+, some -	+
	Spot diameter on sample	$\sim$1 mm	0.1 mm	0.2 x 0.3 mm elliptical
	Sample Irrad. time	150 ns	100 μs, spiking	80 μs, spiking multiple pulses
	Irradiance on sample[3]	$\sim 10^6$ Wcm^{-2}	$\bar{N} \sim 10^7$ Wcm^{-2} $\hat{N} \sim 10^8$ Wcm^{-2}	$\bar{N}$ 10^4- 10^5 Wcm^{-2} $\hat{N}$ 10^5- 10^6 Wcm^{-2}
Mass Spectrometer	Type	double focusing and time of flight (TOF)	double focusing magnetic sector	double focusing
	Detected mass range	$M_r/Z < 500$	$M_r/Z \leqslant 500$	magn. scan range M_r/Z 10
	Transmission[4]	not reported	not reported	not reported
	Resolution	doub.foc.:$M/\Delta M$=500 TOF: $M/\Delta M \leqslant 800$	$M/\Delta M \sim 2000$ at $M_r/Z = 400$	$M/\Delta M$ 1000 at $M_r/Z = 365$
Detector		SEM - ADC	Photoplate, multi-shot exposure	SEM, multishot spectra
	Major substances investig.	Quart. ammonium salts, glychocol. acid	see other FOM column	Sucrose Na_2-AMP Arginine

For footnotes see page 195

Table 1. Experimental parameters of the different groups (continued)

		Univ. of Houston [26,27]		Purdue Univ. W. Lafeyette [8]
Frequency-multiplier Laser	Type	Coumarine dye	Nd-YAG	Nd-YAG
	Time regime	pulsed 25 Hz prf	Q-switched 1-2 kHz prf	Q-switched 10 Hz prf
	Laser energy or power	400 µJ	$\leqslant$ 1 mJ	0.1 J
	Mode pattern	multimode	TEM_{oon}	not reported
	Wavelengths used	483 nm	1060 nm	1060 nm
Sample Stage	Sample geometry[1]	"thin" layer		not reported
	Substrate	metal		metal
	Irrad. geom.[2]	45^0		0^0
	Ion extraction geometry[2]	0^0		90^0
	Ion polarity	+	+ and -	+
	Spot diameter on sample	0.1 mm		2 mm
	Sample Irrad. time	5-7 ns multiple pulses	20-30 ns multiple pulses	10 ns multiple pulses
	Irradiance on sample[3]	10^7 Wcm^{-2}		10^8 Wcm^{-2}
Mass Spectrometer	Type	scanning quadrupole		double focusing reverse geom.
	Detected mass range	M_r/Z < 750		M_r/Z $\leqslant$ 500
	Transmission[4]	not reported		not reported
	Resolution	M/ΔM: 200-400		not reported
Detector		SEM + boxcar integrator or SEM + multichannel analyzer		integrating electrometer
	Major substances invest.	amino acids, peptides, nucleosides, oligosaccharides		sucrose

For footnotes see page 195

Table 1. Experimental parameters of the different groups (continued)

		Institute of Spectroscopy Moscow [48,49]	LAMMA 500 and LAMMA 1000 Universities Düsseldorf,Frankfurt,Pitsburgh LH,Köln: [2,3,9,10,11,28-38,40,41,43,47,56]		
Frequency-multiplier / Laser	Type	N_2, XeCl, KrF	Nd-YAG	Nd-YAG	ruby
	Time regime	pulsed	Q-switched	modelocked	Q-switched
	Laser energy or power	3 µJ – 1 mJ	0.5 J	0.1 J	1.0 J
	Mode pattern	Multimode	TEM$_{oon}$		multimode
	Wavelengths used	337, 308 nm 249 nm	(1060, 532 nm) 353, 265 nm	(1060 nm) 532, 265 nm	(694 nm) 347 nm
Sample Stage	Sample geometry[1]	"thick" and "thin" layers	"thin" layers or particles of $\leqslant$ 2 µm diameter		
	Substrate	metal	LAMMA 500 : none or organic film ($\sim$500 Å) LAMMA 1000: bulk metal		
	Irrad. geom.[2]	83^0	LAMMA 500: 0^0; LAMMA 1000: 30^0		
	Ion extraction geometry[2]	0^0	LAMMA 500: 180^0;LAMMA 1000: 0^0		
	Ion polarity	+	+ and –		
	Spot diameter on sample	$\sim$1 mm	0.5 – 5 µm		
	Sample Irrad. time	$\sim$15 ns mult. pulses	$\sim$10 ns	$\sim$30 ps	$\sim$30 ns
	Irradiance on sample[3]	10^4-10^7 Wcm^{-2}	10^6 - 10^{10} Wcm^{-2}, depending on sample; always 2-10 times threshold for ion gener.		
Mass Spectrometer	Type	time of flight	time of flight LAMMA 500: with reflector LAMMA 1000:with or without refl.		without reflector
	Detected mass range	not reported	M_r/Z < 2200		
	Transmission[4]	not reported	1 - 10 % for ions of low initial energy		
	Resolution	not reported	with reflector: M/ΔM $\leqslant$800 at M_r/Z = 200 without reflector: M/ΔM = 200 at M_r/Z = 200		
Detector		SEM	SEM - ADC (10 ns min. sampling time)		
	Major substances investig.	adenine anthracene peptides	Oligosaccharides, aminoacids, peptides nucleobases, -sides and -tides org. acids and salts		

For footnotes see page 195

- The interaction time range, determined by the irradiation time, the laser pulse width, or the sample evaporation time, from about 30 ps to minutes covers 13 orders of magnitude. At the short end, this time may be considerably shorter than thermal equilibration times for a number of processes (depending on substance parameters, sample- and irradiation geometry, nature of phase changes, etc.), whereas it appears hard to believe in any macroscopic non-equilibrium state for times in the second or minute domain and for the geometrical extents involved.

- The corresponding 10 - 12 orders in magnitude of irradiance will again cover the whole range from only classical absorption at the low irradiance end to a large variety of nonlinear optical processes for irradiances above about 10^6 Wcm^{-2}. Most applications have used irradiances in the transition range between linear and nonlinear absorption, where predictions about the dominating processes are particularly hard to make. Phase- or constitutional changes, induced by the laser radiation, may moreover drastically change the relevant sample properties during the irradiation.

- Sample irradiation and ion extraction geometry as well as the substrates used all have strong influences on the induced processes as evidenced by a number of recent experiments. More investigations are necessary to completely unravel the influence of these factors. Schematic diagrams of the possible arrangements are shown in Fig. 1. Depending on sample layer thickness and the extinction of the substance investigated at the laser wavelength used, the laser energy may be absorbed in the sample or penetrate down to the substrate. In the latter case reflectivity and/or absortivity of the substrate determine the total amount of energy deposited. Peak temperatures reached, temperature profiles during and after irradiation and the temperature-time course at a given sample point will depend on the size of the irradiated area (and therby on the total energy delivered) as well as on the heat capacity and thermal conductivity of the substrate and sample. In repetitive pulsing the average delivered power may futhermore lead to a stationary temperature increase of the sample that influences the results.

---

footnotes to Table 1

1) The Bonn and Baltimore groups prepared their samples in layers of about 0.1 - 0.5 mm thickness. All other groups used samples of less than 1 μm thickness. The "thin" layers must all be considered thin compared to the penetration depth of the laser radiation used (1/e - value). Penetration depth of the CO_2 laser radiation into the oligosaccharides must be expected to be much less than the thickness of all samples prepared in Bonn. Penetration of the CO_2-laser radiation into the Baltimore samples depends on extinction of the substances investigated at the wavelength. Axial temperature profiles can be expected to become stationary during the sample irradiation- or ion generation time for all arrangements discussed.

2) The angles given all refer to the sample surface-normal facing the incoming laser beam.

3) All irradiances given are averaged over the nominal spot size and the whole irradiation time. For the spiking Neodymium lasers an estimate of the peak irradiance ($\hat{N}$) for single spikes is also given.

4) The given numbers are rough estimates of the order of magnitude only.

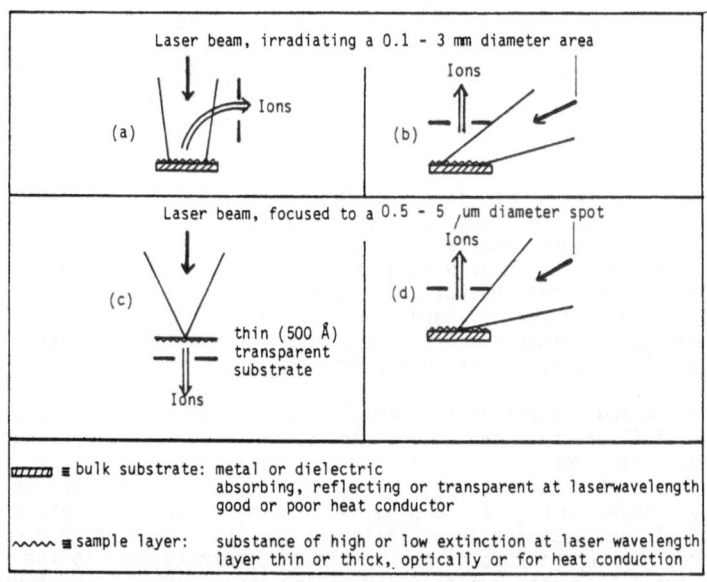

Laser beam, irradiating a 0.1 - 3 mm diameter area

(a) Ions

(b) Ions

Laser beam, focused to a 0.5 - 5 ,um diameter spot

(c) thin (500 Å) transparent substrate Ions

(d) Ions

⬚⬚⬚ ≡ bulk substrate: metal or dielectric
absorbing, reflecting or transparent at laserwavelength
good or poor heat conductor

〜〜〜 ≡ sample layer: substance of high or low extinction at laser wavelength
layer thin or thick, optically or for heat conduction

Fig. 1. Schematic diagrams of sample -, substrate -, irradiation - and ion-
extraction geometries and material properties

- The different mass spectrometers used not only differ widely in transmission,
i.e. absolute sensitivity and dynamic range, they also (together with the
ion extraction geometry) exhibit different selectivity of acceptance as a
function of the distribution of initial energy and momentum of generated
ions. Metastable ions will moreover be recorded to a different degree and
at different apparent mass numbers.

- Sample preparation is another very important parameter. It influences the
relative intensities of substance-specific ion signals (quasimolecular- or
large fragment ions) to those of unspecific background signals and may also
change the relative magnitude of the specific ion signals among each other,
e.g. that of the $(M + H)^+/(M + Alkali)^+$ to $(M - H + 2 Alkali)^+$ or the abundance of cluster ions. In addition, the relative abundance of metastable ions
may be a function of sample preparation [6] .

All samples used have been in the solid state. In most cases sample substances have been dissolved at concentrations in the range of 10^{-6} mol/l to
saturation. Small aliquots of the solution have then be dried onto substrates
in air or vacuum or electrosprayed onto substrates. In some cases aerosols
have been produced from the solution and particles in the range of 0.1 - 3 μm
diameter have been impacted onto substrates. Powders, as supplied by the manufacturers, have also been analyzed directly. Several investigators report reproducible differences in the spectra for samples analyzed directly or prepared from solution [7] . Admixtures of alkali- or ammonium halides influence
the spectra, as expected, but the concentration dependence does not appear to
be the same for the different techniques used.

An impressingly wide variety of organic substances have been analyzed by
laser desorption. They include oligosaccharides, organic acids including amino-

acids and oligopeptides, organic salts particularly quarternary- ammonium salts, poly- and mononucleotides and their building blocks, enzymes and a variety of pharmaceutical products aswell as some organic, polymeric solids. Tab. 1 contains only those molecules which have been investigated by most of the groups and therefore allow intercomparison of results. Alkali halides have been analyzed quite extensively by several groups and, though they are inorganics, the results are important for systematic investigations into the different ion formation processes.

Most of the molecules so far analyzed have relative molecular masses below 500 dalton. More recently an increasing number of spectra with masses up to 1000 dalton have appeared and in rare cases molecules of $M_r/Z = 2000$ and above have been detected.

3. General features of spectra

A number of properties are common not only to laser desorption spectra but also to results obtained with the other so-called desorption techniques. These are in particular:
- polarity is supportive, if not a prerequisite of desorption, in contrast to thermal evaporation where polar groups must, as a rule, be derivatized to render a molecule volatile

- ions are predominantly of the even electron type, protonation and deprotonation rather than electron abstraction or attachment are common processes. Radical ions are only rarely generated from strongly aromatic compounds.

- Cationization by alkali-metal ions is very frequent. Cationization by other metal ions (e.g. Ag, Cu, Mg ,etc.) has also been observed under suitable conditions [8, 9, 10] as has been anionization,e.g. by chlorine [9] .

- Ions of both polarities are generated, usually at comparable abundances. They carry complementary information. For acidic compounds specific ions (e.g. $(M - H)^-$) are found mostly in the negative ion spectra, for basic compounds specific ions (e.g. $(M + H)^+$, $(M + Alkali)^+$) are more easily identified in the positive ion spectra.

- Cluster ions such as $(2M + Na)^+$ are relatively frequent, even for parent molecules with rel. molecular mass of several hundred dalton.

- For the techniques using very short, high irradiance laser pulses a more or less smooth transition to pyrolysis of the sample is observed with increasing irradiance. It appears that at least for the LAMMA technique this transition does not always occur at identical irradiances for positive and negative ions, an observation that deserves attention for further clarification.

- Spectra are qualitativly influenced by the immediate surrounding of the molecules analyzed, e.g. by an organic matrix [11] .

These similarities between all desorption spectra are believed to result from common chemistry that the molecules or ions undergo. It appears that in all desorption techniques there must be a step at which such chemical reactions can occur, most probably as a stabilizing step. Most of this chemistry seems to be governed by much the same molecular properties that are also known to govern most of the chemistry in the gas phase or in solution. It is then obvious that these common features must not necessarily indicate equal excitation and ion formation processes for all the desorption techniques or even among the different laser desorption arrangements. Indeed the rather different mechanisms, discussed in the following section, have only rather recently been recognized, after the more subtle differences in the spectra and other ion properties had been worked out.

4. Ion formation mechanisms

Four ion formation processes can at least in principle be distinguished. They contribute to varying degrees to the spectra obtained with the different techniques :

1. Thermal evaporation of ions from the solid.
2. Thermal evaporation of neutral molecules from the solid followed by ionization in the gas phase.
3. "True" laser desorption.
4. Ion formation in a laser—generated plasma.

The first two processes are called thermal, because they can also be induced by classical Joule- or non-laser radiative heating, usually in conjunction with heat conduction to the sample surface. Thermal evaporation of cations of quarternary ammonium salts and anions of sodium tetraphenylborate has been demonstrated by several groups [12, 13, 14] . Such a thermal evaporation of ions, common for metals and inorganic salts such as alkali halides, had not originally been expected to occur for organics as well. It should be most probable for organic salts with quarternary salts possibly exhibiting a singularly high yield because of their somewhat special structure [12] .

Volatility, together with thermal stability, are the two terms that are used to characterize the chance for thermally evaporating neutral molecules from the condensed - into the gas phase. Based on former experiments employing mostly electron impact ionization most organic molecules, particularly bioorganic ones with relative molecular masses above about 100 dalton were believed to be involatile and thermally unstable. Recently Röllgen et al. [13, 14, 15, 16] as well as Kistemaker et al. [17, 18, 19] have published investigations that strongly evidence a thermal evaporation of neutral molecules of sucrose and a number of other organic substances followed by ionization in the gas phase by molecule reactions with alkali ions. If no separate source for the alkali ions is provided, the existence of separate areas of lower temperature, just high enough for the onset of evaporation of intact neutral molecules (e.g. ∼ 300 - 350° C), and of high enough temperatures (> 750° C) for thermal evaporation of alkali ions are a prerequisite for this process to occur. This in all likelihood is the dominating ion formation process in all the CW - CO_2 - laser experiments. Besides ion-molecule reactions in the gas phase, thermal surface ionization at hot surface areas presumably through decomposition of neutrally evaporated clusters of salts of carboxylic- and sulfonic acids has also been suggested [13] . COTTER et al. [20, 26] have in addition shown that the evaporation of ions and even more of neutrals may persist for times considerably beyond the laser exposure, if temperature decay is slow, because of the ammount of energy deposited and because of limited heat conduction. Several authors have suggested to make more efficent use of these evaporated neutrals through suitable gas phase ion-molecule reactions or, e.g. chemical ionization in specially designed sources, in order to substantially increase the total ion yield.

These results certainly call for future extensive and systematic investigations of the volatility of organic molecules and ions as a function of molecular size and structure. So far such processes seem to be limited to molecules of $M_r/Z < 350$, but there seems to be some evidence that neutral peptides up to $M_r/Z < 1900$ can be evaporated by slow radiative heating of samples on a gold substrate [22] . In view of these observations, the frequent statement that polarity generally supports "desorption" may also need further consideration, at least in cases where thermal evaporation plays a key role in the ion formation.

As discussed above, the yield of thermally generated ions or neutrals will depend strongly on the geometry of the sample/substrate arrangement and their optical and thermal properties. The influence particularly of the substrate has been demonstrated by the Amsterdam group [18] . They also observed that the ion generation ceased completely if metal substrates had been carefully polished, presumably because the increased reflectivity prevents sufficient substrate heating [23] . Some authors claim that "rapid" heating also enhances ion yield or even is a prerequisite for quasimolecular ion generation. Though it appears somewhat unlikely that temporal- or spatial temperature gradients directly influence the evaporation process, they could certainly change the degree of competing chemical processes in the sample or, e.g. maximize concurrent codesorption of molecules and alkali ions for the subsequent ion-molecule gas phase reaction. For the likelihood of the latter process one would also expect an influence by the openness of the whole ion source

The experiments by van der PEYL [18] , COTTER [14, 20] and HERESCH [24, 25] strongly support the assumption that thermal processes also dominate ion generation in some if not most of the experiments in which laser pulses in the ns- or μs-domain have been used, particularly if repetitive pulsing has been applied as by some experimentators, e.g. by HARDIN and VESTAL [26, 27] and the sample/substrate configuration was of the type shown in Fig. 1a or b.

"True" laser desorption must be assumed where the ions generated by laser irradiation of solid samples show properties which exclude a thermal process of generation. Such ion properties are most obvious and best documented for results obtained with LAMMA (Laser Microprobe Mass Analyzer)- instruments [2, 3, 9, 10, 11, 28 through 38, 40, 41, 43, 47, 56] . Though most of the ion properties discussed below have been measured for inorganic ions in order to ease interpretation, there is no reason to expect them to be principally different for organic ones. The most important features are:

1. Even at threshold irradiances for ion generation, ions have initial energies typically above 10 eV often even 50 - 100 eV. Van der PEYL et al. in contrast have determined the initial energy of alkali ions in their instrument, using the 100 ns pulses of the CO_2-laser, to about 0.3 eV, i.e. in the range of thermal energies for the surface temperatures measured. Measurements of the initial energy of organic ions are under way in that group and will certainly add to the evidence of the processes involved [39] .

2. The distributions of cluster ions, obtained, e.g. from pure graphite, show significant contributions of very large clusters which would not be compatible with temperatures needed for thermal evaporation and which are indeed not observed in thermal evaporation experiments [40] .

3. Within a sizable range of laser irradiances above threshold irradiance for ion generation, the contributions of large vs. small clusters increase with increasing irradiance. In a thermal model increasing irradiance should lead to higher temperatures and thereby to smaller clusters.

4. Cluster distributions of alkalihalides cannot be reasonably fitted by an Arrhenius function, assuming physically realistic temperatures, particularly

199

for experiments with samples on substrates, preheated to static temperatures up to 800 K [41] .

An analysis of the peak widths of the ion signals in the LAMMA TOF-spectra and the observation of a qualitative agreement of spectra of many organic subs-tances obtained with a straight time of flight tube or with an ion reflector showed that there were no major contributions to the signals originating from gas phase reactions that would take place at times longer than about 10 ns after ion generation (i.e. after the laser pulse is over) or beyond distances of about 20 μm from the sample surface. This holds for bimolecular including ion-molecule-ractions as well as for monomolecular decay.

The mechanisms that lead to such "true" laser desorption are now believed to be collective, non-equilibrium processes in the condensed phase [38] . In this respect they are closer to processes that must be assumed to lead to ion generation in SIMS and plasma desorption rather than to the thermal laser induced ion generation discussed above, even though the spectra are often in-distinguishable for all different laser techniques. The recently reported ob-servation of metal ion (Cu, Ag, Mg,etc.) attachement for desorption with high power, short pulse lasers [8, 9, 10] also points to the similarity with SIMS.

The reason for the dominance of such "true" desorption processes vs. thermal evaporation is most easily understood for the LAMMA 500^R arrangement, as shown in Fig. 1c. Because of the tight focusing of the laser beam, the total energy delivered is typically about 100 nJ or less. There is moreover no subst-rate that could absorb energy and then conduct it to the sample; both sample and thin-film-substrate are usually "evaporated" simultaneously in the focus during the laser shot. The total affected volume of about<0.1 μm^3 expands at a speed of 10^5 cm s^{-1} or more. As a result, probability of gas phase collisions for a given ion or molecule is very small compared to arrangements with large irradiated areas and/or long irradiation times. Thermal processes are much more likely for the LAMMA 1000^R arrangement as shown in Fig. 1d. FEIGL et al. [42] were able to show that at ion generation threshold (~ 0.1 μJ) from a bulk metal surface, ion generation occurs essentially during the laser pulse time only. At 10-times threshold energy (~ 1 μJ) the strong ion emission during laser irradiation is followed by a emission of ions lasting for several micro-seconds, presumably of thermal origin due to the heated bulk sample.

The details of the formation processes of the observed ions are as yet not well understood. The frequent observation of ions which have undergone substantial structural rearrangements or even chemical reactions, unlikely to occur on a nanosecond timescale in the solid state, suggest that there is an intermediate state of higher mobility, similiar to suggestions made for SIMS. The upper limits given above for times and locations within which such reactions would have to take place would certainly allow such an intermediate state to have particle densities several orders of magnitude below that of solid-state density. Yet average particle distances and the lack of a shield-ing solvent probably prohibits a direct application of liquid- or gas phase chemistry to this intermediate state, particularly as it is of transient nature, such that no chemical and probably even no thermal equilibrium is attained.

At laser irradiances of typically about 10^{10} Wcm^{-2} and above dense plasmas are formed from any solid sample, as is well documented by the large number of laser fusion experiments. In this mode of operation energy is deposited into the solid during the initial phase of the laser pulse only, creating a highly absorbing plasma in front of the surface, shielding it from incoming radiation of the later part of the laser pulse [43] . A substantial amount of energy of the laser beam will thereby be deposited into the plasma, in-creasing its temperature as well as density of ions and electrons. The plasma mode is certainly not suited for organic mass spectrometry. Molecules will be mostly broken down to their atomic constituents, multiply-charged atomic ions will get abundant with increasing laser irradiance and ion initial energies will typically be in the range of kiloelectronvolts. For the analysis of inorganic specimens such as metals or minerals this mode has on the other hand been shown to have decided advantages, because the ion yields become nearly uniform for all elements throughout the periodic table [44, 45, 46].

Though there seems to be sufficient experimental evidence for the four discussed mechanisms to be active in the various laser desorption experiments, the actual interpretation of experimental results may not be as straightforward. In many cases more than just one process may contribute to the results observed. It is, e.g. quite likely that under a suitable choice of experimental parameters in laser desorption with high power lasers one can simultaneously create a plasma in the center of the non-uniformly irradiated area, get "true" laser desorption from the periphery and even thermal emission due to a heated substrate for times longer than the laser pulse.

5. Discussion and outlook

Laser desorption so far has essentially been confined to the laboratory scale for feasibility demonstration. Because of the greatly differing techniques used and the different ion formation processes involved, only very general statements can be made about the performance data.

The yield of ions relative to neutrals generally seem to be relatively low for all techniques. For the LAMMA experiments the ratio is estimated to be in the range of $10^{-3} - 10^{-4}$ and reaches values of up to 10^{-1} only in rare cases such as the detection of alkali ions out of organic matrices. For laser-induced thermal desorption this ratio most probably is much less, though much will depend on the efficiency of the gas phase ion-molecule reaction which in turn strongly depends on the density of alkali ions generated in one or the other way. The ratio of quasimolecular- or sample specific fragment ions to unspecific background signals is even more difficult to estimate. It depends not only on the technique used and the physical parameters chosen, it is moreover strongly influenced by sample preparation and the environment of the molecules, may it be a pure crystal or, e.g. an organic matrix as in biological specimens. For the detection of inorganic ions out of an organic matrix, such as epoxy in the LAMMA technique, sensitivity differs by as much as a factor of 100 for different elements and different binding energies to the matrix.

Reproducibility also depends strongly on sample preparation and some other experimental parameters that influence, e.g. the total ion emission time in CW-laser experiments. Under careful experimentation and suitable sample preparation reproducibility of major peaks in LAMMA spectra, taken from homogeneous samples, is about 10%-20%.

Relative quantification for a given atomic- or molecular ion is generally possible if enough care is taken to standardize sample preparation and stabilize other experimental parameters. This way concentration profiles of ions along biological structures can, e.g. be obtained with the LAMMA microprobe. Relative quantification among different atomic- or molecular ions is a very difficult task so far and this holds even more for any absolute quantification. The main reason for this is the above-mentioned influence of the surrounding matrix on the ion yield as well as possibly competitive ionization of different species. Some attempts have been made to use external or preferably internal standards [47] but great care must be taken for the standard to really have comparable physical and chemical properties.

It should be clear from this review that laser desorption is still in its infancy and much work remains to be done to clarify the underlying mechanisms and to fully exploit the potentialities of the technique by an optimization of experimental parameters. Investigations into the ion formation mechanisms above all require the design of more "clean" experiments such that contributions by the different processes are more clearly separated. Obviously the design of the sample/substrate stage and the ion source play a key role in this respect besides choice of suitable laser parameters and mass spectrometers. Measurements of the fraction and the properties of metastable ions generated by the different techniques should contribute significantly to the understanding of the mechanisms in the different laser as well as other desorption techniques. Some very interesting results on metastable ions have already been published [26] and more are expected in the near future.

For the techniques based on thermal evaporation the investigation of the volatility of molecules or classes of molecules seems to be the most pressing problem for future research. Besides this, optimal conditions for gas phase ion formation of evaporated neutral molecules should deserve particular attention. For "true" laser desorption much more basic research appears to be necessary for an understanding of the main aspects of this process. To this end a more systematic variation of the physical parameters as well as of the chemical properties of the analyzed samples will be necessary. Only two, rather preliminary reports on the influence of pulse widths in the subnanosecond domain have been published [37, 48]. The reported results on the influence of the laser wavelengths are rather contradictory. For a collective process in the condensed phase no resonance effects would be expected as a function of wavelength apart from the wavelength dependence of energy deposition into the solid via linear (e.g. vibrational excitation in the infrared) or nonlinear optical processes (e.g. via multiphoton absorption by transparent dielectrics in the ultraviolett). ANTONOV et al. [49] have on the other hand reported results that suggest strong resonances through UV single photon electronic excitation of organic molecules in laser ion formation from organic solids. More recently the same authors have reported the selective desorption of the aromatic tryptophane residue from a protected oligopeptide a at wavelength around 260 nm [50]. There are also some reports on resonance-enhanced ion formation for wavelength in the visible and ultraviolet in a field ionization source [51, 52, 53] and on the dependence of the yield of desorbed ions on the wavelength from oligosaccharides and alcohols in the region of the CO_2-laser emission around 10 μm [15, 54, 55]. It can therefore be expected that investigations using tunable lasers would yield very interesting information particularly with respect to the properties of the phase within which at least some important steps of the ion formation presumably take place.

The range for practical applications of laser desorption, particularly the upper limit of relative molecular mass, is another important question that undoubtedly will be pursued in the future. With the present, very limited basis of experimental data and only rudimentary understanding of the processes no sound prediction can be made. There is reason to expect that techniques based on thermal processes will be more limited in mass range than "true" desorption techniques. The recently reported quasimolecular signals of some lipipolysaccharides in the range of relative molecular mass of 1900 dalton by SEYDEL [56] and of an oligopeptide of $M_r/Z \sim 2100$ in the author's group, both obtained with LAMMA instruments, are at least encouraging results. It may well turn out that under optimized conditions the signal to noise ratio will in the end be more of a limiting factor than absolute detection sensitivity, as may be true for other non-laser desorption techniques as well.

This review is based on the published literature but also to a large extent on discussions with colleagues, the work of which has been reviewed and, most

202

importantly, on work and discussions in the author's research group at the University of Frankfurt. It reflects the opinion of the author who takes full responsibility for the content, but the comments, ideas and suggestions of all, who so valuably contributed, is greatly appreciated.

References

1. R.J. Conzemius and J.M. Capellen: Int. J. Mass Spectrom. Ion Phys. $\underline{34}$ (1980) 197

2. E. Unsöld, F. Hillenkamp and R. Nitsche: Analusis $\underline{4}$ (1976) 115

3. E. Unsöld, F. Hillenkamp, G. Renner and R. Nitsche: Adv. Mass Spectr. $\underline{7}$ B (1978) 1425

4. M.A. Posthumus, P.G. Kistemaker, H.L.C. Meuzelaar and M.C. Ten Noever de Brauw: Anal. Chem. $\underline{50}$ (1978) 985

5. R. Stoll and F.W. Röllgen: Org. Mass Spectr. $\underline{14}$ (1979) 642

6. M.L. Vestal: personal communication

7. K.-D. Kupka and D.M. Hercules: personal communication

8. D. Zakett, A.E. Schoen and R.G. Cooks: J. Am. Chem. Soc. $\underline{103}$ (1981) 1295

9. K. Balasanmugam, T.A. Dang, R.J. Day and D.M. Hercules: Anal. Chem. $\underline{53}$ (1981) 2296

10. B. Schueler, P. Feigl, F.R. Krueger and F. Hillenkamp: Org. Mass Spectr. $\underline{16}$ (1981) 502

11. B. Ollmann, diploma thesis: Universität Frankfurt, 1982

12. R. Stoll and F.W. Röllgen: Org. Mass Spectr. $\underline{16}$ (1981) 72

13. U. Schade, R. Stoll and F.W. Röllgen: Org. Mass Spectr. $\underline{16}$ (1981) 441

14. R.B. van Bremen, M. Snow and R.J. Cotter: Int. J. Mass Spectr. Ion Phys. in print

15. R. Stoll and F.W. Röllgen: Z. Naturforsch. $\underline{37a}$ (1982) 9

16. R. Stoll and F.W. Röllgen: J. Chem. Soc. Chem. Comm. (1980) 789

17. G.J.Q. van der Peyl, K. Isa, J. Haverkamp and P.G. Kistemaker: Org. Mass Spectr. $\underline{16}$ (1981) 416

18. G.J.Q. van der Peyl, J. Haverkamp and P.G. Kistemaker: Int. J. Mass Spectr. Ion Phys. $\underline{42}$ (1982) 125

19. G.J.Q. van der Peyl, K. Isa, J. Haverkamp and P.G. Kistemaker: Nuclear Instr. Methods $\underline{198}$ (1982) 125

20. R.J. Cotter: Anal. Chem. $\underline{53}$ (1981) 719

21. R.J. Cotter and A.L. Yergey: Anal. Chem. $\underline{53}$ (1981) 1306

22. E. Constantin and J.F. Muller: Proceedings of the 9th Int. Conf. Mass Spectr., paper 14/14, E.R. Schmid, K. Varmuza and I. Fogy eds. Elsevier Pub. Comp., in print, and E. Constantin, personal communication

23. P.G. Kistemaker: personal communication

24. F. Heresch, E.R. Schmidt and J.F.K. Huber: Anal. Chem. $\underline{52}$ (1980) 1803

25. F. Heresch: Proceedings of the 9th Int. Mass Spectr. Conf., paper 6/9, E.R. Schmid, K. Varmuza and I. Fogy, eds. Elsevier Pub. Comp., in print see also: F. Heresch, this volume

26. E.D. Hardin and M.L. Vestal: Anal. Chem. $\underline{53}$ (1981) 1492

27. E.D. Hardin and M.L. Vestal: 13. Ann. Conf. on Mass Spectr. and Allied Topics, poster WPB 26, Honolulu, 1982

28. F. Hillenkamp, E. Unsöld, R. Kaufmann and R. Nitsche: Appl. Phys. $\underline{8}$ (1975) 341

29. R. Wechsung, F. Hillenkamp, R. Kaufmann, R. Nitsche, E. Unsöld and H. Vogt: Microscopica Acta Suppl. $\underline{2}$, 1978, P. Echlin and R. Kaufmann, eds.

30. H.J. Heinen, S. Meier, H. Vogt and R. Wechsung: Adv. Mass Spectr. $\underline{8}$ (1980) 942

31. K.-D. Kupka, F. Hillenkamp and Ch. Schiller: Adv. Mass Spectr. $\underline{8}$ (1980) 935

32. B. Schueler and F.R. Krueger: Org. Mass Spectr. $\underline{15}$ (1980) 295

33. F. Hillenkamp and R. Kaufmann, eds.: special issue Fresenius Z. Anal. Chem. $\underline{308}$, No. 3 (1981)

34. J.A. Gardella, D.M. Hercules and H.J. Heinen: Spectr. Lett. $\underline{13}$ (1980) 347

35. D.M. Hercules, R.J. Day, K. Balasanmugam, T.A. Dang and C.P. Li: Anal. Chem. $\underline{54}$ (1982) 280 A

36. S.W. Graham, P. Dowd and D.M. Hercules: Anal. Chem. $\underline{54}$ (1982) 649

37. F. Hillenkamp, R. Kaufmann and R. Florian: Proceedings of the 7. Vavilov Conf. Nonlinear Optics, Novosibirsk, 1981, in print

38. B. Jöst, B. Schueler and F.R. Krueger: Z. Naturforsch. $\underline{37a}$ (1982) 18

39. G.J.Q. van der Peyl, K. Bederski, A.J.H. Boerboom and P.G. Kistemaker: Proceedings of the 9th Int. Conf. Mass Spectr., paper 6/3, E.R. Schmid, K. Varmuza and I. Fogy eds., Elsevier Pub. Comp., in print

40. N. Fürstenau and F. Hillenkamp: Int. J. Mass Spectr. Ion Phys. $\underline{37}$ (1981) 135

41. B. Schueler, F.R. Krueger and P. Feigl: Proceedings of the 9th Int. Conf. Mass Spectr., paper 6/2; E.R. Schmid, K. Varmuza and I. Fogy eds., Elsevier Pub. Comp., in print

42. P. Feigl, B. Schueler and F. Hillenkamp: Proceedings of the 9th Int. Conf. Mass Spectr., paper 6/6; E.R. Schmid, K. Varmuza and I. Fogy eds., Elsevier Pub. Comp., in print

43. G.G. Devyatykh, S.V. Gaponov, E.D. Kovalev, N.V. Larin, V.I. Luchin, G.A. Maksimov, L.I. Pontus and A.I. Suchov: Sov. Techn. Phys. Lett. 2 (1976) 356

44. J.A.J. Jansen and A.W. Witmer: Spectrochimica Acta 37 B (1982) 483

45. R.J. Conzemius and H.J. Svec: Anal. Chem. 50 (1978) 1854

46. R.J. Conzemius, F.A. Schmidt and H.J. Svec: Anal. Chem. 53 (1981) 1899

47. W. Schröder, D. Frings and H. Stieve: SEM/1980/II pg. 647

48. V.S. Antonov, V.S. Letokhov, Y.u.A. Matveyets and A.N. Shibanov: Laser in Chemistry (to be published)

49. V.S. Antonov, V.S. Letokhov and A.N. Shibanov: Appl. Phys. 25 (1981) 71

50. V.S. Antonov, V.S. Letokhov and A.N. Shibanov: Appl. Phys. B 28 (1982) 245

51. S. Nishigaki, W. Drachsel and J.H. Block: Surface Science 87 (1979) 389

52. W. Drachsel, S. Nishigaki and J.H. Block: Int. J. Mass Spectr. Ion Phys. 32 (1980) 333

53. W. Drachsel, T. Jentsch and J.H. Block: Proceedings of the 9th Int. Mass Spectr. Conf., paper 4/9 and 4/10, E.R. Schmid, K. Varmuza and I. Fogy eds. Elsevier Publ. Comp., in print

54. M. Mashni and P. Hess: Chem. Phys. Lett. 77 (1981) 541

55. M. Mashni, B. Schäfer and P. Hess: Proceedings of the 9th Int. Mass. Spectr. Conf., paper 6/15; E.R. Schmid, K. Varmuza and I. Fogy eds., Elsevier Publ. Comp., in print

56. U. Seydel: this volume

4.2 Time Resolved Laser Desorption

Robert J. Cotter, Mark Snow, and Michael Colvin

Department of Pharmacology, The Johns Hopkins University,
Baltimore, MD 21205, USA

1. Introduction

The combination of laser ionization and mass spectrometry has been used
for a number of years for elemental analysis of solid surfaces [1] and
the pyrolysis of nonvolatile compounds [2]. However, an article by KISTE-
MAKER et al. [3] in 1978 focussed attention on the use of laser desorption for the analysis of large, nonvolatile bio-organic molecules, using
conditions which produce intact molecular ions. There have been a number
of contributions since [4-9] which have generally indicated that the kinds
of ions produced by laser desorption are similar to those common in spectra produced by field desorption, plasma desorption, secondary ion mass
spectrometry, thermal desorption, and fast atom bombardment (Fig. 1.).

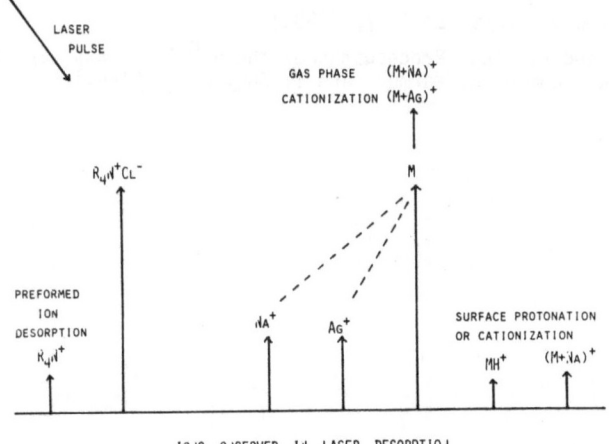

Fig. 1

    Quaternary ammonium ions, R_4N^+, are easily desorbed from their ha-
lide salts using pulsed or continuous wave, CW, lasers [4]. Cationized
molecules, produced by alkali ion attachment (i.e., $(M+Na)^+$ or $(M+K)^+$)
have been observed by nearly all researchers, and have been studied in detail [5] to determine if alkali attachment precedes desorption, or is the
product of an ion-molecule reaction between co-desorbed alkali ions and intact neutral molecules. The conclusion that the latter process is dominant, at least for small molecules such as sucrose, is not surprising. Alkali ions such as Na^+ and K^+ are often observed in greater abundance

than sample ions, even when not deliberately added, and the desorption of neutrals in even larger abundances following a laser pulse has long been recognized [10].

In a variation of this method, COOKS et al. have induced Ag^+ ion attachment by depositing solid samples on silver foil doped with ammonium chloride [8].

Figure 1 represents (schematically) only those processes which lead to intact neutral or ionic molecular species. In addition to these, there are the products of surface decomposition which may also be desorbed as ions, neutrals or cationized/protonated species, as well as gas phase fragmentation ions.

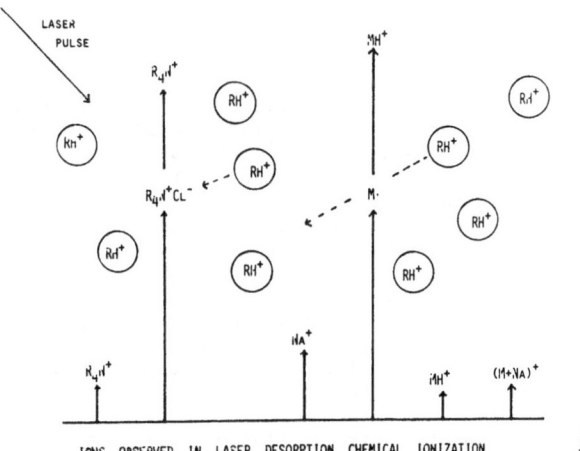

IONS OBSERVED IN LASER DESORPTION CHEMICAL IONIZATION

Fig. 2

In our laboratory we determined that there might also be great analytical utility in exploiting the large numbers of neutrals which are desorbed, and combined laser desorption with chemical ionization [6,7] (Figure 2). Protonated species are plentiful because of the opportunity for multiple collisions with the reagent gas ions in a high-pressure source.

2. Time-Resolved Laser Desorption

Currently a laser desorption, time-of-flight mass spectrometer system is being used to study in greater detail the desorption of molecular ions, neutral molecules, cationized species and pyrolysis products. The instrument has been recently described [11], and uses two techniques to monitor competing processes: electron impact ionization of desorbed neutrals and time resolution. Following a 40n nS laser pulse, the ion withdrawal pulse can be delayed from 0 to 500 μS. When thermal processes are involved ions (and neutrals) will continue to be emitted from the surface irradiated by the laser pulse for some time afterwards, so that the spectra can be "time resolved". The electron beam may be pulsed just prior to extraction of the ions if neutrals are to be measured.

3. Results and discussion

3.1 Alkali ions

When a mixture of alkali halide salts, e.g., NaCl, KCl and CsCl are deposited on the probe tip and desorbed by a laser pulse, the maximum ion currents for Na^+, K^+ and Cs^+ are reached at different times following the laser pulse. The order generally follows that of decreasing ionization potentials(of the neutral atom) so that Na^+ (I.P.$_{Na}$=5.1 e.v.) maximizes around 6 μS and has disappeared at 12 μS, when K^+ ion (I.P.$_k$=4.3 e.v.) reaches a maximum. K^+ ions are emitted for close to 20 μS, and Cs^+ ions are emitted for longer times. Double charged ions, such as Ca^{+2} and Ba^{+2} from their chloride salts, are not easily observed. Instead Ba^{+1} and $BaCl^+$ are observed, and show strong currents less than a microsecond after the laser pulse.

3.2 Quaternary ammonium ions

For a number of reasons, ionization potentials (reflecting a Langmuir-Saha type of thermal behavior) do not represent the best interpretation of desorption phenomena. The Langmuir-Saha equation describes ionization from neutral atoms in an equilibrium situation rather than ions from their salts; and, even if it can be imagined that such a neutral/ion equilibrium is in fact established, it is difficult to describe a corresponding neutral precursor for a quaternary ammonium ion (e.g., $(CH_3)_4N^+$). Lattice energies have been used to predict the desorption of quaternary ammonium ions at lower temperatures than either Na^+ or K^+ ions in thermal desorption experiments [12]. The time-resolved laser desorption results confirm that result, since quaternary ammonium ions are emitted for up to 30 μS after the laser pulse, when the probe tip has cooled down sufficiently that Na^+ and K^+ are no longer observed.

Using the electron beam in conjunction with the laser pulse produces some interesting results. First, the abundance of molecular ions, $(CH_3)_4N^+$, is greatly increased, suggesting that neutral precursors for this ion, such as $[(CH_3)_4N^+Cl^-]_n$ as described by DAVES et al. in flash desorption experiments [13] are abundant in laser desorption. Secondly, the ions $(CH_3)_3N^+$, CH_3Cl^+ and CH_3^+ resulting from electron impact ionization of the neutral products of the thermal (pyrolytic) degradation reaction:

$$(CH_3)_4N^+Cl^- \longrightarrow (CH_3)_3N + CH_3Cl$$

are observed for hundreds of microseconds following the laser pulse, and are clearly thermal effects because of their long emission times.

3.3 Cationization

If cationization proceeds as an ion-molecule reaction of co-desorbed alkali ions and intact neutral molecules [5], then it would seem that the process would be less favorable in the "open" sources generally used on time-of-flight mass spectrometer. This seems to be the case. The time resolved results for a mixture of LiCl and palmitic acid are presented in Figure 3, and show that Li^+ ions and neutral palmitic acid reach maximum desorption at very different times. The intermediate appearance of (palmitic acid + Li)$^+$ may correspond to the optimal co-desorption point.

4. Conclusions

There are several observations and comments that can be made about the mechanisms involved in laser desorption. The first involves the causes for the time resolution. The delay between the laser pulse and maximum ion emission has several origins. Using a thermal interpretation [14], the maximum substrate temperature of the probe (in this case, made of Vespel) may be reached rather quickly, while dissipation of the heat to the organic layer may persist for hundred of microseconds to produce the results observed. This is the ideal case for a thermal model, since the heat dissipation results in lower temperatures and differentiates processes requiring different energies. However, the 40 nS laser pulse from the Tachisto laser has an elongated (400 nS-1µS) tail, so that energy is actually pumped into the sample for a somewhat longer period. In addition, the ion residence time in the source, which may be of the order of a few microseconds must be taken into account. This means that a direct (non thermal) mechanism for ions formed during the first few microseconds cannot be entirely discounted. On the other hand, all the types of ions produced have been produced also by purely thermal means [12,15], and the long desorption times of neutral species point clearly to their thermal origin.

Lattice energies represent a convenient way of predicting the relative stability of singly charged ions removed from the solid phase, but do not describe a mechanism, since deposition of samples from solutions may result in solvation and because such thermodynamic predictions assume equilibrium conditions. Thermodynamic stability of preformed ions cannot give a complete picture of the ions which will be detected. The fact that preformed doubly charged ions are not easily observed in most of the

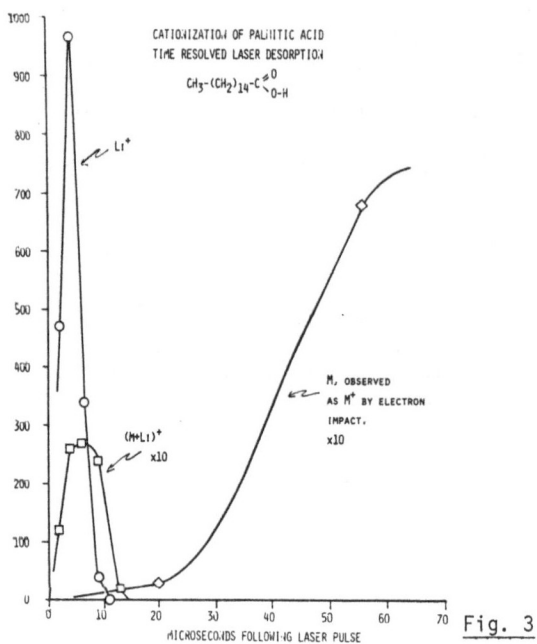

Fig. 3

desorption methods, except field desorption [16], suggests that there is a considerable activation energy involved in their removal from the surface.

Time resolution however does add a dimension to the observation of desorption phenomena, which can distinguish between processes which proceed at different rates. In addition the observation of abundant neutral species, which may be present in other desorption methods and have not yet been exploited, is also emphasized.

ACKNOWLEDGEMENT

This work was supported by grant CHE 80-16440 from the National Science Foundation.

REFERENCES

1. R. J. Conzemius and H. J. Svec, Anal. Chem. 50, 1854 (1978).
2. M. A. Posthumus, N. M. Nibbering, A. J. H. Boerboom and H.-R. Schulten Biomed. Mass Spectrom. 1, 352 (1974).
3. M. A. Posthumus, P. G. Kistemaker, H. L. C. Meuzelaar and M. C. Ten Noever de Brauw, Anal. Chem. 50, 985 (1978).
4. R. Stoll and F. W. Rollgen, Org. Mass Spectrom. 14, 642 (1979).
5. G. J. Q. Von der Peyl, K. Isa, J. Haverkamp and P. G. Kistemaker, Org. Mass Spectrom. 16, 416 (1981).
6. R. J. Cotter, Anal. Chem. 52, 1767 (1980).
7. R. J. Cotter, Anal. Chem. 53, 719 (1981).
8. D. Zackett, A. E. Schoen, P. H. Hemberger and R. G. Cooks, J. Amer. Chem. Soc. 103, 1295 (1981).
9. F. Heresch, E. R. Schmid and J. F. K. Huber, Anal. Chem. 52, 1803 (1980).
10. F. J. Vastola and A. J. Pirone, Adv. Mass Spectrom. 4, 107 (1968).
11. R. B. van Breemen, M. Snow and R. J. Cotter, Int. J. Mass Spectrom. Ion Phys.. in press.
12. R. J. Cotter and A. L. Yergey, J. Am. Chem. Soc. 103, 1596 (1981).
13. T. D. Lee, W. R. Anderson, Jr. and G. D. Daves, Jr., Anal. Chem. 53, 304 (1981).
14. G. J. Q. van der Peyl, J. Haverkamp and P. G. Kistemaker, Int. J. Mass Spectrom. Ion Phys. 42, 125 (1982).
15. R. J. Cotter and A. L. Yergey, Anal. Chem. 53, 1306 (1981).
16. R. J. Cotter, R. van Breemen, D. Heller and J. Yergey, 6th International Conference on Mass Spectrometry, Vienna, 1982.

4.3 New Developments in Laser Pulse Induced Field Desorption

J.H. Block, W. Drachsel, N. Ernst, Th. Jentsch, and S. Nishigaki

Fritz-Haber-Institut der Max-Planck-Gesellschaft, Faradayweg 4-6,
D-1000 Berlin 33, Fed. Rep. of Germany

1.   Introduction

The formation and desorption of ions from solid surfaces can be
achieved by field desorption (FD)[1][2]. Laser photons enhance
the desorption rate due to thermal or electronic excitation[3].
The combination of a field ion microscope (FIM) with a pulsed
laser photon source is used to perform time-correlated time-of-
flight (TOF) measurements. This leads to a new analytical tool
for investigating surface composition and surface reactivity
with an ultimate lateral resolution in the atomic scale [4][5].
Recent experimental improvements [6] are discussed which allow
quantitative measurements. Reactions of hydrogen [7] in adsorption layers are investigated. An attempt is made to understand the generation of multiply charged ions and of cluster
ions [8][9].

2.   Experimental

Laser photon induced FD mass spectrometry is performed in a
FIM-TOF-spectrometer, where FD of admolecules is induced by
laser photon impact, occasionally by possible quantum effects
[3], and generally by the resulting temperature pulse [5].

2.1. The Laser Source

A N_2-laser or a pumped dye laser with an energy density of
~ 0.1 to 1 J/pulse cm^2, a pulse width of 5 ns and repetition
rate of $\lessgtr$ 50 cps is used; after frequency doubling a pulse
width of 2.5 ns could be achieved.

2.2. The Field Ion Microscope

Figure 1 shows the schematic diagram of the TOF-spectrometer.
The specimen (tip) is spotwelded to a heating loop, connected
to W-rods through a glass cooling finger. The tip temperature
is measured by a thermocouple and can be scanned by an electronic programmer. The tip position is adjustable, its voltage
can be programmed (stair case scans). The einzel lens in front
of the tip is simultaneously used for focussing, as counter
electrode and as gas supply (nozzle). Field ions are detected
by a dual channelplate and phosphor screen. For field ion imaging the assembly is shifted to a position 5 cm away from the

tip, in the TOF-mode it is 30 cm away, still displaying part of the field ionization (FI) image. The anode is divided into an outer part and an inner probe area for TOF-measurements.

2.3. The TOF-Measuring Device

The conditioned signals of the photo-cell and from the probed area start and sequentially stop the five fast 15 bit counters in a selected time window so that up to five different masses can be monitored within the 10 ns pulse pair resolution. A voltage controlled oscillator maintains the clock frequency (500-68 MHz), proportional to the square root of the applied tip voltage (within an error of 10^{-3}%). Thus, mass lines will occupy the same channel numbers in the TOF-spectra independent of the applied voltages.

2.4. The Data Handling Device

The use of an LSI-11 computer for data aquisition, storing and evaluation provides much more flexibility than a conventional multichannel analyzer. In our design (Fig.1), with each laser pulse the TOF-data, the total ion counts between pulses, the actual tip voltage, tip temperature and laser intensity are fed into the computer and after data compression are stored in time sequence on a floppy disk. No information is lost from the experiment. A reduced TOF-spectrum is displayed on line.

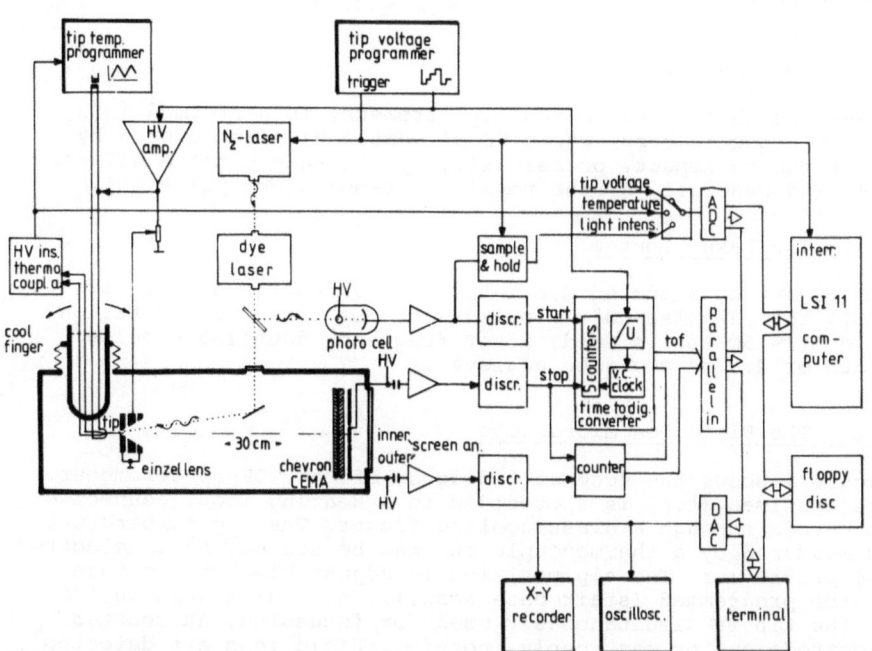

Fig.1. Scheme of the time-of-flight-spectrometer

3. Surface Reactions of Hydrogen

Since the sampling time with the laser induced FD is so short (5 ns), the species measured in the TOF-spectra originate only from the probed surface and definitely not from impinging gas molecules. In Fig.2 the field dependence of H^+-, H_2^+- and H_3^+-intensities is shown for the case of laser-induced FD (Fig.2a) and for normal FI (Fig.2b). In normal FI an energy filter selected only those ions which originated in a zone of $\sim$ 1 nm depth above the surface.

With increasing field strength (Fig.2a) H_2^+-ions are formed above 18 V/nm and reach an intensity maximum at 24 V/nm. H^+ and H_3^+ appear at increased field strengths. The H^+-intensity maximum coincides with the steepest slope of the decreasing H_2^+ abundance. Above 30 V/nm all ion intensities of surface species vanish since the surface is depleted of adsorbed layers due to complete ionization of impinging molecules. The total ion current (dotted line fig.2a), representing the sum of normal FI

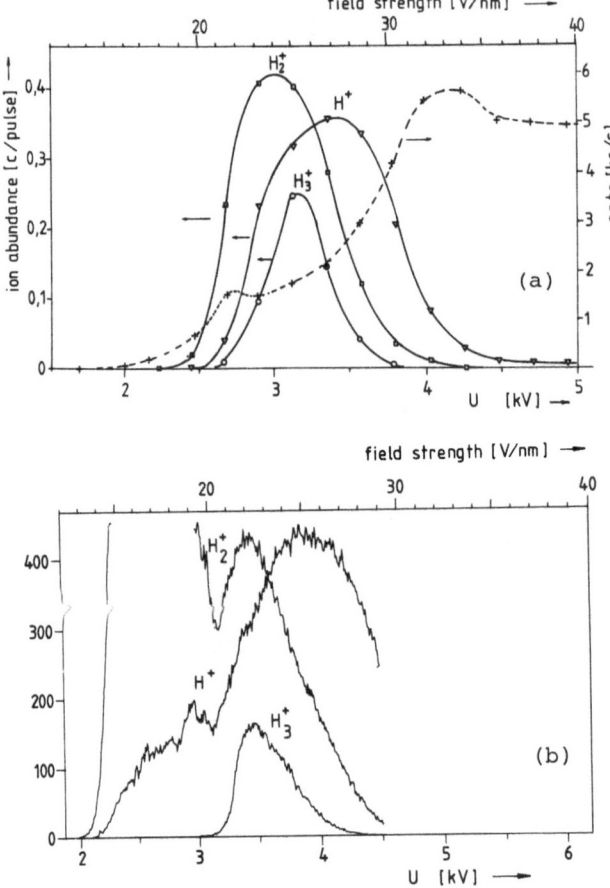

Fig.2. Field dependence of hydrogen ion yield, tungsten tip T_{tip} = 78 K, P_{H_2} = 5 · 10^{-5} Torr

(a) Laser-induced FD. Dashed line total ion current of FD and FI. F-calibrated by BIV of H_2= 22 V/nm

(b) Normal FI and FD. With the retarder voltage δ = - 30 V only a zone of $\sim$1 nm above critical distance is sampled. H_2^+-maximum reaches 1000 c/s and photon-induced desorption shows saturation above 35 V/nm, again, indicating total ionization of impinging molecules in the monitored area.

For normal FI (Fig.2b) at $F > 21$ V/nm H^+-, H_2^+- and H_3^+-intensities resemble those of fig.2a. At $F < 20$ V/nm an extended maximum of H_2^+- and a measurable H^+-intensity are obtained, but no H_3^+-ions. Two regimes are measured in normal FI: At low fields gas phase FI leads to H_2^+ and by fragmentation to H^+. At increased field gas phase FI occurs beyond the detected zone of ~ 1 nm above the surface. The detected intensities at $F > 21$ V/nm are generated by FD of surface species and are therefore comparable with those in fig.2a.

The integral energy distribution displays a continous intensity increase for H^+, a low-energy tail for H_2^+ and a sharp onset signal for H_3^+ [10]. The temperature dependence of intensities shows a decrease with T. Above 350 K H_3^+ intensities vanish. Below 150 K an H/D isotope effect in H_3^+-formation is observed [10][11].

In conclusion, H_2^+ is formed by FI and FD, H^+ is a dissociation product of H_2^+, the FD of atomic H is still not confirmed. H_3^+ is shown to be a surface product, most likely formed from field adsorbed molecular hydrogen and chemisorbed H atoms.

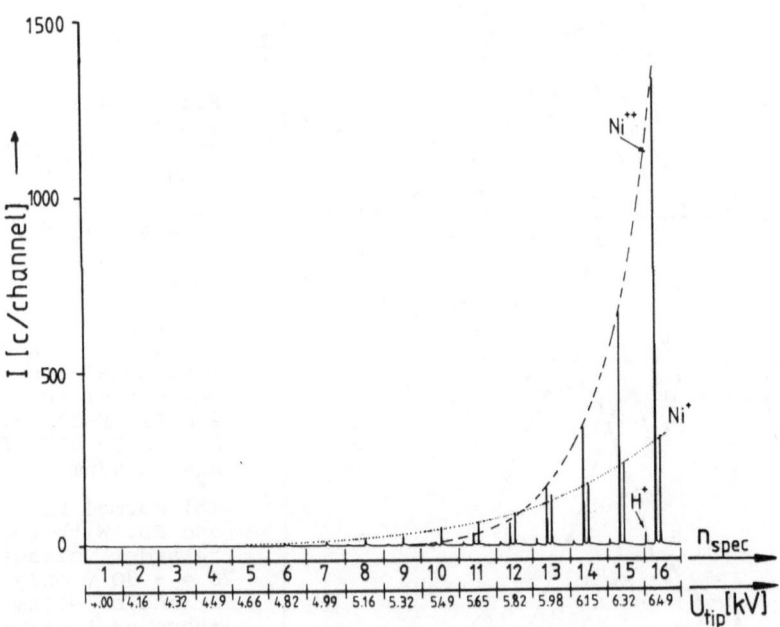

Fig.3. Evaporation of nickel, 16 TOF-spectra with increasing tip voltage at constant laser power, $\lambda = 337$ nm, $T_b = 80$ K, 1 kV $\triangleq$ 4 V/nm

4.   Multiply Charged Metal Ions and Ion Clusters

We can distinguish between two regimes of ion formation: I. At low temperatures and high field strength, singly and multiply charged metal ions are field evaporated. II. At high temperatures and relatively low field strength values the evaporation of metal cluster ions is observed.

As an example for regime I   photon-induced field evaporation at a moderate laser photon density ($\sim$ 0.2 J/pulse cm^2) near the onset field of field evaporation is demonstrated for nickel in fig.3. A sequence of 16 TOF-spectra with increasing tip voltage is presented. The Ni^{2+}/Ni^{1+} intensity ratio increases with the tip voltage, whereas the ratio is independent of the resulting pulse temperature of the tip surface [9]. This behavior can be explained by post-FI of primarily formed Ni^{1+} ions [12][13]. The sudden temperature jump at the tip surface is estimated to be 100-500 K with 80 K tip base temperature. The TOF-peaks during these experiments are as sharp as the laser pulses of 5 ns. Other metals, like Cu, Au or Ag, display a similar behavior.

The regime II is established by lowering the field strength by 30 to 50 % and by applying higher laser power ($\sim$1 J/pulse cm^2). In this mode the PIFI intensities jump to 10^5 ions/pulse. That means a current of $\sim$1 μA during the pulse. The estimated temperature increase amounts to 1000-3000 K, i.e. the melting point of the emitter metal is reached.

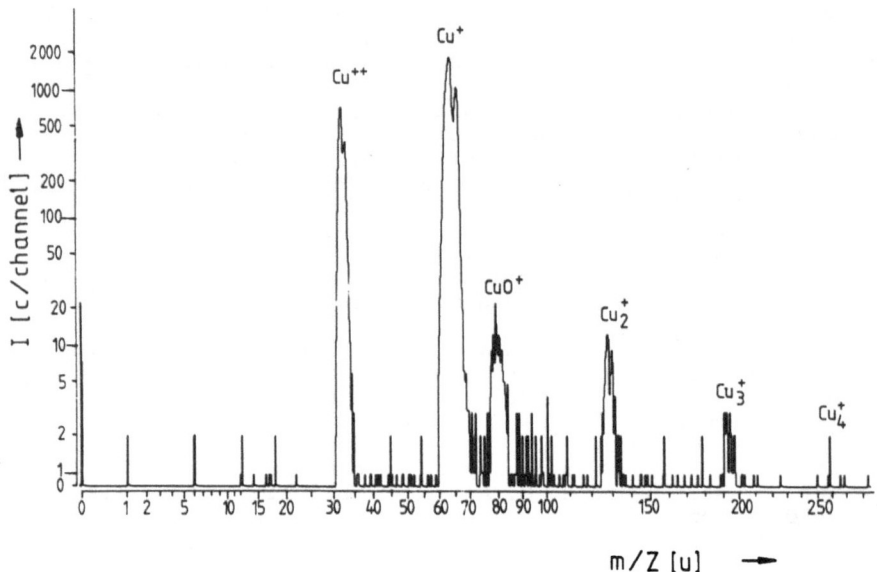

Fig.4. TOF-spectrum of a copper tip, $p_{res} = 10^{-8}$ Torr, $U_{tip} = 2$ kV, (F 8 V/nm), $T_b = 300$ K, $\lambda = 337$ nm

The TOF-spectra now display cluster ions Me_n^+ in addition to the Me^+ and Me^{2+} species, as shown for a copper tip in fig.4 [14]. At these low fields Cu^{2+} ions are not expected by the post FI model. A support for the formation of Cu^{2+} by collisions is found in the strong increase of the Cu^{2+}/Cu^+ intensity ratio with the applied laser power in contrast to the findings in regime I. Another phenomenon emerges with increasing laser power: in front of the Me^+ and Me^{2+} lines there are new peaks appearing,which can only be explained as highly energetic species. This effect is most clearly seen for a monoisotopic metal like gold. The surplus energy of several hundred eV cannot be understood in terms of the present FI models. The fast build-up of a high charge density within the short laser pulse is probably responsible for the energy transfer. The experiments show that the existence of the super-fast Au^+ and Au_2^{3+}, and especially the 4.6 μs-peak (probably Au_2^{3+} with $\Delta E \approx +$ 660 eV) is correlated with high current bursts (induced by high laser intensity). However, the actual mechanism of energy transfer is still obscure.

5.   Acknowledgement

This work was supported by the Deutsche Forschungsgemeinschaft (SFB 6/'81).

References

1   H. D. Beckey in "Principles of Field Ionization and Field Desorption Mass Spectrometry", Pergamon Press, Oxford 1971
2   F. W. Röllgen in "Field Desorption"(Review), this volume
3   S. Nishigaki, W. Drachsel and J.H. Block, Surf. Sci. 87 (1979) 389
4   W. Drachsel, S. Nishigaki and J. H. Block, Int. J. Mass Spectrom. Ion Phys. 32 (1980) 333
5   G. L. Kellogg, J. Appl. Phys. 52 (1981) 5320
6   W. Drachsel, Th. Jentsch and J. H. Block,   Int. J. Mass Spectrom. Ion Phys. 9 (1982) in press
7   W. Drachsel, S. Nishigaki, N. Ernst and J. H. Block, Int. J. Mass Spectrom. Ion Phys. 9 (1982) in press
8   Th. Jentsch in "Feldverdampfung und -desorption unter dem Einfluß von Laserbestrahlung", Dissertation, F.U. Berlin (1982)
9   W. Drachsel, Th. Jentsch and J. H. Block, Proc. 29th Int. Field Emission Symp., Stockholm (Sweden)(1982), eds. H. Nordén and H. O. Andrén, publ. Almquvist & Wiksell International
10   N. Ernst, G. Bozdech, S. H. Allam and J. H. Block, Proc. 29th Int. Field Emission Symp., Stockholm (Sweden)(1982) eds. H. Nordén and H. O. Andrén, publ. Almquvist & Wiksell International
11   N. Ernst and J. H. Block, Surf. Sci., ECOSS V (1982), submitted
12   N. Ernst, Surf. Sci. 87 (1979) 469
13   Th. Jentsch, W. Drachsel and J. H. Block, to be published
14   Th. Jentsch, W. Drachsel and J. H. Block, Int. Mass Spectrom. Ion Phys. 38 (1981) 215

4.4 Thermal Processes in Repetitive Laser Desorption Mass Spectrometry

F. Heresch

Institute for Analytical Chemistry, University of Vienna,
A-1090 Vienna, Austria

Introduction

The laser-induced desorption of organic solids has been shown
[1-4] to be essentially a rapid heating effect resulting in the
flush evaporation of neutrals and ions which subsequently can
undergo gas phase molecule - molecule/ion reactions, e.g.
cationization. Although further types of irradiation interaction
cannot be excluded, thermal effects must be assumed to play a
major role. A recent investigation on supermolecular artefacts
observed with repetitive laser desorption (RLD) [5]   indicated
that saccharides, due to their thermolability, may serve as a
very sensitive probe for such thermal effects. As is known from
carbohydrate chemistry [6] , thermal degradation of saccharides
in general occurs via formation of anhydrosugars, e.g. glucosans,
which - among other reactions - strongly tend to polymerize. The
generation of higher oligosaccharides/glycosides (by successive
addition of anhydrosugar units) should thus be **particularly** indi-
cative of thermal processes, regardless of the desorption
technique applied. In order to investigate this aspect in
more detail, the temperature dependence of the formation of
such artefacts in RLD-experiments was studied.

Experimental

All measurements were made with the RLD technique [7] , i.e.   in
multiple pulse mode using an infrared-laser (Nd-YAG) and a
scanning-type double focusing mass spectrometer. For details see
[7] . Samples were deposited from solution (water, pyridin) onto
a glass substrate to give a bulky layer. In order to provide repetitive and reproducible ion formation from a single sample spot,
the layer was indented as described previously[7].The temperature
of the sample holder which could not be independently heated was
regulated via that of the source. There is evidence [5] that
during measurements, i.e. as a result of repetitive irradiation
(1o pulses/s), a temperature gradient builds up within the
non-conducting sample layer. The temperature near the irradiated
region is estimated to be 1oo to 2oo$^{\circ}$C above the ambient (=source)
temperature, depending on the period of laser irradiation.

Results and Discussion

As reported recently [5] , sucrose forms a series of artefacts
of mass (342 + n·162) and (162 + n·162), n having been observed
up to **7** at a slightly elevated temperature (8o$^{\circ}$C). The relative

intensities of the sodiated and potassiated species are given in Table 1.

Table 1. Relative Intensities of Sucrose Adducts $(M + n \cdot 162)$ C^+

M	C^+	n=0	1	2	3	4	5	6	7
342	Na	<u>1oo</u>	35	17	1	o	o	o	o
162		82	1o9	36	12	6	2	1	o
342	K	<u>1oo</u>	124	65	22	5	3	o	o
162		96	1oo	13o	45	3o	15	9	2

Note: M = basic unit [dalton] , C^+ = attached cation

Fragments resulting from ring ruptures [5] suggest that these products are higher oligosaccharides formed by (linear) addition of an anhydrohexose unit, presumably 1,6-anhydrohexopyranose [6]. Similar observations could be made when glycosidic compounds were mixed to the sucrose as shown in Fig. 1 for an unknown digitalis diglycoside (mol.wt. 742).

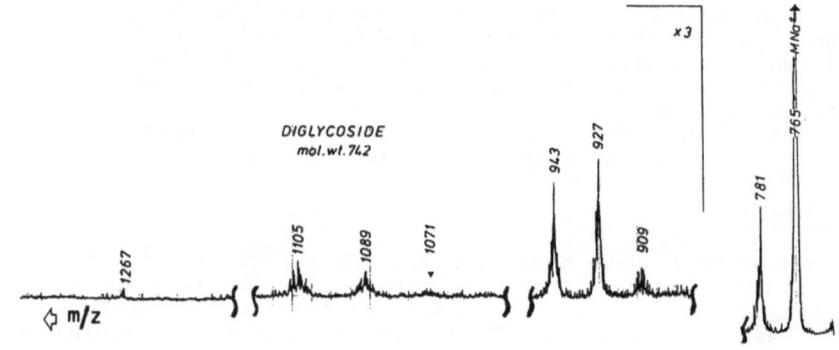

Fig. 1. Supermolecular adducts (+ n·162) of digitalis diglycoside.
Source temperature 5o°C

Effects of temperature

The dependence of artefact formation on temperature was studied by recording partial mass spectra (see Fig. 1) repetitively during a temperature cycle 2o .....8o.....2o°C.

As to be seen in Fig. 2, the concentration of generated higher saccharides/glycosides essentially remains at the highest level achieved so far. Obviously this level cannot be decreased by

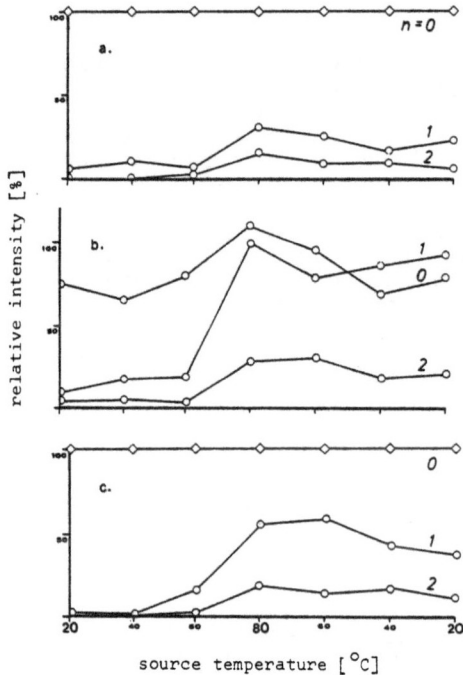

Fig. 2. Artefact ion profiles for temperature cycle.
Ions: a.) sucrose $(342 + n \cdot 162)$ Na^+,
b.) sucrose $(162 + n \cdot 162)$ Na^+,
c.) diglycoside $(742 + n \cdot 162)$ Na^+
Intensities: a) and b) [%] of $(342+Na)^+$-Signal $(\Diamond)$
c) [%] of $(742+Na)^+$-Signal $(\Diamond)$

reducing the temperature. Such results may only be explained by
assuming that the sample itself is irreversibly changed during the
experiment. It must be added that the initial artefact intensity could only be reproduced when changing the irradiated area
(by focusing the laser onto a fresh sample spot). All this strongly
indicates that the reactions involved in the artefact formation,
are thermally induced by both the probe/source heating and the
laser heating.

It should be noted that the intensity distribution of the artefact species appears to depend on other factors, too. As indicated
by the data in Tab. 1, the distribution at a given temperature is
shifted to higher adducts (higher n) when going for instance from
Na^+ to K^+ containing series. This becomes evident by a change in
n for both the most abundant and the highest observable species
(see Tab. 1). The details of these phenomena are still under investigation. With regard to the observed influence of the attached
cation, a steric effect such as the size-dependent formation of
chelates might have to be considered.

Anhydro-saccharides

As to the (162 + n·162)-dalton species, there are two possible
pathways of formation, i) loss of water from the corresponding
oligosaccharide, and ii) polymerization of the anhydrohexose.
From the available data (see Fig. 2b) it cannot be concluded as
to what extent these processes contribute to the observed abundances.

No significant effect of temperature is observed in case of
the monomeric anhydrohexose (n=o) itself. To understand this, it
has to be taken into account that anhydrohexose species are also
formed via unimolecular gas-phase decomposition of sucrose.
Fig. 3 shows the (2ndFFR) metastables for this fragmentation, i.e.
the formation of m/z 185 (=162+Na) from m/z 365 (=MNa) and 347
(=M-H_2O+Na), respectively. (Note: The ripple on top of peaks
results from the pulsed ion formation in RLD.)

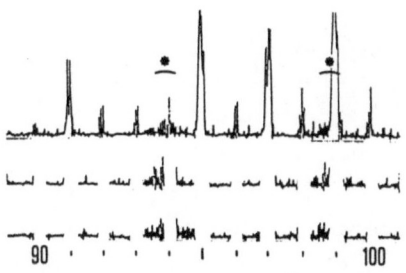

Fig. 3. Partial RLD-Spectrum
of sucrose with metastable
signals (●) at m/z 93.8 and
98.7. (Three recordings, the
lower two with peaks being
omitted.)

This interglycosidic cleavage which was also proven by MS/MS [8] ,
can be expected to be governed mainly by the energy transferred
during desorption and ion attachment, and it should therefore be
rather independent on slight temperature changes. It may also be
assumed that free anhydrohexoses generated in the sample do not
accumulate because of their reactivity under measurement conditions. The observed ion profile, in conclusion, seems to be in
agreement with the proposed thermal degradation mechanism.

Summary

Evidence has been obtained that in case of saccharides thermal
degradation effects are induced by (repetitive) laser irradiation.
The formation of higher oligosaccharide artefacts is shown to be
an irreversible process most likely **occurring in the sample layer.**
The generated species appear to be well suited to be used as a
sensitive probe for thermal effects with desorption techniques,
in general.

Acknowledgement

The mass spectrometer was made available by the Austrian "Fonds zur Förderung der wissenschaftlichen Forschung". Glycoside samples were kindly supplied by Prof.W.Kubelka, Pharmacognostic Institute, University of Vienna.

References

1   G.J.Q. van der Peyl, K. Isa, J. Haverkamp, P.G. Kistemaker: Org. Mass Spectrom. 16, 416 (1981).
2   G.J.Q. van der Peyl, J. Haverkamp, P.G. Kistemaker: Int. J. Mass Spectrom. Ion Phys. 42, 125 (1982).
3   R. Stoll, F.W. Röllgen: Z. Naturforsch. 37a, 9 (1982).
4   R.J. Cotter, A.L. Yergey: Anal. Chem. 53, 13o7 (1981).
5   F. Heresch: Adv. Mass Spectrom. 9, in press.
6   M. Cerny, H. Stanek: In Advances in Carbohydrate Chemistry and Biochemistry, vol. 34, ed. by R.S. Tipson, D. Horton (Academic Press, New York 1977) p. 24.
7   F. Heresch, E.R. Schmid, J.F.K. Huber: Anal. Chem. 52, 18o3 (198o).
8   D. Zakett, A.E. Schoen, R.G. Cooks: presented at the 29th Annual Conference on Mass Spectrometry and Allied Topics; Minneapolis, MN, 1981; p. 27.

4.5 Laser Mass Spectrometry of Organic Compounds

D.M. Hercules, C.D. Parker, K. Balasanmugam, and S.K. Viswanadham

Department of Chemistry, University of Pittsburgh, Pittsburgh, PA 15260, USA

1. Introduction

The present communication presents preliminary results of laser mass spectrometry on four systems. First, we report on production of quasi-molecular cations and anions in amino acids. Second, we have studied methyl group transfer reactions for amino carboxylates indicating involvement of a thermal process. Third, we have studied the negative ion spectra of aromatic hydrocarbons and nitro compounds. Organic nitro-compounds cause substitution reactions via a chemical ionization solid — state source. Fourth, we have studied fragmentation patterns for amino acids contrasted with SIMS and low-energy electron impact.

The laser mass spectra in this study were obtained using a Leybold-Heraeus LAMMA-500^R laser mass spectrometer. Samples were dissolved in an appropriate solvent and evaporated on a Formvar filmed grid to give a thin layer. A detailed description of the instrument can be found elsewhere [1].

2. Pair Production in Amino Acids

High yields of positive and negative quasi-molecular ions, $(M+H)^+$ and $(M-H)^-$ respectively, are observed in the laser mass spectra of organic acids [2]. The yields of $(M+H)^+$ are particularly high for the α-amino acids. Two possibilities for the formation of $(M+H)^+$ are direct proton transfer from another acid molecule, or reaction with a proton donor in the laser plasma. We have been interested in the production of ion pairs in laser MS; earlier we proposed the process of "pair production" to account for ions observed in the LDMS spectra of polymers. We report here experiments which indicate that the concept of pair production can be extended to ionization processes in amino acids. Specifically, we have demonstrated that ionization results from simple intermolecular proton transfer reactions. Similar results have been found in plasma desorption mass spectrometry [3].

Amino acids are known to exist in the solid state as head-to-head dimers. Thus, intermolecular proton transfer induced by the laser can account for ionization in these materials. The overall process can be written as:

LAMMA-500^R is a registered trademark of Leybold-Heraeus, GmbH

$$R-CH\overset{NH_3^+---^-OOC}{\underset{COO^---^+H_3N}{\diagdown}}CH-R \xrightarrow{h\nu} RCH\overset{NH_3^+}{\underset{COOH}{\diagdown}} + RCH\overset{NH_2}{\underset{COO^-}{\diagdown}} \quad . \tag{1}$$

This reaction is referred to as "pair production" because ionization involves transfer of a single proton between two molecules to produce one positive ion and one negative ion.

We have carried out a study in which the labile protons of the amino acids have been exchanged for deuterium. Figure 1 shows the positive LDMS spectra of valine (VAL) and deuterium-exchanged valine (VAL-D). The quasi-molecular peaks occur at m/z 118 for VAL and 122 for VAL-D. Also, note the absence of a significant peak at m/z 121 for VAL-D.

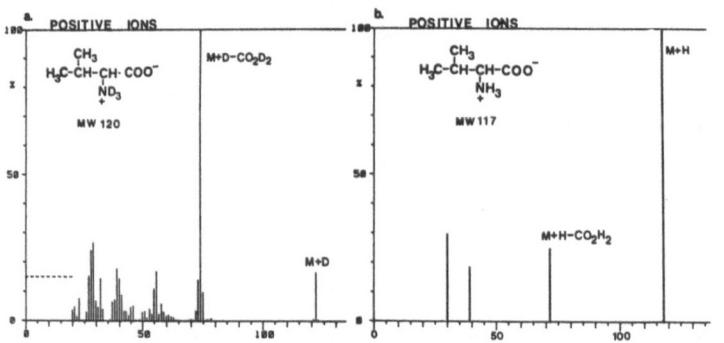

Figure 1. Laser Mass Spectra of a) Deuterated Valine, b) Valine

The above experiment proves conclusively that the protons involved in forming (M+H)$^+$ come from the amino group of VAL and not from the aliphatic chain. This indicates high likelihood that pair production is the major mechanism for ionization of the acid. Similar results have been obtained for other amino acids and simple carboxylic acids.

One could argue that the labile amino protons are involved in plasma protonation reactions. However, at high laser power when H$^+$ and D$^+$ are clearly evident, only (M+D)$^+$ is observed for VAL-D. Ionization could occur via an intermediate species such as an eximer or by volatilization of an ion pair. It is clear, however, that statistical theory [4] supports ion pair formation in laser processes near threshold.

3. Methyl Group Transfer in Amine Carboxylates

Field desorption mass spectrometry (FDMS) has indicated that an alkyl transfer reaction produces a peak at (M+15) (M = intact molecule) in the positive ion FD spectra of quaternary amine carboxylates [5]. This system constitutes another example of possible pair production in laser mass spectrometry. The LD spectra of a series of amino hexanoates are summarized in Table 1.

The peak corresponding to (M+CH$_3$)$^+$ is evident in the positive ion spectra of the four examples studied. (M+H)$^+$ is very abundant in all spectra. Common fragment ion peaks in the positive spectra correspond to loss of CO$_2$ from both (M+H)$^+$ and (M+CH$_3$)$^+$. Loss of the C$_5$ acid chain

Table 1. Laser desorption mass spectra of quaternary amine carboxylates (1)

Positive-Ion Spectrum

Compound	M+CH$_3$[a]	M+H[a]	M+CH$_3$-CO$_2$[a]	M+H-CO$_2$[a]	M+H-C$_5$H$_{11}$COOH[a]	m/z=114
R = CH$_3$	94(188)	39(174)	7(144)	46(130)	100(58)	7
R = C$_8$H$_{17}$	11(186)	44(272)	11(144)	70(118)	44(156)	12
R = C$_{12}$H$_{25}$	17(342)	28(328)	6(298)	38(284)	6(212)	22
R = C$_{24}$H$_{49}$	9(510)	28(496)	3(466)	69(452)	4(380)	34

Negative-Ion Spectrum

Compound	M-CH$_3$	m/z=158	m/z=142	m/z=113	m/z=97	m/z=59	m/z=45
R=CH$_3$	61(158)	61	5	56	100	87	89
R=C$_8$H$_{17}$	52(256)	82	41	100	100	*	*
R=C$_{12}$H$_{25}$	43(312)	34	11	50	86	100	100
R=C$_{24}$H$_{49}$	29(400)	63	17	31	87	26	100

*Not measured in this range

[a]Numbers given are relative intensities with the base peak as 100.
Numbers in parentheses are mass numbers.

accounts for ion 2. A peak at m/z 158 is observed corresponding to loss of RH. The m/z 114 peak is accounted for by loss of CO$_2$ from the m/z 158 peak. The base peak in all spectra is m/z 58 (3). These results correlate with the FD spectra in which alkyl transfer and loss of CO$_2$ from (M+H)$^+$ are observed.

$$R-\overset{\overset{\displaystyle CH_3}{|}}{\underset{\underset{\displaystyle CH_3}{|}}{N}}{}^{\pm}-C_5H_{11}COO^-$$

1

$$R-\overset{\overset{\displaystyle CH_3}{|}}{N}=CH_2$$

2

$$CH_3-\overset{\overset{\displaystyle CH_3}{\diagup}}{\underset{\underset{\displaystyle CH_2}{||}}{N}}{}^+$$

3

$$(CH_3)_2N-C_5H_{10}COO^-$$

4

The highest mass peaks detected in the negative ion LD spectra correspond to (M-CH$_3$)$^-$; all compounds studied show this peak as seen in Table 1. This result coupled with the positive ion spectra provides conclusive evidence for pair production. The peak at m/z 158 (4)is strong in all negative ion LD spectra. The lower mass range of negative ion spectra are nearly identical for all compounds; all ions can be considered as fragments from m/z = 158. These ions are shown in structures 5 - 7 and their relative intensities are shown at the bottom of Table 1. Peaks at m/z = 59 and 45 correspond to acetate and formate ions.

Similar behavior has been observed for a series of zwitterionic compounds known as sultanes (8). Both (M+CH$_3$)$^+$ and (M-CH$_3$)$^-$ are observed in the positive and negative ion spectra respectively, confirming pair production in the sultanes. The fragmentation observed is comparable to the amine carboxylates in both the negative and positive-ion spectra.

$$CH_2=N-C_5H_{10}COO^-$$

5

$$CH_2=CH(CH_2)_3COO^-$$

6

$$CH_2=CH-CH=CHCOO^-$$

7

$$R-\overset{\overset{\displaystyle CH_3}{|}}{\underset{\underset{\displaystyle CH_3}{|}}{N}}{}^{\pm}-(CH_2)_3SO_3^-$$

8

4. Negative Ion Spectra of Hydrocarbons and Nitrocompounds

Formation of odd-electron molecular ions, although uncommon in laser mass spectra, is observed in the positive ion spectra of aromatic hydrocarbons. Compounds containing suitable electron withdrawing groups yield odd-electron molecular anions by electron capture, a case in point is nitrobenzene derivatives studied by EI negative ion mass spectra. In order to determine the extent of molecular anion formation in LDMS, the negative ion mass spectra of aromatic hydrocarbons and nitrocompounds have been investigated.

The formation of odd-electron molecular ions is not observed for aromatic hydrocarbons; a series of cluster ions is observed corresponding to C_n^- and C_nH^-. Interestingly the maximum n-values for the carbon clusters correspond to the number of carbon atoms in the aromatic molecule. The common fragments observed in the negative ion mass spectra of nitrocompounds are CN^-, OCN^-, C_3N^-, and $(M-NO)^-$. In addition, a very intense peak at $(M+15)^-$ was observed for o-dinitrobenzene, 1,8-dinitronaphthalene and 1,3,5-trinitrobenzene. The $(M+15)^-$ ion is the base peak in o-dinitrobenzene.

Formation of the peak at $(M+15)^-$ probably results from a nucleophilic substitution reaction in the laser plasma. One possible mechanism is attack of the aromatic ring by NO_2^- with subsequent elimination of HNO:

$$ArH + NO_2^- \longrightarrow \left[Ar \diagdown^{\nearrow H}_{\searrow NO_2} \right]^- \longrightarrow ArO^- + HNO \ . \qquad (2)$$

It is interesting that the $(M+15)^-$ peak is negligible in the case of m- and p-dinitrobenzenes. Steric interactions between the nitro groups in o-dinitrobenzene and 1,8-dinitronaphthalene obviously aid formation of NO_2^- in the plasma. However, proximity of the two nitro groups is not essential since 1,3,5-trinitrobenzene also produces a high level of $(M+15)^-$.

Formation of adduct ions in the laser plasma from ion molecule reactions constitutes a new kind of chemical ionization source. By running organic compounds which produce adduct ions, one can cause chemical ionization of thermally labile and nonvolatile compounds. The potential for such sources is immense because of the large number of possible adduct ions which can be formed in the laser plasma.

5. Laser Mass Spectra of Amino Acids

Although there have been some LDMS studies of amino acids [6], no detailed evaluation of the fragmentation patterns has been reported. We have begun such a study and report preliminary results here.

Simple, aliphatic amino acids show three major peaks, $(M+H)^+$, $(M+H-formic\ acid)^+$, and $(M-H)^-$. The spectra are free of spurious peaks at low laser power. The fragmentation pattern for these acids is shown in Figure 2.

The aromatic amino acids show features comparable to the aliphatic acids; the nature of the aromatic group can modify fragmentation significantly. Fragmentation of phenylalanine, shown in Figure 3, is

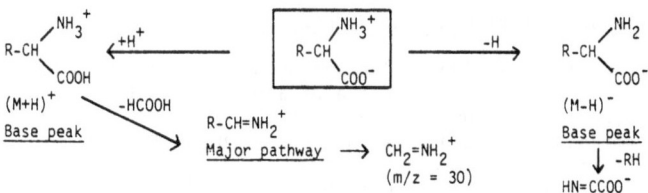

Figure 2. Fragmentation Pattern for Aliphatic Amino Acids

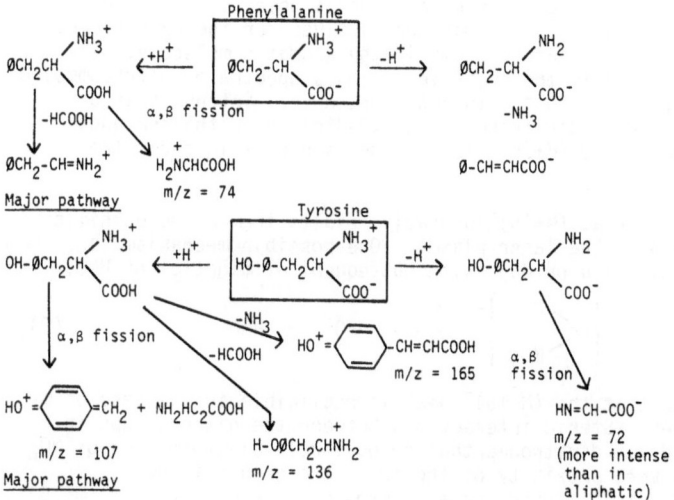

Figure 3. Fragmentation Patterns for Aromatic Amino Acids.

comparable to the aliphatic acids, except that α,β fission and NH₃ loss are more prominent. Figure 3 also shows the fragmentation pattern for tyrosine. The effect of the p-hydroxyl group changes the charge distribution in α,β fission in the positive spectrum. It also initiates this process in the negative spectrum.

The major fragmentation in the α-amino acids can be viewed either as loss of formic acid from (M+H)⁺, or loss of COOH from M⁺. Because M⁺ peaks are never seen in amino acid spectra favors the former interpretation. If loss of formic acid from (M+H)⁺ occurs, the reaction should be influenced by the position of the amino group, since it is the charge center in the ion. Figure 4 shows how the fragmentation pattern varies with position of the amino group for three amino acids. Variation of major fragment elimination with the amino group constitutes strong evidence that one is seeing fragment ions from (M+H)⁺. Reaction probably occurs through a four-member transition state which is not uncommon for even electron ions. Deuteration studies confirm that the proton transferred to the acid comes from the protonated amino group.

226

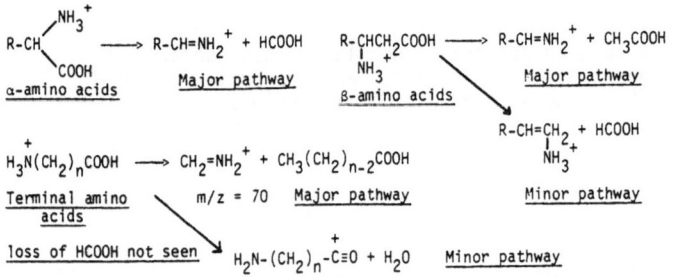

<u>Figure 4.</u> Effect of Amino Group Position on Amino Acid Fragmentation

One of the most interesting aspects of the LDMS study of the amino acids is a comparison with other mass spectral techniques. Table 2 presents such a comparison between LDMS, SIMS [7,8] and low-energy electron impact (LEEI) mass spectrometry [9]. Inspection of the table reveals striking similarity between LDMS and SIMS, but significant differences with LEEI. Even at the low-electron energies used (2-4 eV) LEEI shows much greater fragmentation. These results confirm the widely-held notion that LDMS is a very soft ionization technique compared to electron impact.

<u>Table 2.</u> Comparison of LDMS, SIMS and LEEI

Positive Ions
Aliphatic Amino Acids, (Alanine)

Technique	Major ions				Minor Ions
	M+H	M+	M+H-CO$_2$H$_2$	(M-H$_2$O)+	M-R
LDMS	100	0	92	0	0
SIMS[7]	40	0	100	0	0
LEEI[9]	100	17	0	52	81

Aromatic Amino Acids, (Phenylalanine)

					α-β fission	β-γ fission
LDMS	100	0	100	0	21α	0
SIMS[7] (Benninghoven)	30	0	100	0	56β	37
SIMS[8] (Klöppel)	84	0	100	4	27β	20
LEEI[9]	64	37	94	16	93(α) 100(b)	ND

Negative Ions
Aliphatic Amino Acids (Alanine)

Technique	M-H	M-	M-H-R (m/z 72)	M-H$_2$O	M-R	M-NH$_2$	M-CO$_2$H	R-
LDMS	100	0	0	0	0	0	0	0
SIMS[7]	100	0	0	0	0	0	0	0
LEEI[9]	100	18	0	2	16	49	0	0

Aromatic Amino Acid (Phenylalanine)

	M-H	M-		M-H$_2$O	M-R	M-NH$_2$	M-CO$_2$H	R-
LDMS	100	0	29	0	0	0	0	0
SIMS[7]	100	0	0	0	0	0	0	0
LEEI[9]	100	20	0	0	29	6	3	12

(a) Alanine, β-alanine, glycine are the only aliphatics not to show this ion.

Acknowledgements

We would like to acknowledge support of our laser mass spectrometry research by the National Science Foundation and the Office of Naval Research. We thank Frank Novak for helpful discussions.

References

1. R. Kaufman, F. Hillenkamp, R. Wechsung, Med. Prog. Technol., 6, 121 (1979).
2. a) R. J. Day, A. L. Forbes, D. M. Hercules, Spectrosc. Letters, 14, 703 (1981); b) C. Schiller, K. G. Kupke, F. Hillenkamp, Z. Anal. Chem., 308, 304 (1981).
3. R. D. Macfarlane and D. F. Torgenson, Science, 191, 920 (1976).
4. N. Ohmichi, J. Silbersten, R. D. Levine, J. Phys. Chem., 85, 3369 (1981).
5. T. Keough, A. J. Destefano, R. A. Sanders, Org. Mass Spectrom., 15, 348 (1980).
6. C. Schiller, K. G. Krupke, F. Hillenkamp, Z. Anal. Chem., 308, 304 (1981).
7. A. Benninghoven, D. Jaspers, W. Sichtermann, Appl. Phys., 11, 35 (1976).
8. K. D. Klöppel, G. Von Bünau, Int. J. of Mass Spect. and Ion Phys., 39, 85 (1981).
9. D. Voigt, J. Schmidt, Biomed. Mass Spectrom., 5, 44 (1978).

4.6 LAMMA 1000, a New Reflection Mode Laser Microprobe Mass Analyzer and Its Application to EDTA and Diolen®

H.J. Heinen, S. Meier, and H. Vogt

Leybold-Heraeus GmbH, P.O. Box 51 0760,
D-5000 Köln 51, Fed. Rep. of Germany

Abstract

A new sample chamber has been developed for LAMMA enabling micro area analysis of large samples. The ionizing pulse laser is ultimately focused onto
a selected sample area from the same side from which the ions are extracted
for mass analysis. Details on the construction of the chamber will be
given.

First applications to metal alloys, glasses, semiconductors, organic
monomeric and polymeric compounds show the wide range of materials which
can be successfully analyzed by LAMMA 1000. The spectra of EDTA and Diolen®
will be discussed.

Introduction

LAMMA® 500 [1], working in a linear geometry of laser focusing and ion
extraction, revealed already characteristic features of laser microprobe
mass analysis: the efficient ionization of all kinds of solid matter (con-
ductive as well as non-conductive), the detection of isotopes of all elements, the detection of inorganic and organic compounds, the control of
fragmentation by the applied laser power density, the lateral resolution
in the 1 µm range, the high detection sensitivity and the very rapid
analysis.

However, LAMMA 500 is limited to samples thin enough to be perforated by
the laser: tissue sections, films and powdered materials on thin supporting
films. To overcome this limitation LAMMA 1000 has been developed where
laser irradiation is from the same side as ion extraction. Emphasis is laid
on an efficient ion extraction perpendicular to the sample. A comparison
of thin film analysis with LAMMA 500 indicates that the expansion of the
plasma plume is essentially independent of the direction of laser irradiation [2] .

®LAMMA is a registered trade mark of Leybold-Heraeus

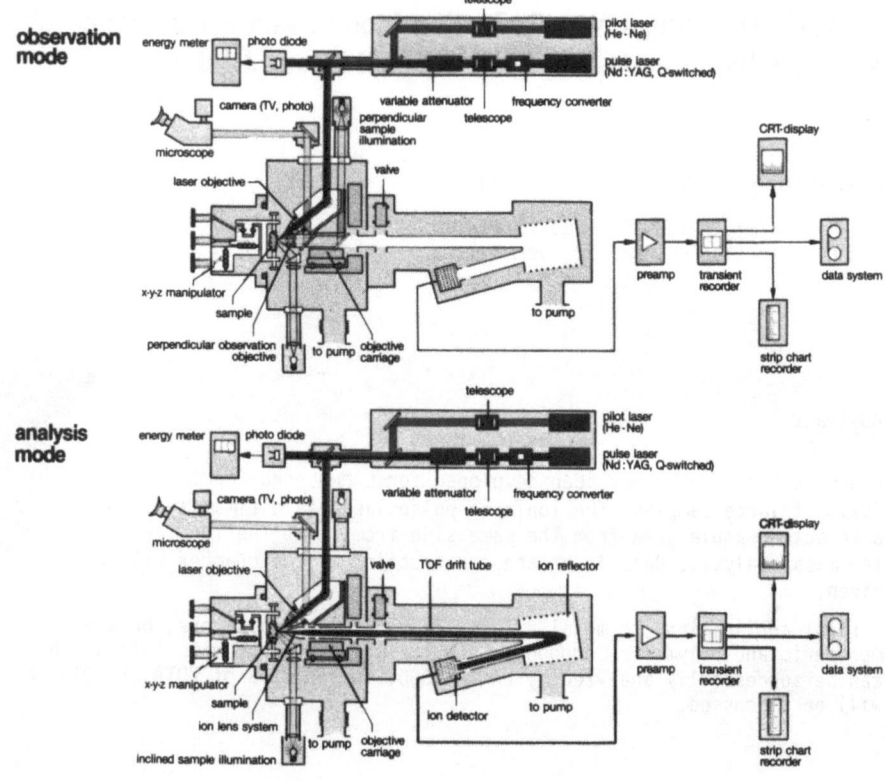

Fig. 1. Schematic diagram of LAMMA 1000

The LAMMA 1000 instrument

Fig. 1 gives a schematic diagram of LAMMA 1000. It was the aim of the development to have a sample observation perpendicular to its surface for a high quality microscopic image as well as an ion extraction perpendicular to the sample surface for a high overall detection sensitivity. To meet this requirement a precise linear motion system is employed carrying side by side the ion extraction system (lower part of fig. 1) and the objective for perpendicular sample observation (upper part of fig. 1). The push – button operated and motor-driven carriage has a mechanical reproducibility of better than 1 μm under vacuum. Position exchange takes about 3 sec. The microscope objective has a numerical aperture of 0.22 permitting a resolution of about 2 μm on the sample. A fixed UV permeable objective of a numerical aperture of 0.2 under 45 degree to the spectrometer axis is used for focusing of the power laser and the colinear He-Ne laser which acts as a focusing aid. The power laser is a frequency quadrupled (λ = 265 nm) Q-switched Nd:YAG laser. The intensity of the power laser can be continuously varied over at least 3 decades by a pair of twisted polarizers. Maximum power per pulse is 100 μJ corresponding to an average power density

230

of 10^{11} W/cm² of the irradiated area on the sample which is estimated to be
an ellipse of about 2 and 3 µm diameter respectively. The craters on the
sample area are usually slightly larger - depending on the sample material
and the laser power density - due to heat conduction effects.

Alternatively to the perpendicular objective the inclined UV objective
can also be used for sample observation. Despite the fact that only a
narrow zone of about 70 µm gives a sharp image,this observation mode is
helpful for precise positioning of the laser at the sample because it is
free of a parallactic effect between inclined laser focusing and perpendicular observation. In the perpendicular observation the viewing field is
about 0.8 mm. Sample illumination for both observation modes is of the
bright field type.

The sample is held on a x-y-z manipulator of 70, 50 and 50 mm carriage
respectively. For an exchange the whole manipulator/sample stage assembly
slides in and out of the analysis chamber on two solid precision guide
bars. The opening in the chamber permits samples of diameters up to 130 mm
to be inserted.

The time-of-flight analyzer has an improved time focusing to accept a
wider kinetic energy distribution of the ions. This became necessary as
ions produced by laser irradiation of thick samples usually have a broader
energy distribution than those from thin samples (LAMMA 500) [2]. The obtained spectra show that the mass resolution known from LAMMA 500 is maintained also for the analysis of bulk metals.

Application to EDTA and Diolen®

Various materials have been investigated with the new instrument: metal
alloys, semiconductors, organic monomeric and polymeric compounds. The analysis of a stainless steel and a glass sample gave singly-charged atomic
ion spectra with high sensitivity [3].

Fig. 2 shows a series of negative ion spectra obtained from an approx.
1 µm thick layer of ethylene diamine tetraacetic acid (EDTA) on a metallic
support. The layer was prepared by drying a droplet of a dilute solution
on the metal. No effort was made towards uniformity of the layer. The
spectra were taken with increasing laser power density from bottom to top
ranging from 8×10^7 to 5×10^8 W/cm². The sensitivity of the detection
system was reduced with increasing total ion currents.

With the lowest power density the deprotonated molecule gives the highest
relative intensity. Weak signals are observed from cleavage of the inner
C-N bond ($[AcNHAc - H]^-$ at m/z 132), loss of three acetic acid fragments
($[AcNHCH_2CH_2NH_2 - H]^-$ at m/z 117) and acetic acid ($[HAc - H]^-$ at m/z 59).
Here Ac stands for $-CH_2COOH$. Hydrogen shows a high mobility during fragmentation and saturates free valencies from bond cleavages.

With increased laser power density also fragments from loss of water
($[M - H_2O - H]^-$ at m/z 273), loss of carbon dioxide ($[M - CO_2 - H]^-$ at
m/z 247), loss of acetic acid together with carbon dioxide
($[M - HAc - CO_2 - H]^-$ at m/z 187) and loss of two acetic acid fragments
($[AcNHCH_2CH_2NHAc - H]^-$ at m/z 175) are observed. The $[AcNHAc - H]^-$ ion

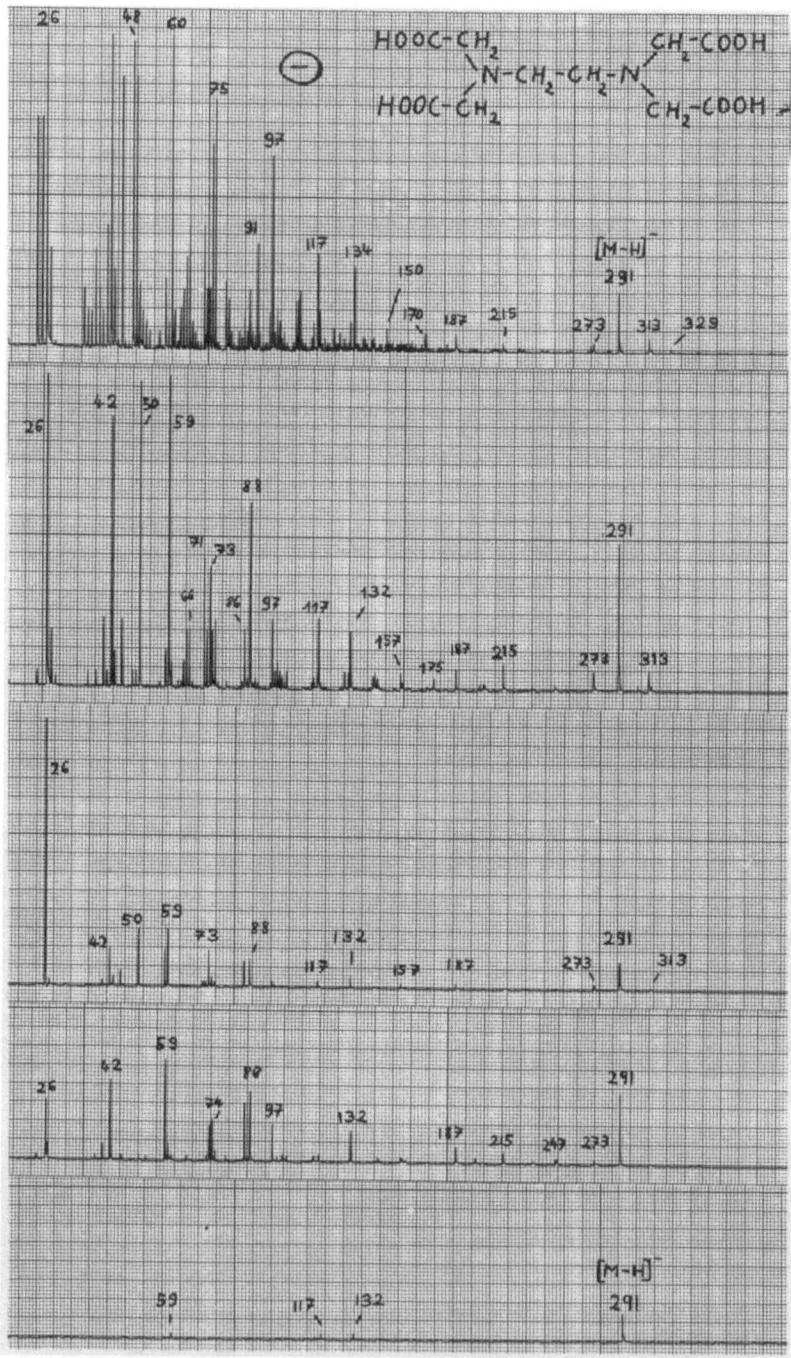

Fig. 2. Negative LAMMA 1000 spectra of EDTA

fragments further to $[CH_3NHAc - H]^-$ and $[CH_2=NAc - H]^-$ at m/z 88 and 86 respectively (loss of CO_2 and HCOOH) and to $[AcNH_2 - H]^-$ at m/z 74. The $[CN]^-$ and $[OCN]^-$ ions at m/z 26 and 42 and, with higher laser power density, the $[C_3N]^-$ and $[OC_3N]^-$ at m/z 50 and 66 are plasma reaction products and frequently observed in negative spectra of molecules containing C, N and O. They are not characteristic for the molecular structure.

With further increasing laser power density, sodium and potassium salt ions $[(M + Na,K - H) - H]^-$ at m/z 313 and 329 respectively are observed due to Na and K impurities during sample preparation. Further low mass fragments are formed. A series of $[C_nH_m]^-$ ions (n = 2,3,4 ...; m = 0,1,2) starting with m/z 24 with a crude C-12 periodicity appears due to plasma reactions of carbon and hydrogen. These ions are frequently observed in high—intensity laser spectra of C, H containing compounds and, like the $[CN]^-$ and $[OCN]^-$ ions, unspecific for the structure of the parent molecule. The still present $[M - H]^-$ peak originates probably from areas surrounding the center of the laser focus.

A polymer is taken as second example. In general, characteristic LAMMA spectra can be expected if, under the action of the laser beam, positive or negative fragment ions of sufficient chemical stability can readily be formed out of the macromolecular structure. During fragmentation hydrogen again shows a high mobility to saturate free bonding sites. For some polymer classes more complicated rearrangement reactions give rise to characteristic spectra [4].

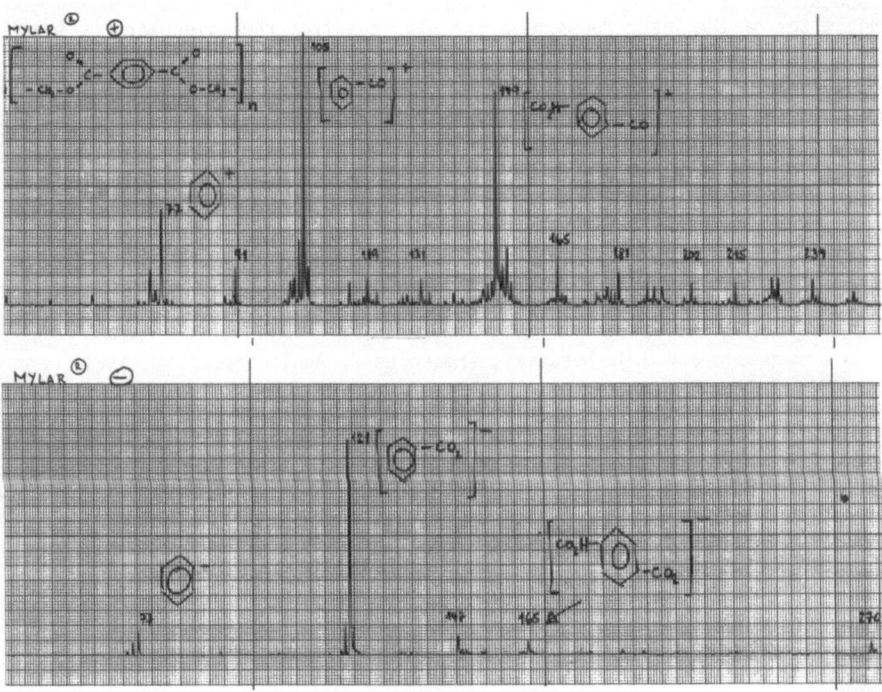

Fig. 3. LAMMA 1000 Spectra of Diolen®

233

Fig. 3 gives the positive and negative LAMMA spectrum of Polyethylene Terephthalate (Diolen®; Mylar®) with a laser power density which was slightly above ionization threshold. Besides the phenyl and tropyllium ions at m/z 77 and 91 intense peaks at m/z 105 and 149 are detected in the positive spectrum. The stable ion structures (fig. 3) can readily be formed out of the polymer structure. The corresponding negative ion structures are 16 masses (oxygen) higher to complete carboxylic groups. The peak at m/z 147 probably arises from water elimination of m/z 165 in the negative as well as m/z 131 from water elimination of m/z 149 in the positive spectrum. The negative spectrum shows also a peak at m/z 270 which might be an addition of m/z 121 and 165 under release of oxygen.

The continuously variable laser power density was found to be very advantageous. The spectra of Fig. 3 could quite reproducibly be recorded within a factor of 1.3 of laser intensity above the threshold of polymer ionization. For higher laser intensities strong pyrolysis reactions and plasma reactions between C, H and O are induced leading to spectra which are unspecific of the compound.

Summarizing, one can say that in the formation of molecular ions the structure of the parent is maintained the more the lower the laser power density is relative to the ion desorption threshold. Fragmentation channels leading to ions of high chemical stability (even electron ions) are generally preferred. The fragmentation level can be controlled by the applied laser power density in a wide range in terms of quasimolecular to fragment ion intensity ratios. With sufficiently high laser power densities a total atomic rearrangement leading to stable plasma reaction products of the involved elements can be induced. From areas surrounding the center of the laser focus, nevertheless, ions from the molecular spectra can be detected besides the plasma reaction products.

REFERENCES

1. H. Vogt, H.J. Heinen, S. Meier and R. Wechsung, Fresenius
   Z. Anal. Chem. 308 (1981) 195 - 200

2. H.J. Heinen, paper in preparation, to be published

3. H.J. Heinen, S. Meier, H. Vogt and R. Wechsung, Proceeding of the
   9th Int. Mass Spectrom. Conf., 30'Aug. - 3. Sept. 1982, Vienna, Austria

4. J.A. Gardella and D.M. Hercules, Fresenius Z. Anal. Chem. 308 (1981)
   297 - 303

4.7 Some Experiments on Laser Induced Cationization of Sucrose *

Paul Wieser and Roland Wurster

Institut für Physik, Universität Hohenheim, Garbenstraße 30,
D-7000 Stuttgart 70, Fed. Rep. of Germany

Introduction

As pointed out by HILLENKAMP [1] quite different instrumental
arrangements are used to desorb "nonvolatile" and thermally
labile substances by laser radiation - organic solid target
interaction. The results obtained are influenced by many experimental parameters and are not yet fully understood [2-4] .

We have started a systematic investigation to see how the
sample geometry can affect the laser—induced molecular evaporation and soft ionization process if the laser microprobe mass
analyzer LAMMA^{R}500 is used. This study is restricted to the alkali ion attachment to sucrose molecules, one of the most convenient soft ionization processes.

It should be noticed that the LAMMA^{R}500 instrument [5] uses a
transmission configuration with the laser beam coming from the
one side and the ions being extracted from the opposite side of
the sample. This means that normally the sample perforation is
a prerequisite for ion detection. In partial contradiction to
this condition on the other hand a sample evaporation in the form
of intact neutrals and/or ionized molecules is wanted. Both conditions may be fulfilled simultaneously, however at different,
spatially separated sites of the sample: the perforation being
created in the central part of the laser spot and the laser induced desorption taking place in the periphery of the laser
irradiated microvolume. To prove this simple idea we analyzed
single particles of different size, supported by different substrates, at systematically varied laser power densities.

Experimental

The laser microprobe mass analyzer LAMMA^{R}500 (LEYBOLD-HERAEUS)
was operated at the following conditions: microscope objective
ZEISS Ultrafluar 32 x giving a laser focus diameter of about 1µm;
pulse width 10 nsec; wavelength 265 nm. Each laser pulse is
energy monitored and the laser energy entering the microscope
can be varied by a set of attenuating filters. The actual laser
irradiance in the laser spot is objected to uncertainties,

---

* Dedicated to Prof. Dr. G. Möllenstedt on the occasion of his
  70th birthday

however. Therefore the monitored laser energy (µJ) multiplied
by the filter transmission factor will be used as a measure for
the laser irradiance.

A compressed air nebulizer[6] is used to produce aerosols of
solid particles. The particles are formed when aqueous solutions
of sucrose , pure alkali halides (NaCl and LiCl) and mixtures of
sucrose and alkali halides are used in the generator reservoir.
The droplets produced are subsequently evaporated, leaving residues which form the solid particle aerosol.

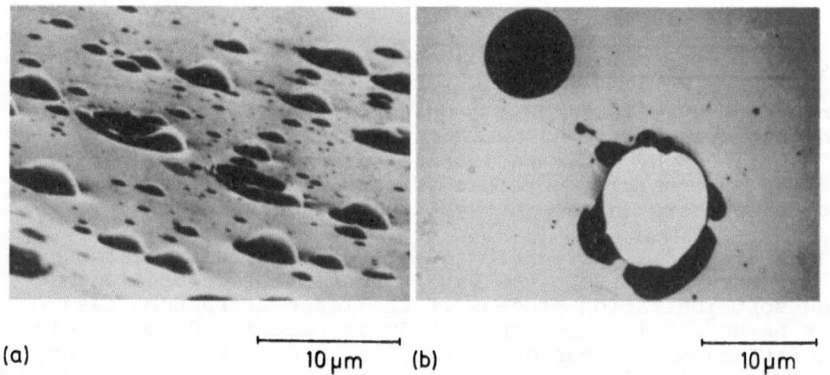

(a) |————————| 10 µm   (b) |————————| 10 µm

Fig.1. (a) Scanning electron micrograph of sucrose particles
different in size, showing the shape of deposited particles
(tilting angle 8o°), (b) transmission scanning electron micrograph of an intact and perforated sucrose particle (tilting
angle 0°)

In a centrifuge[7] these particles are directly deposited on
LAMMA suitable copper grids, which have been coated by either
pioloform foils (thickness about 20 nm) or self-supporting silver
films (thickness about 30 nm). As shown in Fig.1 the particles
are deposited as a highly viscous material. They have a rather
flat shape and can be easily perforated by a single laser shot.
The ratio of particle diameter and height decreases slowly with
decreasing projected area diameter. Therefore the particle heights
of the investigated set of particles range from about 1 µm to
.2 µm, whereas the projected area diameters range from 10 µm to
.8 µm.

Results and Discussion

Figure 2 shows typical time-of-flight mass spectra of positive
ions obtained from NaCl containing sucrose particles deposited
on the pioloform foil and the self-supporting silver film. The
silver supported particle gives rise to a considerable gain of
the quasi-molecular ion [M+Na]$^+$ intensity. In contrast the intensity of sucrose molecules cationized by Ag$^+$ ion attachement is very
small. Due to its better light absorption most of the thin silver
film will be evaporated during the very beginning of the laser

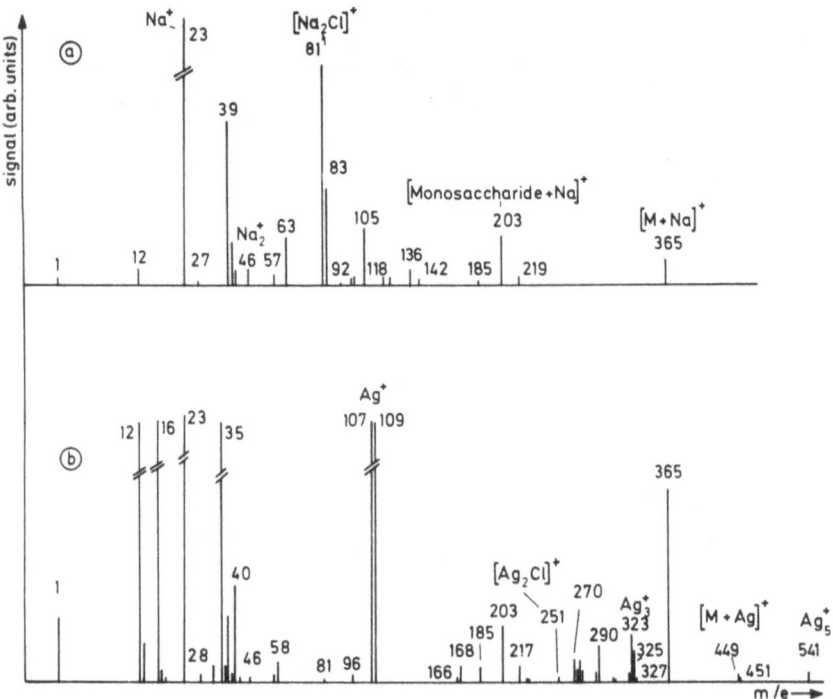

Fig.2. Typical time-of-flight mass spectra of positive ions obtained from sucrose particles containing NaCl (diameter about 10μm) on a pioloform foil (a) and a self supporting silver film (b)

pulse, thus causing a poor synchronization of Ag$^+$ ion and sucrose molecule ejection. On the other hand we conclude from a comparative study on perforation holes in pioloform foils that the thin silver film will improve the energy deposition in the sucrose particles.

The results obtained from more than a thousand mass spectra are as follows:

Both the signal of the cationized sucrose and the cationized fragment-monosaccharide depend on the molar ratio of NaCl and sucrose in the particle material [8]. As shown in Fig.3 the best results are obtained at relatively high NaCl concentrations.

The signal of the quasi-molecular ion depends on the laser irradiance. This dependency seems to be quite different if the pioloform foil is replaced by the silver film (Tab.1).

At nearly optimal laser irradiance the signals of both the cationized sucrose molecule and the cationized monosaccharide strongly depend on the particle size. Though the evaporated mass

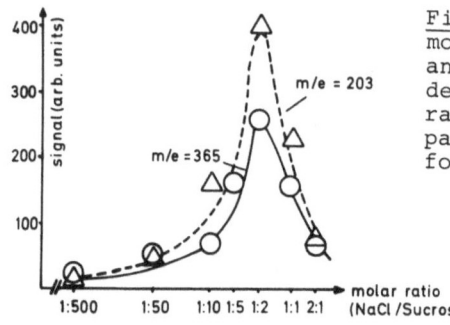

Fig.3. Intensities of cationized monosaccharide molecules (m/e=203) and sucrose molecules (m/e=365) depending on molar NaCl-sucrose-ratio (laser energy about 10µJ, particle size about 10µm, pioloform substrate)

is nearly constant, the signal of the quasimolecular ion decreases by one order of magnitude if the particle diameter is varied from 6.2µm to 3.7µm (Tab.2). In addition from Tab.2 we clearly see the influence of the silver film on the measured ion intensities. The reproducibility of the mass spectra dramatically decreases with decreasing particle size. In the size range of about 1µm only a few percent of the mass spectra show the mass lines in question. The reproducibility of the values reported in Tab.1 and Tab.2 is about 30%. The fragmentation probability increases with decreasing particle size. These facts may confirm the suggestion that the high sodium ion concentration produced in the central part of the laser spot and the simultaneous desorption of intact molecules at the rim of the laser perforation give rise to an optimal cationization yield (about 10^{-5}).

High quasi-molecular ion signals are detected if a great amount of residual, obviously molten particle material is transported to the rim of the perforation during the laser beam-particle interaction (Fig.1b).

The intensities of Ag_3^+, Na_2Cl^+ and Ag_2Cl^+ ions show quite different dependencies on particle size and laser irradiance. This may be due to quite different processes and conditions under which these ions are formed. The Ag_3^+ ions show a significant shortening of their time of arrival (about 100 nsec) when being produced at the rear side of a sucrose particle compared to the pure silver film.

Table 1. Signal (m/e=365) depending on the laser irradiance (arbit. units). Substrate: Pioloform, silver film

particle size (µm)	6.2		4.8		3.7	
	p	s	p	s	p	s
m/e = 365	8.6	14	2.5	14.5	0.9	7
m/e = 203	13.4	10.4	9.3	16.1	2.5	15.9

Table 2. Signals (m/e=356, 203) depending on the particle diameter (arbit. units). p: piologorm, s: silver substrate

laser energy (µJ)	0-2	2-4	4-8	8-12	12-16	16-20
pioloform	3.3	10	13.3	9	7.3	3.2
silver	0.6	7.1	13.7	18.9	18.5	12

238

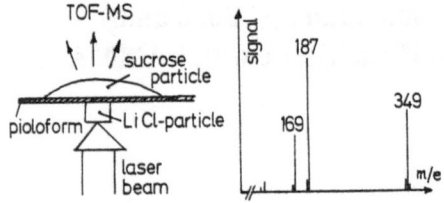

Fig.4. Sample con-
sisting of clearly
separated sucrose-
and LiCl particle
with the correspon-
ding mass spectrum

As shown in Fig.4 the cationization by Li^+ attachement also takes place if the sucrose and LiCl are spatially clearly separated. Under these conditions the cationization process prerequisites an instantaneous and sufficient fast mixing process, which may occur in the so-called transitions region of the solid and gaseous phase.

The experiments have shown the cationization of sucrose molecules by alkali ion attachement to be dependent on laser energy, particle size, alkali halide concentration, and on particle substrate material. The optimal conditions with respect to the investigated parameter range suggest a cationization process which is based on at least partly separated sources of alkali ions and intact sucrose molecules. Moreover some experimental indications exist that the liquid phase and fast mixing processes play an important role concerning the molecular evaporation and soft ionization. There are also experimental indications that remarkable amounts of the irradiated material leave the interaction volume in the liquid state.

Acknowledgement

The authors acknowledge the support given by Prof. Dr. H.Seiler and the technical assistance of Mrs. R. Jacobi.

References

1. F. Hillenkamp, Laser Desorption, this volume
2. P.G. Kistemaker, G.J.Q.van der Peyl and J. Haverkamp, Advances in Mass Spectrometry - Soft Ionization, Heyden, London 1981
3. R. Stoll and F.W. Röllgen, Org.Mass Spectrometry,14, 642 (1979)
4. See R.J. Conzemius and J.M. Capellen, Int. J. Mass Spectrom. Ion Phys., 34, 197 (1980)
5. H. Vogt, H.J. Heinen, S. Meier and R. Wechsung, Fresenius Z. Anal. Chem. 308, 195 (1981)
6. M. Corn and N.A. Esmen, Aerosol Generation, in Handbook on Aerosols, R. Dennis ed., Oakridge 1976
7. W. Stöber and H. Flachsbart, Environ.Sci.Technol.,3,1280 (1969)
8. J. Gruber, Inst. f. Physik, Universität Hohenheim, unpublished

4.8 Mass Spectrometry of Organic Compounds (≲ 2000 amu) and Tracing of Organic Molecules in Plant Tissue with LAMMA

Ulrich Seydel and Buko Lindner

Biophysics Division, Research Institute Borstel,
D-2061 Borstel, Fed. Rep. of Germany

1. Introduction

The laser microprobe mass analysis as realized in the commercially available LAMMA®500-instrument is applied in our laboratory mainly for microbiological investigations on single bacterial cells to characterize their physiological state from intracellular cation concentrations [1,2].

Here, two other applications are reported: 1. The pure laser desorption mass spectrometry (LD-MS) for testing the possibility of getting information on structure and molecular weight of rather complex organic compounds (components of the outer membrane of gram-negative bacteria, especially lipid A). In a first step these investigations are performed on synthetic compounds with known chemical structure corresponding to the fundamental lipid A structure, then on a natural molecule isolated from Rhodomicrobium vaniellii. 2. A microprobe application for the direct tracing of an organic molecule in plant tissue (the phytoalexin glyceollin) which soybean seedlings accumulate upon infection with certain fungi.

2. Experimental

2.1 Instrumental Parameters

For a detailed description of the LAMMA®500-instrument (Leybold-Heraeus Comp., Köln, F.R.G.) see,e.g. [3]. The parameter settings for the present investigations were: laser output wavelength 265 nm, laser power density at the sample surface 10^{10}-10^{12} W/cm^2, laser pulse duration 15 ns, diameter of the focused laser beam ≳1 µm, photomultiplier voltage 3500 V, and high input sensitivity of the transient recorder.

2.2 Samples and Preparation

The synthetic lipid A-like compound was a generous gift of T. Shiba and S. Kusumoto, Osaka University, Japan [4]. The natural lipid A was isolated from Rhodomicrobium vaniellii by O. Holst, Max-Planck-Institute, Freiburg, F.R.G.. Both substances

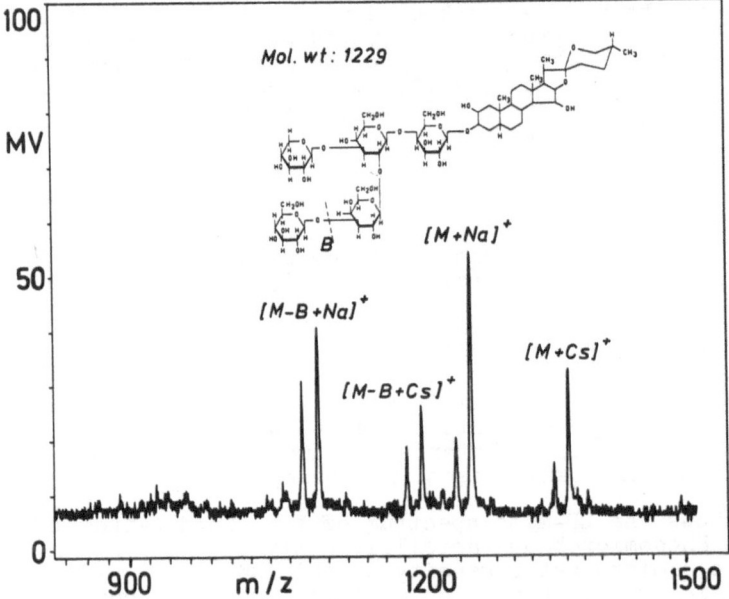

Fig.1. Structure and LD-mass spectrum of digitonin (mol. wt. 1229). Na-,K-,CsCl were added to a total of 5:1 w/w

were given at hand as crystalline powders and were prepared for LAMMA-analysis by admixture of appropiate amounts (5:1 w/w) of one or more alkali salts. The mixtures were brought onto Formvar-filmed Cu-grids.

For the tracing of glyceollin 10 μm freeze microtome sections from soybean cotyledons were freeze-dried on Cu-grids. The thin sections were cut perpendicular to the border of infection, allowing the analysis of the transition region from uninfected to infected tissue. The identification of glyceollin in the tissue was accomplished by comparison with mass spectra of the pure substance isolated from infected tissue. This work was done in cooperation with the Biological Institute II, University of Freiburg, F.R.G. [5].

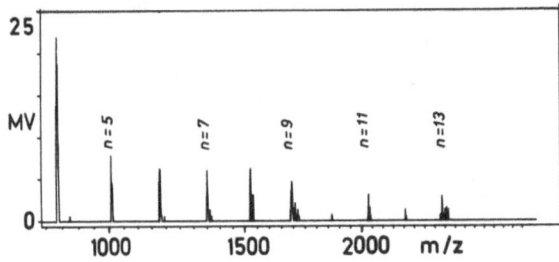

Fig.2. LD-mass spectrum of CsCl, showing cluster formation of the type $[Cs(CsCl)_n]^+$

Mass scale calibration was achieved via digitonin and CsCl. The LAMMA-spectrum of digitonin - prepared as above - shows a simple, easily identifiable pattern in the mass region from approximately 1100 to 1400 amu (Fig.1). CsCl leads to the formation of clusters of the type $[Cs(CsCl)_n]^+$ which could be identified up to n=13 $\triangleq$ 2317 amu (Fig.2).

3. Results and Discussion

Figure 3 shows the structure and the LD-mass spectrum of a synthetic lipid A-like substance (mol. wt. 1890). Only the upper part of the total spectrum is given here, showing little fragmentation and very pronounced quasi-molecular peaks from cation attachment ($[M+Na]^+$, $[M+K]^+$, $[M+Cs]^+$). It is worth mentioning that from very similar compounds - differing only in that the hydrogen marked with an asterisk in Fig.3 is substituted by an organic phosphate group - no interpretable spectra could be obtained. These zwitterionic substances produce a large number of low-intensity fragment peaks resembling background noise only.

Figure 4 shows the LD-mass spectrum of the lipid A-fraction isolated from Rhodomicrobium vaniellii. Fatty acid determination points to a micro-heterogeneity of this special lipid A molecule explaining the absence of a single prominent quasi-molecular peak (only NaJ was added) and the very low overall intensity. Peak identification is still to be done.

It should be pointed out that optimal results for all compounds were obtained only at highest laser power density and directing the laser beam on particles much larger than its diameter. When particles of the same size as the laser beam dia-

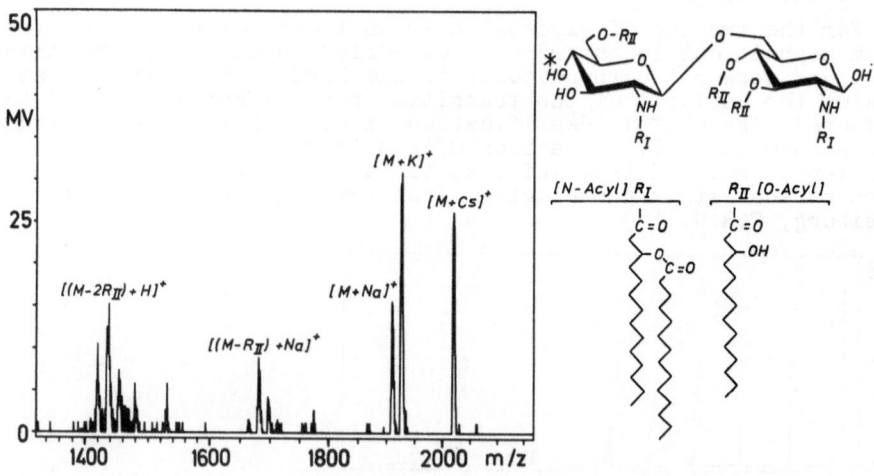

Fig.3. Structure and LD-mass spectrum of a synthetic lipid A-like compound (mol. wt. 1890). Na-, K-, CsCl were added to a total of 5:1 w/w

meter or smaller are irradiated under otherwise identical conditions neither molecular nor higher fragment peaks can be observed. These observations are in agreement with the following rough model for the ionization mechanism: Upon the interaction of the laser pulse ($\approx 10^{-8}$ s) with the solid sample a shock wave is generated. Its energy is at least partially thermalized by the induction of collisional processes of the molecules ($\approx 10^{-13}$ s). As in these short times dissipation of heat from the irradiated microvolume can be neglected, superheating and subsequent 'phase explosion' [6] - the explosive decomposition of the superheated liquid into a 2-phase system (wet vapour) in thermodynamic equilibrium - is achieved. This leads to the sudden release of the surrounding cold sample material, among others also as intact molecules. The failure in producing interpretable spectra of phosphate-containing lipid A's should also be explainable by this model. The introduction of a phosphate group leads to a zwitterionic state of the molecule which might disturb the superheating and releasing processes due to higher 'cohesive forces' in the solid.

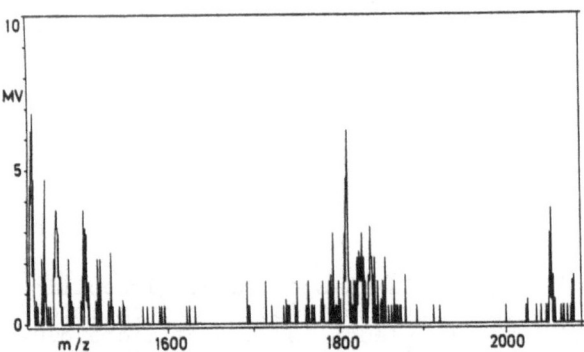

Fig.4. LD-mass spectrum of a mixture of lipid A - isolated from Rhodomicrobium vaniellii - and NaJ (5:1 w/w)

In Fig.5a-c the results for tracing of glyceollin are presented, showing the spectra for the pure substance (a) in comparison with spectra taken from infected (b) and uninfected (c) tissue, respectively. Clearly, the $[M-OH]^+$-peak can be identified in the spectrum of the infected region but is completely absent in the spectrum of the uninfected tissue. Evaluation of 150 microprobe analyses revealed a steep rise in glyceollin content at the borderline of infection [5].

This is the first time that such highly localized glyceollin accumulation has been shown at the cellular level. One can be hopeful that the laser microprobe mass analysis could be a promising approach for other cases in which organic molecules have to be traced in biological tissues and that with high lateral resolution.

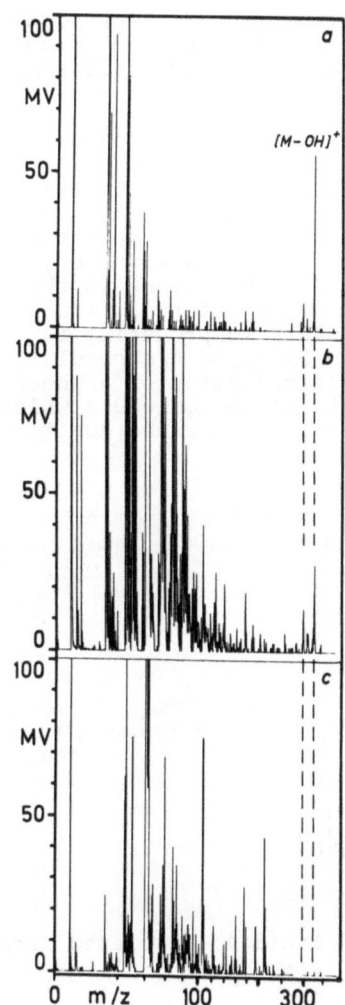

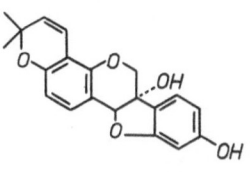

Glyceollin

Fig.5. LAMMA-spectra of glyceollin (mol. wt. 338) isolated from infected cotyledon (a) and of microvolumes in infected (b) and uninfected (c) cotyledon tissue of soybean. The dashed lines indicate pronounced fragment peaks of glyceollin

References

1. U.Seydel, B Lindner: Fresenius Z. Anal. Chem. 308, 253(1981)

2. B.Lindner, U. Seydel: J. Gen. Microbiol. (in press)

3. H.J.Heinen, S.Meier, H. Vogt, R.Wechsung: In 'Advances Mass Spectrometry', ed. by A.Quayle (Heyden, London 1980) p.942

4. M.Inage, H. Chaki, S.Kusumoto, T Shiba: Tetrahedron Lett. 21, 3889(1980)

5. M.M.Martynyuk: Sov. Phys.-Techn. Phys. 21, 430(1976)

6. P.Moesta, U.Seydel, B.Lindner, H.Grisebach: Z. Naturforsch. (in press)

Part 5

Other Ion Formation Processes

5.1 Ion Emissions from Liquids (Review)

Marvin L. Vestal

Department of Chemistry, University of Houston, Houston, TX 77004, USA

Introduction

The production of charged particles by spraying or otherwise disrupting
liquid surfaces has been studied for more than a century [1] but only during
the past twenty years has ion emission from liquids been used as a source of
ions for mass spectrometry. Applications to organic mass spectrometry are
of even more recent origin. Several studies have shown that it is possible
to extract ions from solution and inject them into a mass spectrometer
without significantly increasing their internal energy; as a result,
desorption of ions from liquid solutions provides a potentially attractive
approach to applying mass spectrometry to large, nonvolatile, or thermally
labile compounds not accessible by conventional techniques.

The various techniques which have been developed for desorbing molecular
ions from liquid solutions differ primarily in the means employed for
disrupting the liquid surface. In the electrohydrodynamic ionization (EH)
technique developed by EVANS and co-workers [2-5] a strong electric field is
applied to the surface of the liquid in a vacuum and the emitted ion
clusters are accelerated directly to the mass spectrometer. The electrospray technique (ES) of DOLE and co-workers [6,7] applies a strong electric
field to the liquid surface at atmospheric pressure. The charged droplets
produced are then desolvated by passing them through dry gas at atmospheric
pressure and the resulting mixture of gas, ions, and particles is passed
through a nozzle to produce a supersonic beam which is sampled by a mass
spectrometer or energy analyzer. Recently, THOMSON and IRIBARNE [8,9] have
used a somewhat similar approach but with some important differences.
Solutions of electrolytes are pneumatically sprayed into the air, the
resulting droplets are charged by induction, and an electric field across
the plume of evaporating spray extracts small ions which are drawn through a
small orifice into the vacuum chamber of the mass spectrometer. A dry air
"curtain" is used to suppress clustering during the free jet expansion
downstream of the sampling orifice. The thermospray (TS) technique
developed by BLAKLEY and VESTAL [10-13], in contrast to the other methods,
employs no electrical fields in the ion source. Solutions of electrolytes
are forced through a heated capillary nozzle to produce a supersonic jet
consisting of vapor and superheated charged particles. The charged
particles continue to vaporize as they pass through a few centimeters of
vapor maintained at reduced pressure and elevated temperature. Molecular
ions or small clusters are desorbed from the vaporizing particles and are
sampled into the mass spectrometer in a similar fashion as in a chemical
ionization source.

All of these techniques produce spectra which are similar to those obtained by field desorption from samples deposited on solid emitters. None

of these techniques produce significant fragmentation of the molecular ions. The major differences in the spectra are due to the varying amount of residual solvation of the ions observed; this may be due to the differences in the way the ions are sampled into the mass spectrometer rather than to major differences in the ion production mechanisms. It seems clear that all of these techniques, including conventional field desorption [14], involve direct desorption of ions from electrically-charged liquid surfaces.

In this review we present the major operational and performance features of these techniques and their apparent strengths and weaknesses from the point of view of both analytical applications and fundamental studies. A simplified model of the ionization mechanism is presented together with some speculations about the importance of this mechanism in other "soft" ionization techniques where desorption from liquid surfaces is not so obviously involved.

Electrohydrodynamic Ionization

Electrohydrodynamic ionization results from the interaction of a strong electric field with a liquid meniscus at the end of a capillary tube. This technique was originally developed for potential use in spacecraft thrusters where charged liquid metal droplets [15] and ions [16,17] as well as droplets from organic liquids [18-21] have been studied. EVANS and HENDRICKS coupled an EH source to a double focussing mass spectrometer and initially employed this system in studies on production of atomic and molecular ions from liquid metals [2]. More recently this instrument has been used in investigations of EH as a soft ionization technique for thermally unstable organic compounds [3-5].

The ion-source used for EH ionization is shown schematically in Fig.1. The sample is dissolved in the host fluid and loaded into the syringe which is then mounted in the source housing and the system is evacuated to a pressure of 10^{-7} torr. Typically, the needle supply voltage is set to +9 kV and the extration voltage to -2 kV. The needle plunger is then advanced until spraying commences as indicated by a positive reading on the needle current meter and also by an increase in source pressure to 5×10^{-7} torr. Typical total emission current are on the order of 10 μA resulting in a 500 V drop across the series resistor.

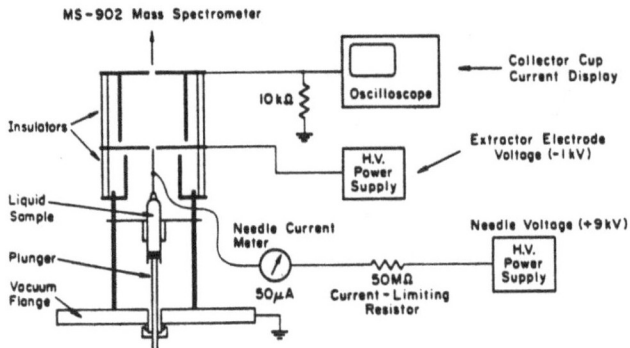

Fig.1. Electrohydrodynamic ion source and electronics. (From ref. [3] with permission)

When liquid metals are sprayed by the EH process, ion emission occurs at the tip of a self-formed cone of the liquid [17]. Organic liquids behave in a qualitatively different manner as described by EVANS et al. [3]. "When a low voltage is applied to the needle, a cone of liquid is formed with a stream of macroscopic droplets emanating from the apex. As the needle voltage is increased, the emission point moves to the rim and the liquid meniscus retracts into the needle. Increasing the voltage still further produces more emission sites on the rim and decreases the diameter of the droplet jets until they are no longer visible. This last stage of spraying occurs at needle voltages above 6 kV and produces beams with substantial ion content when NaI-glycerol is the working fluid."

The proper choice of host fluid is very important to successful operation of the EH source. The requirements include low vapor pressure, minimal corrosive behavior, and high solution capability, both for samples of interest and for suitable electrolytes. Early studies showed that the average mass of the droplets produced by EH spraying strongly correlates with the electrical conductivity of the liquid. Higher conductivities yield droplets of lower mass [19,20]. Glycerol doped with suitable electrolytes, usually alkali halides, is the host fluid normally used. The electrical conductivity of pure glycerol is 5×10^{-7} ohm^{-1} cm^1 and under typical EH spraying conditions the average droplet has m/e values of over 10^6 (amu/unit charge). However, glycerol is a good solvent for a number of ionic salts and its conductivity can be increased by their additions. With a solution of NaI-glycerol at a concentration of 300 mg/ml the conductivity is 2×10^{-4} ohm^{-1} cm^{-1} and the average m/e values for the ions produced are typically in the range of 500-1000. An example of an EH mass spectrum of NaI-glycerol solution is shown in Fig.2.

With the EH source, ions are produced with a broad range of kinetic energies ranging from an upper limit near the needle potential to a lower limit about one-half of the needle voltage. Measurements of energy spectra by HUBERMAN [22] have shown that this energy spread is primarily due to

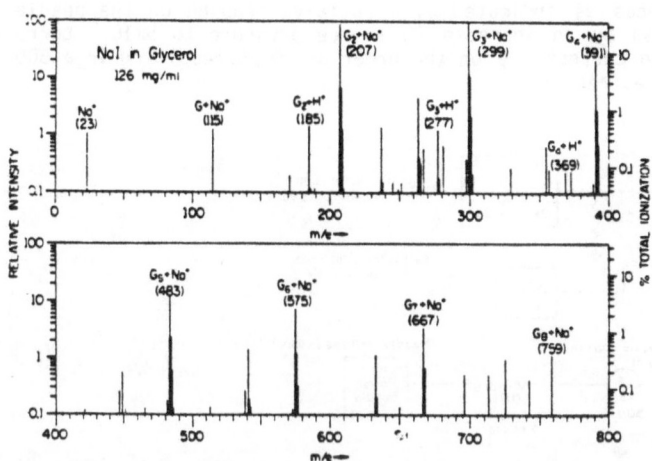

Fig.2. EH spectrum of NaI in Glycerol at a concentration of 126 mg/ml. G = Glycerol, $C_3H_8O_3$. (From ref. [3] with permission)

in-flight evaporation of glycerol molecules from the cluster ions with the true dissipative energy loss in the spraying process being on the order of 400 eV. The spectrum shown in Fig.2 was obtained at an ESA voltage corresponding to 5 KeV ions; thus, the observed ions correspond to clusters which have lost on the average about one glycerol molecule in flight. By changing the ESA voltage the portion of the ion decomposition spectrum sampled can be varied from the stable portion of the clusters produced directly from the liquid (at ca. 8 KeV) to sampling only the final products of decomposition (at ca. 4 KeV). In most of the applications of EH to nonvolatile organic molecules the ESA has been operated so as to sample primarily the directly formed ions prior to any in-flight decompositions; however, the total intensity is higher if a somewhat lower energy is sampled, indicating that a majority of the ions decompose in flight.

Results have been obtained on a variety of compounds of low volatility including proline, sugars (glucose, sucrose, raffinose), nucleosides (adenosine, thymidine, uridine), a tripeptide (glutathione) and an amino-cyclitol antibiotic (neomycine) [4], and low molecular weight polymers of ethylene glycol [23]. All samples yield several ions containing one or more intact molecules of the sample clustered with solvent molecules on either a proton or an alkali cation. Fragmentation of sample molecules has not been observed.

An extremely important result of the EH ionization studies is that they demonstrated for the first time that ions present in solution could be desorbed intact and efficiently transmitted to a mass spectrometer. The technique is potentially applicable to the continuous analysis of non-volatile solute; however, a number of the characteristics of the source may seriously limit its applicability. In particular, the sprayed beam is composed of massive, charged droplets as well as analytically useful ions and it would be desirable to remove the droplet component. Ion current stability is marginal, with typical fluctuations being on the order of 20% over a five-minute period. The fluctuations are probably due to erosion of the needle rim which affects the local electric field. Needle erosion can also cause contamination of the spectrum. The EH source is presently only operable with glycerol as the solvent which is a very severe limitation on its application. Perhaps the most serious limitation from the point of view of analytical applications is the broad energy spectrum of the ions produced. This requires the use of an expensive double focussing mass analyzer and even then the ion transmission efficiency is necessarily very low.

Electrospray

DOLE and co-workers have reported the results of a series of experiments [6,7] aimed at obtaining a molecular beam of macroions derived from high molecular weight polymers. A schematic diagram of their apparatus is shown in Fig.3. In their experiments, solutions of polystyrenes (0.001 to 0.1 wt.%) in a benzene-acetone solvent (3:2 by volume) were used. The solution flowed through a hypodermic needle (0.2 mm ID by 0.4 mm OD with a sharply tapered tip) at a typical rate of 0.3 ml/min. For producing negative ions the needle was typically operated at -10 kV. Positive ions were produced by merely reversing the polarity of the needle supply. Nitrogen gas flushed through the spray chamber at room temperature and atmospheric pressure with flows in the range of 5 to 15 l/min. The nozzle and skimmer system was similar to that described by KANTROWITZ and GREY [24]. Both the skimmer and nozzle aperture were 0.1 mm in diameter and the nozzle to skimmer spacing was 0.75 mm. Pressure downstream of the skimmer was about 10^{-4} torr with

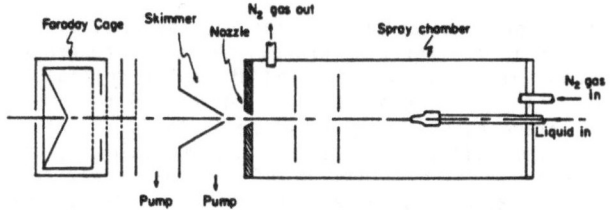

Fig.3. Electrospray apparatus. (From ref. [6] with permission)

the pumping system employed. Current to the first collimating plate was typically about 0.2 µA and current to the Faraday cup detector was typically in the 10^{-14} A range.

This apparatus was applied to studies of polystyrene solutions ranging in nominal molecular weight from 600 to 411,000 amu. Retarding potential energy spectra were obtained which were interpreted as indicating the presence of desolvated polystyrene macroions in the beam, but attempts at time-of-flight mass analysis of these ions were unsuccessful [7]. In retrospect, it appears that the spectrum of ions produced in the desolvation chamber must have been drastically altered by clustering reactions in the adiabatic expansion downstream of the nozzle. The utility of this particular technique for producing analytically useful ions has not yet been adequately established, and unfortunately this work seems to have been discontinued.

Ion Evaporation from Aqueous Droplets

IRIBARNE and THOMSON have reported some experiments which show that small ions evaporate from evaporating liquid droplets carrying electrical charges. Their initial experiments [8] employed a time-of-flight mobility analyzer to reproduce some much earlier results of CHAPMAN [25] indicating that singly-charged ions containing at most a few solvent molecules could be obtained by pneumatically spraying aqueous solutions at atmospheric pressure and room temperature. More recently [9] the ions were analyzed with a quadrupole mass spectrometer sampling directly from the atmosphere. A schematic diagram of the latter apparatus is shown in Fig.4. A pneumatic sprayer was used to atomize aqueous solutions of electrolytes into the air, and ions

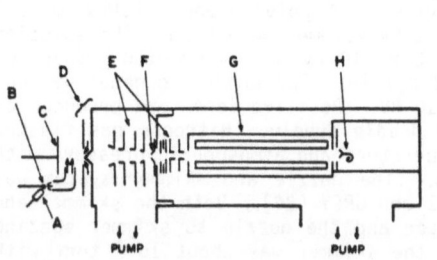

Fig.4. Experimental apparatus for studying ion evaporation from liquid droplets. A, induction electrode; B, sprayer nozzle; C, deflecting electrode; D, air current with confining plate and 25 m orifice; E, ion lenses; F, 2 mm differential pumping orifice; G, quadrupole mass filter; H, electron multiplier detector. (From ref. [9] with permission)

which evaporated from the droplets were drawn by an electric field toward a 25 mm orifice in the vacuum chamber. Ions and neutral molecules carried through the orifice were separated by focussing the ions with a series of electrostatic lenses and pumping away the neutrals with a high-speed oil diffusion pump. Differential pumping was used to maintain the first chamber at 10^{-4} torr and the quadrupole chamber at $<10^{-6}$ torr, under a total inflow of 1 torr 1/sec of air.

A unique feature of this apparatus is the "dry air curtain" developed to control clustering in the free jet expansion; this problem is particularly severe when sampling from a region of high humidity at atmospheric pressure and temperature. When the ions were sampled directly from the spray, many of them grew to a size beyond the range of the mass spectrometer (500 amu), and the rest were observed with 10 or more water molecules attached, making identification of the core ion difficult. The clustering was suppressed to a large degree by introducing a curtain of air (dried and purified through a charcoal trap in liquid nitrogen) between the orifice and the region of the evaporating spray. The curtain repelled the moist air while allowing the ions to be drawn through it by the electric field, and ensuring that very little water vapor was present in the free jet to condense on the ions. Evidence for the success of this technique is the fact that the cluster distributions observed are similar to those expected for equilibrium in dry air.

The droplets produced by the spraying process were charged primarily by the symmetric charging mechanism described by DODD [26]; however, since the ion currents were rather weak an induction electrode was placed near the sprayer head to increase the charge on the droplets. With ±3500 V (depending on the polarity of drop charge desired) on the electrode situated approximately 1 cm from the sprayer head, the net current from the sprayer increased by up to two orders of magnitude. After directing the spray through a 90° copper pipe elbow in order to remove the largest drops, the plume was passed across the orifice plate, far enough away not to wet it. A field of 700 V/cm was used to draw the ions toward the orifice, and under the most favorable conditions an ion current of 4×10^{-11} A could be measured inside the vacuum chamber. This current is very sensitive to the conditions of the air flow around the orifice area.

Results have been obtained from spraying a number of inorganic and organic electrolytes in aqueous solution, typically at concentrations of about 10^{-3} moles/liter. The spectra generally show that the ions expected to exist in solution are observed with approximately the degree of hydration expected for equilibrium in moist air at room temperature and atmospheric

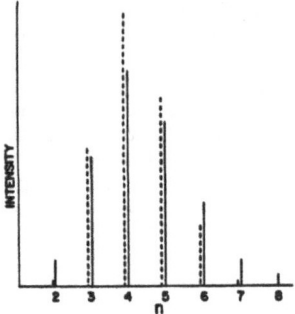

Fig.5. Comparison of the measured distribution of $Cs^+(H_2O)_n$ (full bars) and the calculated equilibrium distribution at 10.5 torr water pressure (dashed bars). n = number of H_2O molecules. (From ref. [9] with permission)

pressure. Results for Cs$^+$ are shown in Fig.5 where they are compared with the calculated equilibrium distribution. The observed spectra may be strongly affected by certain trace impurities; for example, the positive ion spectra generally show substantial contributions from hydrated ammonium ions due to trace amounts of ammonia present either in the solutions or in the air. Many of the negative ion spectra show clusters containing CO_2, HNO_2, and NO_2 also presumably due to trace impurities. Only monovalent ions are observed by this technique. Some transition metal ions which might be expected to be readily vaporizable were not observed. These include Ag$^+$, Tl$^+$, and Cu$^+$.

Recently this approach has been applied to some larger nonvolatile organic molecules which are ionized in aqueous solution [27]. Positive ion spectra have been reported for glutathione, arginine, tetrabutyl ammonium chloride, and dopamine, while the negative ion spectrum is reported for adenosine triphosphate disodium salt. Generally, compounds which have acidic hydrogens are observed as the deprotonated negative ion while compounds which have basic functions are observed in the positive ion mode as the protonated molecule. Little or no fragmentation is observed. In some larger molecules with multiple ionizable groups, multiply charged molecular ions are observed (for example $(M-2)^{2-}$ in ATP).

The mass spectrometer system employed in this work has been described in detail and is now commercially available [27].

Thermospray

The thermospray technique has emerged from efforts to develop an LC-MS interface suitable for efficiently analyzing samples dissolved in aqueous mobile phases at typical analytical flow rates on the order of 1 ml/min. Early efforts employed a focussed CO_2 laser to vaporize the solvent [10]. Later, the laser was replaced by an array of oxy-hydrogen torches [11-13]. In the present version of the apparatus an inexpensive electrical heater is employed. The original approach involved the production of a molecular beam of the vaporized effluent which could be directed into an EI or CI source of more-or-less conventional design. In the course of this work it was found that, under certain conditions, ions were produced even though the hot filament normally used to produce the primary ionizing beam was turned off [13]. Initial measurements of mass spectra produced from nonvolatile compounds such as peptides, nucleosides, and nucleotides showed that the spectra were quite different from those obtained by chemical ionization and were, in fact, most similar to those obtained by field desorption [28].

Since this initial discovery, a number of studies have been undertaken in an effort to elucidate the mechanism of the ionization process and to improve the analytical utility of the technique. A schematic diagram of the "thermospray" system presently in use is shown in Fig.6. This interface and ion source are much simpler than the earlier versions and requires only minor modification of a commercial quadrupole mass spectrometer. The present version of the vaporizer consists of a few centimeters of 0.015 mm ID × 1.5 mm OD stainless steel tubing which is brazed at one end into a copper block. The block is heated by two commercial 100 watt cartridge heaters normally operated at substantially below their rated power. The length of tubing immersed in the copper block is not particularly critical, but lengths on the order of 3 cm are presently used. It is very important that the stainless tube be in good thermal contact with the copper block; the brazed joint is essential. The heater temperature is controlled by a

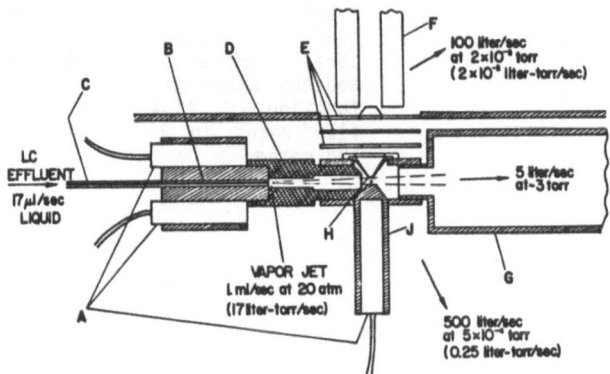

Fig.6. Schematic diagram of thermospray apparatus using electrically heated vaporizer. A, 100 watt cartridge heater; B, copper block brazed to stainless steel capillary; C, 1.5 mm OD × 0.15 mm ID SS capillary; D, thick-walled copper tube; E, ion lenses; F, quadrupole mass filter; G, pump line connected to mechanical vacuum pump; H, sampling aperture; J, source heater

proportioning temperature controller. When properly controlled, the system is capable of stably vaporizing up to 2 ml/min of aqueous effluent.

In the thermospray system the electrolyte is forced through the vaporizer by a high-pressure pumping system and sufficient thermal energy is coupled into the liquid to cause nearly complete vaporization at the rate the liquid is supplied. In the present system relatively high concentrations of ions in solution (typically 10^{-3} to 10^{-1} moles/liter) are required to achieve useful ion currents; however, the sample may be present at much lower levels. Thermospray ionization has been used successfully with a number of solvents including methanol, acetonitrile, etnanol, and propanol, and mixtures of these with water, and it appears that almost any solvent system can be used so long as the concentration of ions in solution is sufficient.

Most of the work to date has involved aqueous buffers at flow rates in the range from 0.5 to 2 ml/min, but it appears that both higher and lower flow rates can be accommodated. The performance at lower flow rates is improved by using smaller capillaries. At higher flow rates, larger heaters and faster pumping systems are required. The sample molecules need not be ionized in solution since they may be ionized by gas phase ion molecule reactions which apparently cannot be vaporized as intact neutral molecules must be present as ions in solution for intact pseudomolecular ions to be observed.

The ion current depends very strongly on both the flow rate and the power input to the vaporizer heater as shown in Fig.7. These results may be understood in terms of the steady—state vaporization model described earlier [29]. The optimum vaporizer temperature for a given flow rate appears to correspond very closely to that required for achieving vaporization of the liquid just as it reaches the exit of the capillary tube. At higher temperatures the heater can supply more than enough power for steady—state vaporization and the boundary between liquid and vapor recedes into the tube; at lower temperatures complete vaporization is not achieved. Visual observations of the thermospray vaporization process in air at atmospheric

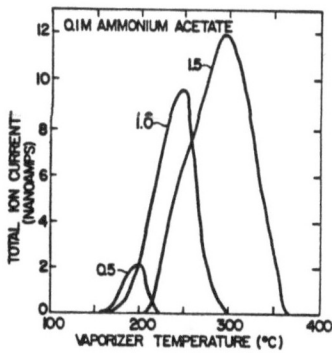

<u>Fig.7.</u> Total ion current entering the quadrupole produced by thermospray ionization of 0.1 M aqueous ammonium acetate as a function of flow rate. The flow rates in ml/min are as indicated

pressure are in agreement with this model. At temperatures and flow rates corresponding to the ion current maxima in Fig.7, the jet issuing from the vaporizer contains visible particles carried in an intense jet of vapor mostly confined to a cone with a half-angle of about 3 degrees, and the jet does not condense on a room temperature surface thrust into the jet. At higher temperatures the jet becomes invisible as complete vaporization is apparently achieved inside the capillary. At lower temperatures the jet becomes progressively weaker and wetter.

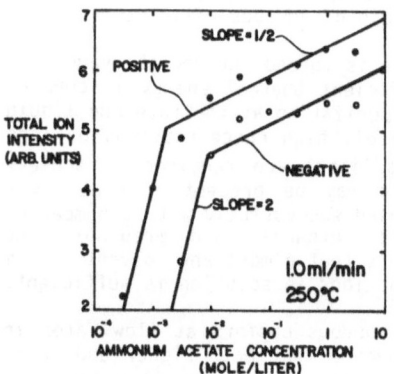

<u>Fig.8.</u> Total positive and negative ion intensities as a function of ammonium acetate concentration for thermospray ionization at a vaporizer temperature of 250°C and a flow rate of 1 ml/min. These results were obtained from measurements using the electron multiplier detector and the apparent differences in intensity between positive and negative ions is primarily due to differences in multiplier detection efficiency

The dependence of total ion current on concentration of ammonium acetate in solution is shown in Fig.8. At higher concentrations (ca. above 3×10^{-3} moles/liter) both the positive and negative currents vary approximately as the square root of the concentration of ions in solution. The apparent difference in positive and negative ion intensities at high concentrations is due primarily to the fact that the electron multiplier detector is somewhat less sensitive for negative ions. Absolute measurements of positive and negative ion intensities indicate that they are generally very nearly the same. Between 10^{-2} and 10^{-3} molar the ion intensities drop sharply with decreasing concentration in solution. The relative intensities of the major positive ions in the mass spectrum are shown as a function of ammonium acetate concentration in Fig.9. At low concentrations the spectrum consists only of NH_4^+ and its hydrates. At higher concentrations water is replaced by acetic acid and ammonia in the clusters. At the highest concentrations, masses 95 and 96 (not shown in Fig.9) are observed which

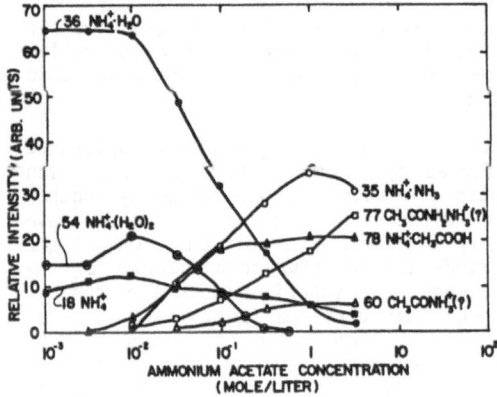

Fig.9. Major peaks in the positive ion thermospray mass spectrum of ammonium acetate as a function of ammonium acetate concentration. Conditions are the same as in the previous figure

correspond to addition of acetic acid to mass 35 and hydration of mass 78. The ions of m/e 77 and 60 do not correspond to simple cluster ions and may be formed by loss of water from the acetic acid-ammonia clusters on NH_4^+ at m/e 95 and 78, respectively. The negative ion spectrum is much simpler, consisting primarily of the acetate ion and its dimer with acetic acid. Very little hydration is observed unless the ion source temperature is much lower than the normal value of ca. 200-250°C. At higher concentrations the acetic acid trimer is also observed.

From results such as those presented above it is quite clear that the primary ions in the mass spectrum are those present in solution; however, the mass spectrum contains significant contributions from secondary ions produced by gas phase ion molecule reactions in addition to the primary ions. It appears that the primary ionization process produces the ionic species present in solution (e.g., MH^+) with little fragmentation, but the gas phase processes may produce significant fragmentation as well as ionization of neutral fragments produced thermally. In some cases quite useful fragmentation patterns are obtained. For example, the spectrum of the dinucleotide ApU, shown in Fig.10, allows the structure of the dinucleotide to be unambiguously determined. Very similar results on dinucleotides have been obtained by SCHULTEN using field desorption [28].

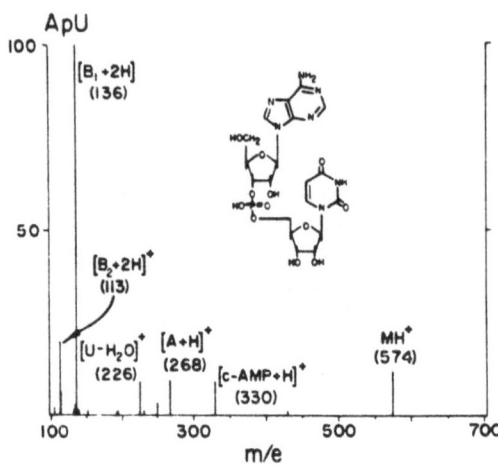

Fig.10.
Positive ion thermospray mass spectrum of the dinucleotide ApU obtained using 0.2 M formic acid at a flow rate of 0.5 ml/min. Obtained from a single scan from 100 to 700 amu in 4 sec. (From ref. [13])

The degree of fragmentation in thermospray ionization depends rather strongly on the operating conditions and very large, poorly controlled variations in fragmentation are sometimes observed. This behavior also appears to be similar to that observed in FD.

One of the reasons for the observed variation in fragmentation appears to be that the fragments are mostly formed by one mechanism (gas phase ion molecule reaction) while the protonated molecular ions are formed by another (direct desorption from the liquid), and the relative contributions of these two mechanisms depends rather strongly on operating conditions. One of the parameters which can be varied to study the relative contributions of the two mechanisms is the distance between the vaporizer and the conical sampling aperture. As shown in Fig.6, this distance is typically several centimeters, but by decreasing it drastically the ratio of primary ions to secondary ions can be increased since the available reaction time is decreased. Some results obtained with the vaporizer located 1 cm from the sampling cone are shown in Figs.11 and 12. For these experiments a thermocouple was placed near the axis of the jet just downstream from the sampling aperture and the temperature indicated by that thermocouple is shown along the top of the figures. The data shown in Fig.11 was obtained using 0.1 M ammonium acetate. The dashed line corresponds to the total intensity of the solvent ions (NH_4^+ and its clusters) while the solid lines correspond to the MH^+ intensities produced by injection of 10^{-4} molar solutions of leucine and arginine, respectively, into the ammonium acetate solution. It is clear that leucine (which is neutral in solution under these conditions) is being ionized primarily by gas phase reactions, while MH^+ ions from arginine are produced solely by direct desorption from solution. It should be noted that fragments of arginine are observed at the lower flow rates (and higher vapor temperatures) and the intensities of the fragment ions behave similarly to the MH^+ ion of leucine. Fig.12 shows the results on arginine in pure water. A higher concentration of arginine is required to produce significant ionization since no other source of ions is present. At high flow rates and low vapor temperatures, the spectrum contains only the MH^+ ion, while at lower flow rates and higher vapor temperatures significant fragmentation occurs. For example, at a flow of 1.3 ml/min the MH^+ ion is about 20% of the total ionization. This fragmentation is presumably the result of excess thermal energy in the desorbed MH^+ ions, although we cannot rule out the possibility of some contribution from thermal desorption of neutral arginine molecules or fragments followed by gas phase reactions with MH^+ since the concentration of arginine used in this experiment was rather high.

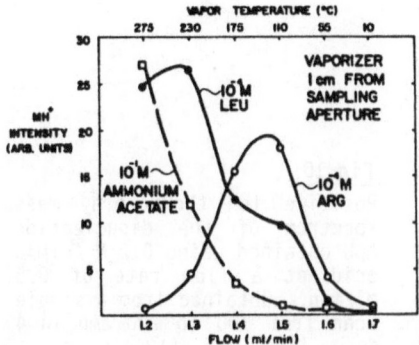

Fig.11. Thermospray ionization of arginine and leucine in 0.1 M ammonium acetate obtained with the copper tube (D in Fig.6) removed and the vaporizer installed directly in the ion source about 1 cm from the sampling aperture. The vapor temperature indicated at the top of the figure was measured with a thermocouple placed near the center of the jet about 0.1 cm downstream of the sampling aperture

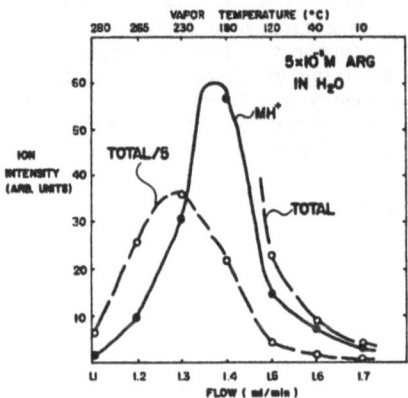

Fig.12. Thermospray ionization of arginine in pure water. Experimental conditions are the same as in the previous figure

The development of the thermospray technique has reached the point that it appears to be a useful analytical tool both for producing spectra of nonvolatile, thermally-labile molecules and also for use in an on-line LC-MS system. An example of the LC-MS performance is shown in Fig.13. Similar results have been obtained on a variety of other classes of compounds including amino acids, small peptides, nucleosides, and antibiotics. Despite these successes, some additional research is required to develop a system which is routinely applicable to a broad range of analytical problems. Many details of the mechanism of ionization are not yet well understood and the optimum configuration of the vaporizer and ion sampling has not been established. Major limitations on the technique at present are the difficulties with day-to-day reproducibility of ionization efficiency and fragmentation patterns and the fairly severe dependence of the ion intensities on the liquid flow rate.

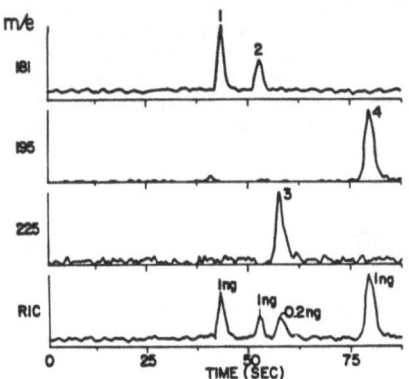

Fig.13. LC-MS analysis of a mixture of xanthine derivatives; 1, theobromine; 2, theophylline, 3, β-hydroxyethyltheophylline; 4, caffeine. Obtained using a 3 μm Ultrasphere (Altex) ODS column and mobile phase consisting of 12% acetonitrile in 0.1 M ammonium acetate at a flow rate of 1.5 ml/min. MS obtained using thermospray ionization of the total LC effluent; sample quantities are indicated on the figure

Mechanism of Ion Desorption from Liquids

The rate of vaporization of molecules from a surface can be represented approximately by an Arrhenius equation of the form

$$Z = vC_s \, e^{-Q/kT} \quad , \tag{1}$$

where Q is the activation energy for vaporization, C_s is the surface concentration of the molecule of interest, v is a frequency factor, T is the absolute temperature, and k is Boltzmann's constant. For application to neutral molecules, Q can be taken as the enthalpy of vaporization and the product vC_s can be estimated from kinetic theory and thermodynamics. For example, if we neglect the temperature dependence of the enthalpy of vaporization, the rate of evaporation of a pure liquid is given by

$$Z = \frac{P_v(T)}{(2\Pi mkT)^{\frac{1}{2}}} = \frac{P_v(T_0)}{(2\Pi mkT)^{\frac{1}{2}}} e^{\Delta H_v/kT_0} \cdot e^{-\Delta H_v/kT} \quad . \tag{2}$$

Thus

$$vC_s = \frac{P_v(T_0)}{(2\Pi mkT)^{\frac{1}{2}}} e^{\Delta H_v/kT_0} \tag{3}$$

where T_0 is a convenient reference temperature, such as the boiling point. If we independently estimate C_s from the molecular diameter, then the frequency factor can be calculated and is generally found to correspond to a typical vibrational frequency of ca. 10^{14} sec^{-1}.

Equation (1) appears to provide a satisfactory approximation to the rate of vaporization of any substance provided the necessary parameter values are available. The pre-exponential factors can be estimated satisfactorily from the molecular diameter and mass, and errors introduced by uncertainties in these values will generally be negligible compared to the overriding effect of the exponential. The activation energy for evaporation of a positive ion from the surface of a pure metal is given by

$$Q = e(V_I + \lambda - \phi) \tag{4}$$

where V_I is the ionization potential, λ the enthalpy of vaporization and ϕ the work function of the metal. For metal surfaces the major short range force between an ion and the surface can be represented by the coulombic attraction between the ion and its image in the surface. If we consider only the simplest case in which only the image force is present, then the effect of applying an electric field to accelerate ions away from the surface is the Schottky lowering of the potential barrier by an amount proportional to the square root of the field [30]. This gives for the activation energy

$$Q = e(V_I + \lambda - \phi - \sqrt{eE/4\Pi\varepsilon_0}) = e(\lambda_I - \sqrt{eE/4\Pi\varepsilon_0}) \quad . \tag{5}$$

As illustrated above, the activation energy from evaporation of metal ions from a metal surface can be obtained from well-known properties of the metal. Unfortunately, in the case of evaporation of ions from liquid solution, the parameter values are not known with accuracy. Furthermore, serious questions may be raised concerning the applicability of this model to the physical situation involved.

IRIBARNE and THOMSON [8] have discussed the energetics of ion evaporation from aqueous solution. Enthalpies and free energies of vaporization of ions can be calculated using thermochemical cycles from tabulated thermochemical data, provided the absolute values for solvation of the proton are known. These have been calculated, but there is apparently little experimental verification of the values. A further complication is that the ions

departing the liquid surface are almost certainly partially solvated since
the energy barrier is much lower for a small cluster ion than for the bare
ion. Using thermochemical data and the measurements of KEBARLE [31],
IRIBARNE and THOMSON [8] have estimated values for several small ions such
as H_3O^+, OH^- and the alkalis and halides. These values appear to fall
primarily in the range between 2 and 3 eV.

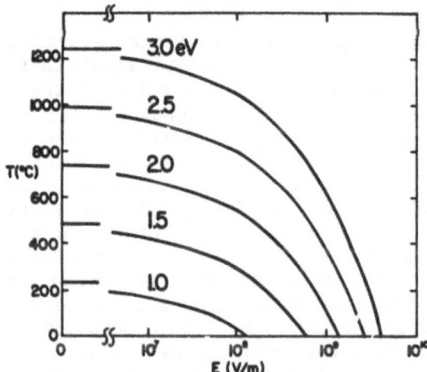

Fig.14. Calculated temperature-field loci for vaporizing ions with rate
constant of 10^4 sec^{-1}. Parameter is the enthalpy of vaporization in
eV. The frequency factor, v, is taken as 10^{14} sec^{-1}

The rate equation can be used to predict the temperature and field
required to evaporate ions at a sufficient rate to be observed in a
particular experiment. An example of such a calculation is shown in Fig.14,
where results are given for obtaining a vaporization rate constant of 10^4
sec^{-1} for activation energies in the range between 1 and 3 eV. From these
results it is apparent that whether a particular vaporization process is
considered field desorption or thermal desorption depends on the experimental conditions as well as the activation energy for desorption.
Activation energies for evaporation of metal ions are typically 3 eV or
higher and temperatures in excess of 1200°C (at low field) or fields in
excess of 3×10^9 V/m (at low temperatures) are required to produce significant rates of desorption. For the desorption energies estimated for small
ions in aqueous solution (ca. 2-3 eV), fields on the order of 10^9 volts/m
are required to produce significant rates of desorption at room temperature
according to this calculation. Fields on the order of 10^8 V/m lower the
temperature required by about 200°C, but this reduction appears insufficient
to significantly enhance the rates of desorption of ions with activation
energies above 2 eV.

The results of these calculations appear to be inconsistent with recent
experimental results on desorption of ions from liquid surfaces. For
example, GIESSMANN and RÖLLGEN [14] have determined threshold fields in the
range of 10^7 to 10^8 V/m for field desorption from aqueous solutions. There
are several possible explanations for this apparent discrepancy. The
activation energies for desorption of ions from liquid solutions are not
known accurately, but perhaps more importantly, some aspects of the
classical field desorption model may not be realistic for desorption of
solvated ions from liquids. One important difference between solids and
liquids, not taken into account in the model, is the local enhancement of

the electrical field which occurs due to field-induced distortion of the liquid surface. Also, it appears that the image force approximation involved in the field-induced lowering of the potential barrier may not be simply applicable to liquid surfaces.

For an ion imbedded in a liquid surface the effect of an external electrical field is substantially different from that for an ion adsorbed on a clean metal surface. In the case of the metal, the image force is clearly a good approximation to the important short range interaction of the ion with the surface, in that it accurately represents the force between the charge and the induced charge distribution on the smooth equipotential surface of the metal. The only approximation is that the metal is treated as a perfectly smooth surface of essentially infinite conductivity; the fact that it is composed of discrete atoms is ignored. However, in the case of a liquid surface distorted by an applied electrical field, the important short-range force on an ion at or near the surface appears to be the liquid cohesive forces which can be represented macroscopically by the surface tension. Until the ion separates from the liquid surface the image force does not exist. After the ion or small cluster has separated from the surface, it will feel an attractive force similar to the image force, but generally smaller in magnitude than in the case of solid metals. Except for the effects of field-induced distortion of the surface, liquid metals behave similarly to solids in that the mobile free electrons allow the surface to remain an equipotential throughout the ion desorption process. However, ions in aqueous solution are much less mobile and it appears likely that the desorbed ion will depart from the surface before the surface equipotential can be re-established. As a result, the image force which is of dominant importance in ion desorption from metals may be weak and unimportant for desorption from poor conductors such as solutions and insulators.

It presently appears that the effect of electric field on desorption of ions from solutions is not correctly predicted by the use of (5) in the rate expression. Rather, it appears that the ion emission process may be initiated by field-induced surface instabilities resulting in Taylor cones [32] on the surface. Since these local projections may significantly enhance the local field so that ions readily evaporate, the threshold for desorption may be determined primarily by the field strength required to initiate the surface instabilities rather than that required for direct emission. From the recent work of GIESSMANN and RÖLLGEN [14] it appears that fields of only 10^7 to 10^8 V/m are required to initiate this process.

Ion Desorption from Liquid Droplets

The mechanism for ion desorption from liquid surface seems to be essentially the same whether the liquid is attached to a solid support as in FD or EH ionization or is present as free, charged droplets. The major difference in the latter case is that the field is self-induced due to the charge on the isolated droplet. When a liquid surface is disrupted by spraying or bubbling, the droplets produced are often electrically charged. For liquids containing substantial concentrations of dissolved ions the charging mechanism (in the absence of an applied electrical field) appears to be dominated by the statistical mechanism described by DODD [26]. If the separation of the droplet from the bulk liquid occurs sufficiently rapidly that conduction in the liquid can be neglected, then the net charge on the droplets is determined by statistical fluctuations. In this case the resulting charge distribution is Gaussian with zero mean and a standard deviation equal to the square root of total number of ions within the volume of the droplet. The absolute mean charge (of either sign) is given by

$$\langle q \rangle = (\frac{2}{\pi})^{\frac{1}{2}}\sigma = (\frac{4VN}{\pi})^{\frac{1}{2}} \quad , \tag{6}$$

where V is the volume of the droplet and N is the number of ions of each sign per unit volume.

There are several limits on the charge which a liquid droplet can accommodate. The Rayleigh limit on the charge on a droplet of radius r is given by

$$n_R = \frac{8\pi(\gamma\varepsilon_0)^{\frac{1}{2}}}{e} r^{3/2} \quad , \tag{7}$$

where γ is the coefficient of surface tension and ε_0 is the permitivity of free space. If the charge on the droplet exceeds this limit it is unstable with respect to subdivision into smaller droplets. Another limit on the droplet charge is provided by the field desorption processes discussed above. This limit is given by

$$n_E = \frac{4\pi\varepsilon_0}{e} E_I r^2 \quad , \tag{8}$$

where E_I is the minimum electrical field required for ion emission at an observable rate. An absolute limit for stability of the droplet relative to ion emission is given by

$$n_G = \frac{4\pi\varepsilon_0}{e^2} \Delta G_I r \quad , \tag{9}$$

where ΔG_I is the change in free energy associated with desorption of the minimum energy cluster ion. Because of the different radial dependence of the limits given in (7) through (9), the important limit depends on the droplet size. Except at very small radii, the lower limit on droplet charge is given by (9); however, the important limit practically is likely to be the ion emission limit (8) since even though the particle is unstable with the rate of emission is likely to be very low unless (8) is exceeded also.

These limits for desorption of a typical ion from an aqueous droplet are illustrated in Fig.15. In this illustration the minimum field for ion emission has been taken somewhat arbitrarily as 10^8 V/m. The dashed line in Fig.16 corresponds to the mean charge expected for spraying a 0.1 M solution and the point corresponds to an initial sprayed droplet with a 1 μm radius. Although this droplet as initially formed is unstable with respect to ion emission, the initial rate of ion evaporation will be low compared to the rate of neutral evaporation. As a result the droplet will vaporize at essentially constant charge until the path on Fig.15 crosses the field desorption limit. If the droplet temperature is maintained sufficiently high, the droplet will vaporize further emitting both neutral molecules and ions with the charge decreasing as a function of radius along the limiting line given by (8), until the energy limit (9) is reached or until essentially all of the solvent has been evaporated and the droplet becomes a solid particle of nonvolatile solute.

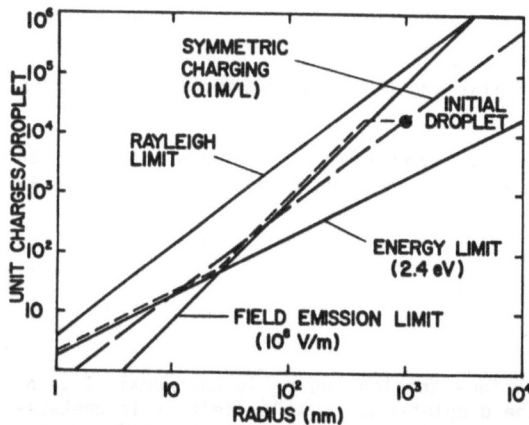

<u>Fig.15.</u> Calculated limits on charge that can be accommodated by a water droplet. The short, dashed line shows the charge as a function of droplet radius for an evaporating droplet initially of 1 μm radius

This model appears to account, at least qualitatively, for the ions observed in the experiments of THOMSON and IRIBARNE [8,9] as well as our observations of thermospray ionization; however, insufficient data are available for reliable quantitative conclusions to be drawn at this time. The extent to which this model applies to desorption of molecular ions from solid surfaces by particle impact or laser irradiation is not yet established. These techniques appear to initially produce a highly excited, partially ionized "clump" of high-density material. It appears that the net charge on this material as it separates from the bulk surface may be determined primarily by the symmetric charging mechanism described above for liquid droplets, and if the net charge is sufficiently high, evaporation of molecular ions can occur. While quite speculative at present, this mechanism appears to provide a possible explanation for some puzzling aspects of molecular ion desorption from solids. In particular, it appears consistent with observations that higher energy densities often lead to more efficient production of molecular ions, and it may also provide a mechanism for efficient transfer of electronic excitation (ionization) into translational energy of desorbed ions.

This work was supported by the Institute of General Medical Sciences (NIH) under Grant GM 24031. We thank Ms. C. R. Hsieh for preparing the figures.

References
1. For references to the earlier literature see, Leonard B. Loeb, <u>Static Electrification</u>, Springer-Verlag, Berlin, 1958.
2. C. A. Evans, Jr. and C. D. Hendricks, Rev. Sci. Instr. <u>43</u>, 1527 (1972).
3. D. S. Simons, B. N. Colby, and C. A. Evans, Jr., Int. J. Mass Spectrom. Ion Phys. <u>15</u>, 291 (1974).
4. B. P. Stimpson and C. A. Evans, Jr., Biomed. Mass Spectrom. <u>5</u>, 52 (1978).
5. B. P. Stimpson, D. S. Simons, and C. A. Evans, Jr., J. Phys. Chem. <u>82</u>, 660 (1978).

6. M. Dole, L. L. Mach, R. L. Hines, R. C. Mobley, L. D. Ferguson, and M. B. Alice, J. Chem. Phys. 49, 2240 (1968).
7. L. L. Mack, P. Kralick, A. Rheude, and M. Dole, J. Chem. Phys. 52, 4977 (1970).
8. J. V. Iribarne and B. A. Thomson, J. Chem. Phys. 64, 2287 (1976).
9. B. A. Thomson and J. V. Iribarne, J. Chem. Phys. 71, 4451 (1979).
10. C. R. Blakley, M. J. McAdams, and M. L. Vestal, J. Chromatogr. 158, 264 (1978).
11. C. R. Blakley, J. J. Carmody, and M. L. Vestal, Anal. Chem. 52, 1636 (1980).
12. C. R. Blakley, J. J. Carmody, and M. L. Vestal, Clin. Chem. 26, 1467 (1980).
13. C. R. Blakley, J. J. Carmody, and M. L. Vestal, J. Am. Chem. Soc. 102, 5931 (1980).
14. U. Giessmann and F. W. Röllgen, Int. J. Mass Spectrom. Ion Phys. 38, 267 (1981).
15. V. E. Krohn, Jr., Progr. Astronaut. Rocketry 5, 73 (1961); J. Appl. Phys. 45, 1144 (1974).
16. J. F. Mahoney, A. Y. Yahiku, H. L. Daley, R. D. Moore, and J. Perel, J. Appl. Phys. 40, 1501 (1969).
17. D. S. Swatik and C. D. Hendricks, AIAA J. 6, 1596 (1968); C. D. Hendricks and D. S. Swatik, Astronaut. Acta, 18, 295 (1973).
18. V. E. Krohn, Jr., Progr. Astronaut. Rocketry 9, 435 (1963).
19. P. W. Kidd, J. Spacecr. Rockets 5, 1034 (1968).
20. R. J. Pfeifer and C. D. Hendricks, AIAA J. 6, 496 (1968).
21. J. Perel, J. F. Mahoney, R. D. Moore, and A. Y. Yahiku, AIAA J. 7, 507 (1969).
22. M. N. Huberman, J. Appl. Phys. 41, 578 (1970).
23. S.-T. F. Lai, K. W. Chan, and K. D. Cook, Macromolecules 13, 953 (1980).
24. A. Kantrowitz and J. Grey, Rev. Sci. Instr. 22, 328 (1951).
25. S. Chapman, Phys. Rev. 52, 184 (1937); 54, 520 (1938); 54, 528 (1938).
26. E. E. Dodd, J. Appl. Phys. 24, 73 (1953).
27. Sciex Application Note 7782-A, Sciex Corporation, Thornhill, Ontario, Canada.
28. H.-R. Schulten and H. M. Scheibel, Nucleic Acids Res. 3, 2027 (1976).
29. C. R. Blakley and M. L. Vestal, Anal. Chem., in press (1982).
30. For an excellent review of field-induced and thermal desorption from metal surfaces see N. I. Ionov, "Surface Ionization and its Applications" pp. 237-354, Progress in Surface Science, Vol. I, S. G. Davison, Ed., Pergamon, Oxford, 1972.
31. P. Kebarle, in Ion-Molecule Reactions. J. L. Franklin, Ed., Plenum, New York, 1972, Vol. 1, Chap. 7, p. 315.
32. G. Taylor, Proc. Roy. Soc. London A280, 383 (1964).

5.2 Fast Dust Particles as Primaries – Comparison of Ion Formation with Other Techniques

Franz R. Krueger

Max-Planck-Institut für Kernphysik, Abt. Kosmophysik,
D-6900 Heidelberg, Fed. Rep. of Germany

1. Introduction

Rapid dissipation of energy is a prerequisite common to most of the modern techniques of ion formation from organic solids. With the particle impact methods like SIMS and fast atom bombardment (FAB) on the one hand and fission-fragment induced desorption (FFID), or more generally with respect to the type of primary ion, fast heavy ion-induced desorption (HIID) on the other hand, energy is brought into the lattice very rapidly either directly by collision cascades or via electronic excitation and subsequent polaron inter-action. By means of Q-switched laser pulses electronic excitation takes place causing a rapid heating of the lattice, too. As a consequence, prompt ion formation from the surface occurs. Neither ion formation rates nor ion-to-neutral ratios, neither secondary ion type distributions nor primary ion/laser parameter dependencies are compatible with any quasiequilibrium proces-ses [1]. It is assumed that these phenomena are to be treated in terms of a non-equilibrium surface-gas phase transition theory.

In order to enlarge the scope of the methods and learn more about these rapid dissipation-ion formation phenomena, the impact of Fe-dust particles (0.1 - 30 μm ∅) within the velocity range of 1 to 60 km/sec on metal foils coated with thin (~.1 μm) layers of organics and salts was studied.

At least with the higher particle velocities complete destruction of the projectile and an appropriate target regime has been expected forming a fully atomized plasma [2], whereas particles of lower velocities should produce atomic alkali ions only [3]. Neither the calculated distribution of plasma ions has been found so far, nor was ion formation thus simple with lower ve-locities. Instead quasimolecular and cluster ion formation as known from the techniques mentioned above has been found and investigated in more detail.

The greater scope of particle impact ion formation is related to the "Giotto" space flight mission to comet Halley in 1986: then a time-of-flight (TOF) mass spectrometer will analyze the particles of the comet tail with a velocity relative to the space-craft of 70 km/sec by impacting a metal foil. In order to interpret the TOF-mass spectra properly the ion formation pro-cesses should be understood.

2. Experimental conditions

The generation of swift particles is maintained by a van de Graaf accelerator and a special dust source as described elsewhere [4]. The particles are then selected according to their velocity range by an electric switch triggered by a time-of-flight Δt-delayed coincidence signal obtained with two position detectors (Δs distance) in the flight tube. Their exact velocity $v = \Delta s/\Delta t$ is measured the same way. Their mass m is given by $m = 2qU_o/v^2$ and may be obtained by measuring the charge q via induction (U_o: acceleration voltage). The switch signal is also used as trigger signal for the Biomation transient recorder, which measures the time-dependent secondary ion current at the TOF-detector. This TOF arrangement is very much comparable with those used with the well-known LAMMA®-technique. However, the dynamic range for ion detection is more than three decades due to the logarithmic amplifier used. So it is justified to compare the mass spectra of both techniques, as the area of energy dissipation is related to the focal diameter d_1 of the laser pulse or the particle diameter d_p, respectively, with $d_1 = 0.6 \dots 3.0$ μm, and $d_p = 0.2 \dots 20$. μm. However, the time period of dissipation is related to the pulse length τ_1 of the laser ($\tau_1 = 10 \dots 30$ nsec, or ~.1 nsec for "pico-second" lasers), but is in the order of 2 nsec for particle dissipation (for 40 km/sec) to about 5 nsec for lower velocities [2]. Consequently, the energy flux densities are not thus comparable. However, it has been calculated and observed by means of crater formation [5], that the thermodynamic state of the shocked material during the impingement of the particle ($v_p = 70$ km/sec) on the target is simulated by laser irradiation (=1 nsec) of a flux of 10^{13} W/cm², whereas similar ion formation is observed with some 10^9 W/cm², often without any cratering. The above-mentioned shocked material is believed to expand into a very hot plasma with a correlated ion flow of a velocity v_{ion} approximately $v_p/2$. (1 keV for Fe [2]) If this would be true, no TOF-measurements of target and/or projectile atomic ions would be possible due to severe line broadening. However, well resolved lines are observed. It seems that the ejected material is almost neutral, only causing secondary ionization at the chamber walls.

With FFID and HIID the dissipation occurs in much shorter time intervals (polaron lifetimes ~10^{-12} sec) in smaller surface areas (some shielding lengths). So these TOF mass spectra may be even less comparable. Due to the mainly electronic interaction of swift ions, generally no target metal ions are observed. However, charged surface states are excited either, and one should expect the same transient ion formation phenomena with dielectrics, at least qualitatively.

The samples have been electrosprayed [6] on silver or iron targets. Only positive ion mass spectra have been taken so far.

Table 1. Approx. relative intensities of the main lines of the positive ion mass spectra of tetrabutylammonium iodide [7]

Meth\m/z	611	368	242	184/6	142	100	57	41/2	28-30
FFID	20	20	75	10	65	7	32	74	150
SIMS	?	1	100	8	30	10	20	45	40
LAMMA^R	2	-	100	10	15	4	8	5	15
PaID 1.8	-	-	100	5	25	3	10	10	20
6.	1	-	100	10	30	5	10	10	20
30.	1	-	100	40	80	60	70	50	20
[km/sec]									

m/z

611: M_2J^+

368: $MJ\pm1^+$

242: M^+

M=But$_4$N

3. Secondary ion distributions

3.1 Tetrabutylammonium Iodide on Iron

The mass spectra obtained by the dust technique are qualitatively the same as known from FFID, SIMS, and LAMMA®. The relative intensities of the molecular lines are shown in Table 1. for three different particle velocities.

3.2 Adenine on Silver

The only intense mass lines with adenine on silver have been due to Na^+, K^+, Ag^+, Ag_2^+; NH_4^+, $C_2NH_6^+$, $(M+H)^+$, $(M+Na)^+$, and $(M+Ag)^+$. The yield of fragment ions was thus remarkably low. This is in good agreement with the LAMMA 1000 ®-technique and SIMS [8] , but differs from the results obtained with FFID and LAMMA 500®.

3.3 Alkali Iodides on Silver

A mixture of $Na/K/Rb/Cs-J$ $10^3:10^2:10:1$ in molar concentrations has been investigated, due to the fact known from other techniques [1] , that the ion species related to heavier alkali atoms are overrepresented in most spectra, esp. K vs. Na, and Cs vs. all other. The identified lines (nearly all have

m/z	Chemical Symbols	Relat. Intens.
23	Na	90
39	K	100
86	Rb	95
108	Ag	20
133	Cs	75
173	Na_2J	30
189	NaKJ	30
205	K_2J	40
216	Ag_2	2
236	NaRbJ	10
252	KRbJ	15
258	NaJAg	1
274	KJAg	1
283	NaCsJ	4
299	KCs/Rb_2-J	8
321/4	$RbJAg, Ag_3$ / Na_3J_2	9
339	Na_2KJ_2	4
346	RbCsJ	2
355	NaK_2J_2	3

Table 2. Relative intensities of ions, AlkJ on silver; Fe projectile particle $v_p=9$ km/sec.

(For sake of simplicity $M_{Rb}=86$ and $M_{Ag}=108$ was taken, but correct summation over the isotopes has been performed)

m/z	Ch.S.	Rel.I.
371	K_3J_2	1
386	Na_2RbJ_2	1
393	Cs_2J	1
402	$NaKRbJ_2$	1

been identified) are due to the types Alk^+, Ag_n^+ (n=1,2,3), Alk_2J^+, $AlkJAg^+$, and $Alk_3J_2^+$. The relative yields of the four Alk^+ lines (Rb^+ twin line) are constant over the entire velocity range, whereas the cluster ion intensities decrease slightly with increasing velocity. Heating up the target to about 500^oC results in disappearing of the cluster ions and the Rb^+ and Cs^+ ion lines. After cooling to room temperature again, the Rb^+ and the cluster ion lines due to Na/K-J recover partly.

4. Ion Formation Processes

The data are inconsistent with plasma ion formation models as calculated by MALAMA and LEONAS [2] as well as those and ion desorption models by DRAPATZ and MICHEL [3] concerning the impact processes.

When assuming a plasma, it has to be very cold ($T \ll 3000$ K) and of equal temperature over the entire velocity range, due to the known differences of alkali iodide ion formation enthalpies and the results from Sect. 3.3.

Almost no radical ions have been found, thus electron impact mechanisms in the gas phase cannot play a major role. Furthermore gas phase collisions should be rare, due to the large amounts of intact organic cations observed, at least for particle velocities below 15 km/sec.

No line broadening due to plasma- (Debye-) density or correlated flow phenomena has been observed, although expected by the calculations of [2]. There are only tails of the Alk^+-lines, apparently due to thermal evaporation after the impact.

Any (plasma- or surface evaporation-) "temperature" derived from the prompt ion distributions should be a varying function of the impact velocity at all ranging from $5 \cdot 10^3$ K (low v_p) to $2 \cdot 10^5$ K (high v_p), a clear disagreement with the experimental fact that there is almost no dependence of relative ion intensities vs. projectile velocity. Other authors (Cosmophysics, Moscow) claim the "plasma" to be a cloud of mainly neutrals, admitting that ions may be produced from the neighbor surface and some sputtering. This view is consistent with the above measurements.

The often-discussed ion formation mechanism being due to a strong sudden perturbation (here caused by the shock wave) thus leading to a far from equilibrium surface-gas phase transition where kinetic time scales are more important than enthalpic energy scales, is believed to be also responsible mainly for these impact phenomena. Independence on the primary perturbation conditions over a wide range of parameters and methods, but strong dependence on the chemical surface composition and sample temperature are common phenomena related to those fast non-equilibrium phase transitions as observed with all these fast dissipation techniques [10].

I should like to thank Mr. W. Knabe and Dr. J. Kissel for their kind support.

References

1. B.Jöst, B.Schueler, F.R.Krueger, Z.f.Naturf. 37a (1982), p. 18

2. V.B.Leonas, Yu.G.Malama, preprint presented at the Halley investigators meeting (Moscow 1982)

3. S.Drapatz, K.W.Michel, Z.f.Naturf. 29a (1974), p. 870

4. H.Fechtig, E.Grün, J.Kissel in "Cosmic Dust" ed. by J.A.M.McDonnell (Wiley 1978)

5. French group; Halley investigators meeting (Moscow 1982)

6. F.R.Krueger, Chromatographia 10 (1977), p. 151

7. B.Schueler, F.R.Krueger, Org. Mass Spectrom. 14 (1979), p. 439

8. A.Eicke, W.Sichtermann, A.Benninghoven, Org. Mass Spectr. 15 (1980), 289

9. B.Schueler, F.R. Krueger, Org. Mass Spectrom. 15 (1980), p. 295

10. F.R.Krueger, Appl. Surf. Science 11 (1982), p. 819

Index of Contributors

E.A. Silinsh

Organic Molecular Crystals

Their Electronic States

Translated from the Russian by J. Eiduss in collaboration with the author
1980. 135 figures, 54 tables. XVII, 389 pages
(Springer Series in Solid-State Sciences, Volume 16)
ISBN 3-540-10053-9

Contents: Introduction: Characteristic Features of Organic Molecular Crystals. – Electronic States of an Ideal Molecular Crystal. – Role of Structural Defects in the Formation of Local Electronic States in Molecular Crystals. – Local Trapping Centers for Excitons in Molecular Crystals. – Local Trapping States for Charge Carriers in Molecular Crystals. – Summing Up and Looking Ahead. – References. – Additional References with Titles. – Subject Index.

W. Press

Single-Particle Rotations in Molecular Crystals

1981. 53 figures. IX, 129 pages
(Springer Tracts in Modern Physics, Volume 92)
ISBN 3-540-10897-1

Contents: Introduction. – Interaction and Rotational Potentials. – Neutron Scattering. – Stochastic Rotational Motion. – Rotational Excitations at Low Temperatures I. Principles. – Rotational Excitations at Low Temperatures II. Examples. – Rotational Excitations at Low Temperatures III. Special Features. – Appendix: Calculation of Transition Matrix Elements. – List of Symbols. – References. – Subject Index.

Mössbauer Spectroscopy II

The Exotic Side of the Method

Editor: U. Gonser
1981. 67 figures. XII, 196 pages. (Topics in Current Physics, Volume 25).
ISBN 3-540-10519-0

Contents: U. Gonser: Introduction. – R.L. Mössbauer, F. Parak, W. Hoppe: A Solution of the Phase Problem in the Structure Determination of Biological Macromolecules. – R.V. Pound: The Gravitational Red-Shift. – V.I. Goldanskii, R.N. Kuzmin, V.A. Namiot: Trends in the Development of the Gamma Laser. – R.L. Cohen: Nuclear Resonance Experiments Using Synchrotron Sources. – U. Gonser, H. Fischer: Resonance γ-Ray Polarimetry. – B.D. Sawicka, J.A. Sawicki: Iron-Ion Implantation Studied by Conversion-Electron Mössbauer Spectroscopy. – R.S. Preston, U. Gonser: Selected "Exotic" Applications. – S.S. Hanna: The Discovery of the Magnetic Hyperfine Interaction in the Mössbauer Effect of ^{57}Fe.

Crystal Cohesion and Conformational Energies

Editor: R.M. Metzger
1981. 55 figures. XI, 154 pages. (Topics in Current Physics, Volume 26).
ISBN 3-540-10520-4

Contents: R.M. Metzger: Introduction. – D.E. Williams: Transferable Empirical Nonbonded Potential Functions. – F.A. Momany: Conformational Analysis and Polypeptide Drug Design. – R.M. Metzger: Cohesion and Ionicity in Organic Semiconductors and Metals. – B.D. Silverman: Slipped Versus Eclipsed Stacking of Tetrathiafulvalene (TTF) and Tetracyanoquinodimethane (TCNQ) Dimers.

Springer-Verlag Berlin Heidelberg New York Tokyo

Applied Physics A
Solids and Surfaces

Applied Physics A "Solids and Surfaces" is devoted to concise accounts of experimental and theoretical investigations that contribute new knowledge or understanding of phenomena, principles or methods of applied research.

Emphasis is placed on the following fields:

Solid-State Physics
Semiconductor Physics: **H. J. Queisser**, MPI Stuttgart
Amorphous Semiconductors: **M. H. Brodsky**, IBM Yorktown Heights
Magnetism (Materials, Phenomena): **H. P. J. Wijn**, Philips Eindhoven
Metals and Alloys, Solid-State Electron Microscopy: **S. Amelinckx**, Mol
Positron Annihilation: **P. Hautojärvi**, Espoo
Solid-State Ionics **W. Weppner**, MPI Stuttgart

Surface Science
Surface Analysis: **H. Ibach**, KFA Jülich
Surface Physics: **D. Mills**, UC Irvine
Chemisorption: **R. Gomer**, U. Chicago

Surface Engineering
Ion Implantation and Sputtering: **H. H. Andersen**, U. Aarhus
Laser Annealing: **G. Eckhardt**, Hughes Malibu
Integrated Optics, Fiber Optics, Acoustic Surface Waves: **R. Ulrich**, TU Hamburg

Coordinating Editor: **H. K. V. Lotsch**, Heidelberg

Special Features:
– Rapid publication (3–4 months)
– No page charges for concise reports
– 50 complimentary offprints
– Microform edition available

Subscription information and/or **sample copies** are available from your bookseller or directly from Springer-Verlag, Journal Promotion Dept., P.O.Box 105 280, D-6900 Heidelberg, FRG

Springer-Verlag
Berlin
Heidelberg
New York
Tokyo